Human Factors and Ergonomics in Practice

Improving System Performance
and Human Well-Being
in the Real World

Human Factors
and **Ergonomics**
in Practice

Improving System Performance
and Human Well-Being
in the Real World

Edited by

Steven **Shorrock**
Claire **Williams**

CRC Press
Taylor & Francis Group
Boca Raton London New York

CRC Press is an imprint of the
Taylor & Francis Group, an **informa** business

CRC Press
Taylor & Francis Group
6000 Broken Sound Parkway NW, Suite 300
Boca Raton, FL 33487-2742

© 2017 by Taylor & Francis Group, LLC
CRC Press is an imprint of Taylor & Francis Group, an Informa business

Printed on acid-free paper
Version Date: 20160815

International Standard Book Number-13: 978-1-4724-3925-3 (Paperback)

Library of Congress Cataloging-in-Publication Data

Names: Shorrock, Steven T., editor.
Title: Human factors and ergonomics in practice : improving performance and well-being in the real world.
Description: New York : Routledge, 2016.
Identifiers: LCCN 2016025148 | ISBN 9781472439253 (pbk. : alk. paper) | ISBN 9781315587332 (ebook)
Subjects: LCSH: Human engineering. | Work design.
Classification: LCC TA166 .H7843 2016 | DDC 620.8/2--dc23
LC record available at https://lccn.loc.gov/2016025148

Visit the Taylor & Francis Web site at
http://www.taylorandfrancis.com

and the CRC Press Web site at
http://www.crcpress.com

Dedication

*Dedicated to the memory of
Professor John R. Wilson (1951–2013)*

Contents

PART I   Reflections on the Profession

PART II   Fundamental Issues for Practitioners

PART III Domain-Specific Issues

PART IV Communicating about Human Factors and Ergonomics

List of Figures

List of Tables

Foreword

I *do* human factors.

As an airline pilot, I sit at the front of an aircraft that passes through the upper atmosphere at close to the speed of sound. My immediate thoughts are about 100 miles ahead. My longer-term thoughts may be many thousands of miles ahead. All the time I can only do what I do because of an amazingly complex aviation system that somehow "understands" the nature of humans within it. It understands how to make my job easier. It understands how to present me with information. It understands how not to distract me up until I need distracting. It understands, in simple terms, how to make it easy to do the right things.

I am supported by colleagues who understand how to make it easy for us to do the right things, as well as understanding the ways to get the best out of us and the best out of the system. And if it all starts to go wrong, they, I, and the system as a whole, understand how we can manage ourselves and all the complexity in a way that ultimately is more likely to deliver us safely back to our homes.

Do the passengers I fly understand human factors and ergonomics (HF/E)? I'm not sure, but I bet they'd really appreciate the preceding paragraphs. Do I understand human factors and ergonomics? I'm not sure either, but I still *do* it and live with it every day of my life.

Sadly, I was to discover whom not all safety-critical work in other domains has benefitted from understanding human factors and ergonomics. In 2005 my wife was admitted to a hospital for a routine elective procedure. It took just over 20 minutes for people and a system that didn't *do* human factors to leave my wife brain dead. It would be another 13 days before she really was dead.

I was shocked, not just by the tragedy that had befallen me and my children, but as an independent report and inquest revealed, the system that had inadvertently killed my wife seemed to be so far behind in its practices. When it came to safety and human factors, it was as if it was stuck in the 1930s.

Since then I've worked, in a voluntary capacity, to try and understand the healthcare system. I discovered there had been some theoretical discussion around the topic at senior levels, and a few very local frontline projects, but in a workforce of 1.4 million people in the United Kingdom alone, HF/E was regarded as a niche research topic that wasn't terribly relevant to the real world.

It became clear very early on that I was out of my depth, I needed specialists in HF/E to work with. Since then I've been delighted to have received the support of so many in your community, far too many to name.

As an outsider, I've been enveloped in the HF/E profession, and I've begun to appreciate just how much HF/E can do for the world. The simple pleasing design of my phone that allows me milliseconds faster access to my contacts to the safety critical seconds that a good system might buy me, which can make the difference between life and death. HF/E is just too important to confine to a lab or research article. Whatever research articles might exist on the issues relevant to the events on that day, they are wasted unless HF/E touches practice—in the broadest sense of the term.

As clinicians the world over have reviewed my late wife's case, in a quiet break-room perhaps, they have all, with very few exceptions, stated clearly: "I wouldn't have done what they did." Yet place those same people in a simulated scenario with the same real-world disorder, which deteriorates into the same challenging moment, and most actually do. This gap illustrates the difference between human performance as imagined and human performance in the real world. Of course this gap exists in other industries as well.

To help close this gap requires the sort of insight that you can provide.

I have no doubt that the most influential HF/E specialists are those who can "mix it" with the frontline employees, CEOs, senior policy makers, even politicians. These people understand the complexity of HF/E and can also explain it in relatively simple terms. They can balance the need to do things in the "best" way, with the real financial, practical, and cognitive limitations of those who have to implement ideas into reality.

You might be surprised that I talk about cognitive limitations when we're discussing a target audience who make up some of the most intelligent people in the world. But I've also observed that in complex systems nurture wins over nature almost always, and, introducing even the simplest concept, that "human error" is normal, is for many a stretch of the imagination a little too far.

This has been my challenge, but here's a challenge for you:

Do you move the profession forward or do you move the world forward?

Moving the world forward involves some trade-offs and compromises in order to include the wider world. The danger of the purist academic, theoretical world is that it is neat and defined. This is not the messy real world occupied by the frontline in any industry. To change the real world, you have to be in the real world, whatever part you play.

This book has been expertly crafted by Steven Shorrock and Claire Williams. It features many human factors experts and ergonomists who have exposed their work to the real world and who have become my own heroes. I know it will inspire you to make a difference for the benefit of not just those of us at the frontline, but everyone who depends on us.

Martin Bromiley
Airline Captain & Founder of the Clinical Human
Factors Group (CHFG), UK

Preface

Around 10 years ago, I found myself caught between two worlds. I had spent several years in practice in various industries, as an internal consultant to an air traffic service provider and as an external consultant for an international consultancy. Then I moved to academia, and set up as a sole trader on the side. I loved both research and practice, and was intrigued by the crossover, and often, the lack of it. I found that the two worlds had different goals, values, resources, constraints, ways of working, and methods of communication. Most of the books and journal articles on human factors and ergonomics (HF/E) were written by people situated in academia. But much of the writing on HF/E did not speak directly to the worlds in which I found myself as a practitioner, or to some of the issues that seemed to really matter in practice. The late John Wilson pointed out that one should not talk of HF/E being a science but a discipline, and—crucially—one that blends craft, science, and engineering. The craft side of HF/E, while a strong focus for practitioners and deeply embedded in context, has received much less attention in the literature.

There seemed to be a gap for a text that speaks more directly to practitioners of HF/E (specialists or those in related professions), written predominately by those embedded in industry or with strong industry links. HF/E, like work, is fundamentally contextual and situated. There are texts on practice and professional issues for other disciplines and professions, often written by practitioners (e.g., school psychologists, teachers, therapists, management consultants, and helping professions generally). These books do not provide description and instruction on theory and methodology as such, but rather address "life as a practitioner," including the context, conditions, constraints, challenges, and compromises that make up the real world of practice. This differs from "textbook HF/E," which may be conceptually and methodologically thorough and well-written, but is devoid of the "mess" in the contexts that people, including HF/E specialists, work with and within, and many of the contextual and situational factors that really make a difference, such as practitioner characteristics and the relationships through which HF/E works.

In April 2005 and April 2007, at the conference of the (now) Chartered Institute of Human Factors and Ergonomics, I was presenting papers on HF/E drawing from counselling on the notions of skilled helping and empathy in practice. At the conferences, I met Claire Williams, a practitioner who was also doing a PhD on HF/E practitioner expertise and competencies. We stayed in touch and some years later decided to try to bring about a book on issues relevant to practitioners. Over our careers to date, we have both worked in practice and research, in industry and academia, and value similar things about the discipline and profession, in particular the diversity of HF/E and how we adjust to the messy worlds in which we find ourselves. These are themes that gradually emerge in this book. Over the past decade or so, we have also become trusted friends and mutual sounding boards. There are few other people with whom I can imagine undertaking something like this book; we did it because we thought it was quite important and because we would learn a lot. Like most of the authors, we also wrote and edited in our "spare" time—on trains and planes, in hotels, during evenings, weekends, and holidays. Sometimes, it seemed impossibly ambitious. Coordinating and editing a book with so many experienced practitioners is in many ways harder than writing a book.

So, this book attempts, in some small way, to focus more on the craft side of HF/E, as well as the application of engineering and science in real contexts. It is not a textbook or a "tools and methods book." Instead, it is about "being a practitioner," and a celebration of the practice of HF/E, in all its diversity. We are very grateful to the practitioners who have made time to contribute to the book. We have learned from you all.

Steven Shorrock

I have been working as an HF/E consultant since 1996, initially providing advice to industry about how to manage musculoskeletal disorders (MSDs) though gradually broadening my focus to look at health and safety more widely, particularly the cultural and behavioral aspects. Over the course of these 20 years, and about the time the ideas for this book were being formulated, I began to question my own expertise, and that of colleagues more and less qualified than I am. Issues around whether HF/E interventions work; why they work; what it takes to make them work; and who is capable of being successful, became important to me.

These questions became paramount in 2003, when I was promoted to Head of Ergonomics in an established occupational health consultancy. At this time I became responsible for assuring the quality of my own and others' work, and for facilitating the professional development of my ergonomics colleagues. It is against this backdrop that I started a PhD looking at expertise amongst HF/E advisors (2004) and first met and started to talk to Steve about these issues.

During my PhD (2004–2008) I aimed to further the understanding of what it meant to be "expert" as an HF/E advisor. Steve and I talked regularly about issues of successful practice; whether the focus of the training courses for HF/E was too technical, too analysis-driven, and insufficiently solution-oriented or "softer-skills" based. This became particularly important when I took on an academic teaching post on the Ergonomics master's at the University of Derby in 2009. My students needed to be academically rigorous, but ours is an applied subject. Most of them would enter practice, and I wanted to be sure they were as prepared as possible to make that step.

The need for a book that talked about practice, in all its various forms for our multifaceted profession, gradually crystallized. Steve first spoke the idea out loud and I said "yes" without thinking about what it might actually mean. The 30 or so chapters that follow are that book; full of the wisdom and expertise of a great many practitioners who have forgotten more than I'll ever know about practice and for which I am incredibly grateful. I have learnt a great deal from reading their writing and it is my firm hope that anyone embarking on or well established in their practice will also reap great benefit from the contributions in this book.

Claire Williams

Acknowledgments

This book has taken over three years to put together, from conception to publication. We are grateful to the many authors who have written and reviewed chapters, and to many others who have reflected on chapters during this time. We are especially grateful to our partners and families who have supported us along the way. Last but not least, we would like to acknowledge the thousands of human factors and ergonomics specialists who work, often in the background, to help improve well-being and performance in almost every industry around the world.

Editors

Steven Shorrock is a chartered ergonomist and human factors specialist, and a chartered psychologist, with a background in internal and external consultancy in human factors and safety management in several industries, government, and as a researcher and educator in academia. He holds a BSc in applied psychology, an MSc (Eng) in work design and ergonomics, and a PhD in human factors in air traffic control. Steven is currently a safety and human factors specialist and European safety culture program leader at EUROCONTROL where he works in countries throughout Europe, and an adjunct senior lecturer at the University of New South Wales, School of Aviation, Sydney, Australia.

Claire Williams is a chartered ergonomist and human factors specialist (C.ErgHF) providing advice and training in risk management to a wide range of industrial and government organizations. She holds a BSc in biological sciences, an MSc in ergonomics, and a PhD in ergonomics expertise. She is a senior human factors ergonomics consultant at Human Applications and a visiting research fellow in human factors and behavior change at Derby University, Derby, United Kingdom.

Contributors

John Allspaw is the Etsy's chief technology officer (CTO), leading the product engineering, infrastructure, and operations teams. For over 17 years, he has worked in system operations in domains such as biotech, government, online media, social networking, and e-commerce. Unsatisfied with shallow explanations of the performance he had seen in teams of engineers operating more and more complex software, he began his human factors research (alongside his day job) in the master's program for Human Factors and Systems Safety at Lund University, Sweden. John is the author of *The Art of Capacity Planning* and *Web Operations: Keeping the Data on Time*, both published by O'Reilly Media.

David Antle, BSc, MSc, PhD, CCPE, is the research director and a senior consultant at EWI Works, Edmonton, Alberta, Canada. David joined EWI Works in 2012, and has since received his doctorate in kinesiology, specializing in occupational biomechanics and ergonomics, from McGill University, Montreal, Quebec, Canada. David works with both industrial and office settings to provide individual assessments, development, and delivery of training, and assist with expansion/development of clients' in-house ergonomics programs.

Andrew Baird is a lecturer at the University of Derby, Derby, United Kingdom, involved with the MSc Ergonomics/Human Factors and MSc Behavior Change programs. A graduate of Loughborough University, Leicestershire, United Kingdom, Andrew spent 12 years in consultancy before rejoining academia and has worked across a range of sectors with a particular specialism in MSDs. Alongside his teaching Andrew continues to practice in relation to MSDs and behavior change. His research interests are focused primarily around musculoskeletal pain and he has published and presented at various conferences on the subject. He also has a developing interest in safety climate research.

Fiona Bird has been applying human factors in the rail industry since 1999, with considerable experience in the research, regulation, development, and verification of complex systems. Fiona is a senior human factors engineer at Bombardier Transportation and is responsible for the design, assurance, and verification of rolling stock and their subsystems through all phases of the lifecycle from the tender phase to the end of warranty period including modification programs.

Martin Bromiley is an airline captain for a major UK airline and an aerobatic instructor. In 2005 his first wife died during a routine hospital operation, and a subsequent review identified numerous systemic and human factors–related issues. As a result, he founded the Clinical Human Factors Group (www.chfg.org) a charitable trust that aims to promote an understanding of human factors in health care. He continues as the chair of the CHFG in a voluntary capacity and has contributed to more human factors initiatives in health care than he can remember. In the 2016 New Year Honours list, he was awarded an OBE for his work.

David Caple has been the director of an independent ergonomics consulting company for the last 33 years and he is also an adjunct professor at Latrobe University, Melbourne, Australia. His company provides consulting services to industry groups, large companies, and government particularly relating to macroergonomics issues. Projects have been conducted in many countries including Singapore, Papua New Guinea, Malaysia, United Kingdom, and New Zealand. Research projects have been funded in a wide range of areas including wheelchair anthropometry, supply chains, and government communication with small business. David has been the facilitator of consultation processes for the Australian Work Health and Safety Strategy. He was also the president of the International Ergonomics Association.

Ken Catchpole is a human factors practitioner whose mission is to improve safety and performance in health care. Initially working at Great Ormond Street Hospital in London, United Kingdom, in congenital heart surgery, he spent six years at Oxford University, Oxford United Kingdom, exploring HF interventions in surgical and acute care, before moving to Cedars-Sinai in Los Angeles, California, to work on trauma care and robotic surgery. He has also contributed to health-care research and improvement at hospitals in the Netherlands, Norway, Australia, and New Zealand. In 2016, he took an endowed chair position at the intersection of research and safety practice at the Medical University of South Carolina, Charleston, South Carolina.

Bernie Catterall was the founder director of the UK-based ergonomics and risk management consultancy Human Applications. Now "part-retired," Bernie nevertheless continues to work internationally as a consultant for numerous large manufacturing clients. With over 30 years of practical consultancy experience in both the public and private sectors worldwide, Bernie evolved as a specialist dealing with Boards and Senior Management Teams in larger organizations. His consultancy work remains based around molding ergonomics principles with risk management practice in order to provide practical improvement solutions both for industrial workplace safety and job design—and for achieving behavioral changes, especially at higher managerial levels.

Ed Chandler is the principal UX consultant at User Vision with over 12 years' experience in user-centered design and leads their accessibility services. Chandler worked at RNIB applying accessibility best practice. Having worked in a variety of sectors, including financial services, ecommerce and retail, travel and tourism, and consumer electronics, Ed specializes in creating and shaping user-centered design strategies, and building longer-term business engagement on UX and accessibility. Ed has a master's degree in human factors from the University of Nottingham and is a member of the Chartered Institute of Ergonomics and Human Factors.

Amy Chung is a registered psychologist with 7 years of experience in human factors research projects, including a 3-year ARC Linkage Project to assess the involvement of human factors in adverse events related to patient safety, a 3-year ARC Discovery Project investigating the nature of skilled performance, and a number of projects in transport safety for Roads & Maritime Services, New South Wales, Sydney, Australia. Amy is currently working as a human factors consultant and an academic editor while completing her PhD studies on the research–practice relationship in human factors and ergonomics at the University of New South Wales, Sydney, Australia (expected graduation in 2016).

Ben Cook is a former military transport pilot, flying instructor, and low-level aerobatics display pilot with formal instructional, training, and human factors qualifications. Ben has over 16 years of safety management experience, with the majority of that time spent as an applied human factors specialist. He has helped organizations and teams to develop and deliver innovative projects to achieve behavioral and cultural change. He is well known for the development of contemporary safety and error management strategies, which have enhanced system resilience to human error. Ben continues to provide pragmatic, applied interventions to help achieve operational excellence.

Ryan Cooper is a human factors specialist within the Australian Defence Force, Canberra, Australia, and provides expertise to support a range of projects targeting the enhancement of human capability at the individual, group, and system levels. He is a registered psychologist with qualifications spanning the fields of psychology, management, human factors, safety, and accident investigation. Ryan was formerly employed as a military psychologist within the Australian Army and held a variety of positions covering a broad spectrum of psychological services and operational settings. His professional interests include organizational/safety climate assessment and interventions as well as human performance in high-risk environments.

Phil Day manages a team of usability, accessibility, and interaction design specialists for NCR Corporation. As the lead on usability, ergonomics, and accessibility issues, he has responsibility across the entire product range for NCR Corporation including software and hardware. Prior to working at NCR, Phil, originally trained as a software engineer, studied human–computer interaction (MSc and then PhD) and then worked in human factors research in several academic research roles.

Sidney Dekker, PhD, is professor of humanities and social science at Griffith University in Brisbane, Australia, where he runs the Safety Science Innovation Lab. Previously, he was professor of human factors and system safety at Lund University in Sweden. After becoming full professor, he learned to fly the Boeing 737, working part-time as an airline pilot out of Copenhagen. He has won worldwide acclaim for his groundbreaking work in human factors and safety, and is the best-selling author of, most recently, *The Field Guide to Understanding "Human Error"* (2014), *Second Victim* (2013), *Just Culture* (2012), *Drift into Failure* (2011), and *Patient Safety* (2011). His latest book is *Safety Differently* (2015). More at sidneydekker.com.

Lisa Duddington is the owner of the multi-award-winning User Experience agency, Keep It Usable. A recognized UX expert, her company works with household brands to understand customers' digital experiences, applying psychology to improve those experiences and increase sales. Her main interests are consumer psychology and behavior. Lisa holds an MSc in Human Factors and Ergonomics and a BSc (Hons) in Computing and Psychology and originally began her career as a mobile usability specialist within Sony Ericsson. Keep It Usable's mobile work has been showcased at 10 Downing Street as an example of how technology can be used to improve a nation's health and well-being.

Alan Ferris is a retired fellow of the Chartered Institute of Ergonomics and Human Factors (CIEHF), Loughborough, United Kingdom. He has worked in HF/E in small and large consultancies including CCD, AIT, FA, ITT, and EMI. His most recent work prior to retirement was on rail systems for Network Rail and Metronet/LUL. Before that it was financial software systems, telecoms, and telecoms standards. He has held a variety of positions on CIEHF's panels and committees and has been part of the team steering CIEHF through to achieving Chartered status over the last decade. He is currently chair of CIEHF's Professional Affairs Board.

Margo Fraser holds a BSc (Co-op) and MSc in kinesiology specializing in ergonomics from the University of Waterloo, Waterloo, Canada, as well as a certificate in occupational health and safety from Ryerson Polytechnic Institute, Toronto, Canada. She is a Canadian Certified Professional Ergonomist (CCPE) and a past-president and current executive director of the Association of Canadian Ergonomists/Association canadienne d'ergonomie. Margo is a practicing kinesiologist with the British Columbia Association of Kinesiologists and has acted as vice-president and secretary general of the International Ergonomics Association (IEA) and member of the IEA technical committees on Ergonomics for Children and Educational Environments (ECEE), and Organizational Design and Management (ODAM).

Dominic Furniss is a researcher co-investigator at University College London, United Kingdom. His research focus is on understanding the design and use of medical devices in sociotechnical systems. He has authored/edited three books on qualitative research and fieldwork for healthcare. He has a keen interest in public and patient involvement. In 2014 he received a UCL Public Engagement Award for his contribution to this area.

Ron Gantt is a vice president of SCM. He has over a decade experience as a safety leader and consultant in a variety of industries, such as construction, utilities, and the chemical industry, to help people see safety differently. Ron has a graduate degree in Advanced Safety Engineering and Management as well as undergraduate degrees in Occupational Safety and Health and Psychology. He is currently pursuing his PhD, studying organizational learning and drift. Ron is a certified safety professional, a certified environmental, safety, and health trainer, and an associate in risk management. He is also coeditor for SafetyDifferently.com.

Don Harris is a professor of human factors in the Faculty of Engineering and Computing at Coventry University, Coventry, United Kingdom. He is a fellow of CIEHF and a chartered psychologist with 30 years of experience of teaching and research in the aviation, defense, and automotive sectors. He has been involved in the design and certification of flight decks, worked in the safety assessment of helicopter operations, and was an accident investigator on call to the Division of Army Aviation. His principal research interests lie in the areas of advanced concepts for flight deck design and aviation safety. He recently published *Writing Human Factors Research Papers* (Ashgate).

Brent Hayward is a registered psychologist (AU) and the MD of Dédale Asia Pacific, Albert Park, Australia, with more than 35 years of experience working with human factors, safety promotion, and safety investigation in civil and military aviation and other safety-critical industries, including rail, maritime, resource mining, and nuclear power production. He has developed and delivered training in aviation psychology, human factors, CRM, and safety investigation techniques for a range of organizations in Australia, Africa, Asia, Europe, the Middle East, the Pacific, and the Americas. This has included international training courses and workshops on behalf of the European Association for Aviation Psychology, for EMBRAER, for EUROCONTROL, and for the Singapore Aviation Academy.

Nigel Heaton is the director of Human Applications. He has worked for BT Research Laboratories, GEC Research Laboratories, HM Treasury, and Loughborough University of Technology. Nigel's main interests are in engaging senior management teams in improving health, safety, and welfare. He has worked worldwide with organizations ranging from large government departments, multinational companies, and NGOs. Recently he has been working on ways to deliver leadership and behavior change programs to support organizations in getting better. Nigel is a chartered fellow of the Chartered Institute of Ergonomics and Human Factors and a chartered member of the Institution of Occupational Health and Safety. He acts as a consultant to organizations who are interested in designing and implementing simple risk management systems that take account of people. Nigel is a reverent Leicester City fan who now believes that anything is possible!

Erik Hollnagel is a professor at the Institute of Regional Health Research, University of Southern Denmark (DK), Denmark, chief consultant at the Centre for Quality, Region of Southern Denmark, Denmark, and professor emeritus at the Department of Computer Science, University of Linköping (S), Linköping, Sweden. He has worked at universities, research centers, and industries in several countries and with problems from many domains including nuclear power generation, aerospace and aviation, software engineering, land-based traffic, and health care. Erik's professional interests include industrial safety, resilience engineering, patient safety, accident investigation, and modeling large-scale sociotechnical systems. He has published widely and is the author/editor of 20 books, as well as a large number of papers and book chapters.

Daniel Hummerdal is the director of Safety Innovation at Art of Work, Australia. After an initial career as commercial pilot, Daniel studied psychology. Since then, he has worked as accident investigator with the Swedish Civil Aviation Administration, as a human factors consultant with Dedale (France), safety innovation leader in engineering, construction and mining, and been engaged in industrial safety research in Sweden, France and Australia. Daniel is also the founder of www. safetydifferently.com

Shelly Jeffcott is a native Aussie human factors specialist with 15+ years' experience in patient safety research and implementation. She is currently the human factors lead for one of Scotland's biggest Health Boards, working with frontline operating theatre teams to design and test safety interventions. Shelly has a degree in psychology from York University, a PhD in human computer interaction from Glasgow University, and did her postdoctoral research in the Engineering Design Centre at Cambridge University. She was a senior research fellow in the Department of Epidemiology and Preventive Medicine at Monash University in Melbourne for 5 years and spent 2 years as a national improvement advisor in NHS Scotland.

Dan Jenkins leads Human Factors and Usability team at DCA Design International, Warwick, United Kingdom, working across four sectors (medical, consumer, transport, and industrial). Dan started his career as an automotive engineer where he developed a keen interest in ergonomics. In 2005, he returned to Brunel University taking up the role of research fellow, studying part-time for his PhD in human factors and interaction design. Dan has worked across a wide range of domains including medical, consumer products, defense, automotive, rail, maritime, aviation, nuclear facilities, and control room design. He has coauthored ten books and over 50 peer-reviewed journal papers, alongside numerous conference articles and book chapters.

Rob Miles is the technical director of Hu-Tech Risk Management Services. Rob has over 25 years' experience in human factors, principally in offshore oil and gas but also rail, defense, and aviation. He was the HF topic specialist on the UK Deepwater Horizon working group and has contributed HF sections in many guidance documents. As chair of the Energy Institute's Human and Organizational Factors Working Group, Rob oversees the production of HF guidance for the UK Energy sector. He is an HF advisor and independent member of CIRAS, the UK's confidential reporting system for rail workers. He is currently developing barrier-based analyses in health care to improve patient safety.

Linda Miller, OT [D], OTD, MEDes, CCPE, CPE, is the president at EWI Works International Inc. and has practiced in ergonomics for 25 years. Linda established EWI Works in 1991 and has extensive experience in large-scale national and international ergonomic projects involving program development and training. Industries include health care, utilities, forestry, mining, and manufacturing. For two decades, she has presented at health and safety conferences and teaches at the University of Alberta. She is a certified ergonomist in Canada and the United States.

Dave Moore is a New Zealander originally from London with 25 years experience specifically in HF/E, who has lived almost exactly half his life in each place. He was codirector of a private consultancy, South Pacific Ergonomics Ltd., Auckland, New Zealand, worked as a research ergonomist for a government-owned institute, and subsequently teaches at AUT University in Auckland. Posts held include the IEA newsletter editor, HFESNZ president, and IEA Council member for NZ.

Ben O'Flanagan is a human factors specialist at Sydney Trains, Sydney, Australia, with a focus on developing resilient systems and improving operational performance. He is a chartered ergonomist and human factors specialist (C.ErgHF) and has a background in human factors consultancy. He has a master's degree in ergonomics from Loughborough University, Leicestershire, United Kingdom.

Dave O'Neill has over 40 years' experience as an ergonomics researcher, practitioner, and manager. Having started his career as an aerospace engineer (mechanical) he moved into agricultural engineering after completing an MSc in ergonomics (UCL, 1974). After 30 years at a government research institute (National Institute of Agricultural Engineering, which became Silsoe Research Institute), Dave set up his own Consultancy (Dave O'Neill Associates) from which he took some time out to be the chief executive of the Ergonomics Society, now the Chartered Institute of Ergonomics and Human Factors (2008–2014). Dave is a chartered ergonomist and a European ergonomist (Eur. Erg.) as well as having affiliations to the Institution of Mechanical Engineers and the Institution of Agricultural Engineers.

Guy Osmond has been in business for over 40 years and in workplace ergonomics for more than 20 years. During that time, he has developed a broad understanding of the variable approaches to workplace ergonomics arising from different cultures and legislative backdrops. He has also been involved in the sourcing, development, and UK introduction of a number of new devices. He has continually pioneered innovative products and services and is especially interested in the use of technology to improve processes and communication.

Jean Pariès is the president of the DEDALE company, Paris, France, and Melbourne, Australia. He is an internationally recognized expert in the field of human and organizational factors of safety. After a career with the French Civil Aviation Authority, then with the French Air Accident Investigation Bureau, he participated in the founding of Dédale, active in the fields of aviation, nuclear power, rail transportation, energy distribution, patient safety, industry, and road safety. Since the early 2000s, he has been actively participating in the research movement of resilience, and he has recently been elected as the president of the Resilience Engineering Association.

Clare Pollard is the director of a human factors consultancy, Clear Pool Consulting Limited, Hampshire, United Kingdom, providing technical assessments, training, and peer review. Over the past 15 years, Clare has been supporting projects for nuclear licensees across the United Kingdom including reactor sites, defense facilities, and research establishments. A chartered fellow of the Institute of Ergonomics and Human Factors, Clare has particular interests in security behaviors, operator error, and the human aspects of decommissioning.

Ian Randle, BSc, MSc, PhD, MIEHF, CErgHF, is the managing director of Hu-Tech Human Factors Consultancy, London, United Kingdom, and the president of the Chartered Institute of Ergonomics and Human Factors, Loughborough, United Kingdom (2016–2017). He has over 30 years' experience in human factors consultancy, research, and teaching. Ian has worked in a broad range of sectors including manufacturing, defense, aviation, nuclear, pharmaceuticals, and health care. However, over the past 15 years Ian has mostly worked in major projects on oil and gas. His work has ranged from managing the human factors integration process through the stages of a project, to undertaking human error analysis of safety critical tasks. As the MD of Hu-Tech, he has overseen human factors support in over 100 oil/gas and maritime capital projects around the world.

Paul Salmon is a professor in human factors at the University of the Sunshine Coast, Queensland, Australia, and is the director of the Accident Research team (USCAR). His research to date has involved the application of human factors theory and methods for understanding and enhancing performance in safety critical systems such as road safety, aviation, defense, the workplace, and outdoor recreation. A significant component of his current research program focuses on bridging the gap between research and practice and on the integration of human factors research outputs and methodologies into real-world practice.

Caroline Sayce is a human factors engineer at Rolls-Royce, currently supporting the safety case for future submarine designs. This work has focused on operability and task analysis, the development and testing of HMIs, and support to the design of systems and components. Previous experience has been gained in the rail industry, in the design of rolling stock and highways industry. She has a bachelor's degree in ergonomics, is a chartered member of the Institute of Ergonomics and Human Factors (C.Erg.HF), and a member of the Institute of Engineering and Technology.

Richard Scaife is a director of The Keil Centre Limited, Edinburgh, United Kingdom, a chartered occupational psychologist, and a chartered ergonomics and human factors specialist. He has 27 years' experience in a range of industry sectors including aviation, nuclear, construction, process, and pharmaceuticals, providing consultancy expertise and training. Richard works in all aspects of human factors, particularly human safety analysis (including human error) and the design and evaluation of equipment to meet user requirements. Richard's current work includes training incident investigators, safety culture assessment, and behavioral safety. Richard holds a BSc (Tech) in occupational psychology and an MSc in occupational psychology.

Graham Seeley is a human factors specialist with over 10 years' experience managing human factors integration in the rail industry. He has a keen interest in the practical application of human-centered design principles, to help deliver safety improvements and to realize business objectives. Graham's design experience covers a broad range of systems, interfaces, and infrastructure; his current focus is the optimization of safe working systems used to protect track workers in the rail corridor. Graham holds a BSc degree in ergonomics from Loughborough University, Leicestershire, United Kingdom.

Sarah Sharples is a professor of human factors and associate pro-vice-chancellor for research and knowledge exchange in the Faculty of Engineering at the University of Nottingham, Nottingham, United Kingdom. She is a chartered ergonomist and human factors specialist, and her main areas of interest and expertise are human–computer interaction, cognitive ergonomics, and quantitative and qualitative research methodologies for examination of interaction with innovative technologies in complex systems. Her work has been based in the domains of transport, health care, and manufacturing and mobile technologies. She was the president of the Chartered Institute of Ergonomics and Human Factors (2015–2016), and coeditor of *Evaluation of Human Work*, 4th edition (2015).

Stuart Shirreff is a consultant with Human Applications, and former head of Safety, Health and Environment for KDC Contractors Ltd., Manchester, United Kingdom. Stuart has worked in the UK construction sector for 33 years, the last 10 years in high-hazard environments. After initially working as a craft tradesman, he studied and qualified as a chartered safety practitioner before specializing in human factors/ergonomics. Stuart is particularly driven to demonstrate that through practical application of HF/E, safety and health can be evidenced as core to any business. Stuart is also a coordinator for the CIEHF specialist interest group, Occupational Safety and Health Ergonomics Network (OSHEN).

Steven Shorrock is a chartered ergonomist and human factors specialist, and a chartered psychologist, with a background in internal and external consultancy in human factors and safety management in several industries, government, and as a researcher and educator in academia. He holds a BSc in applied psychology, an MSc (Eng) in work design and ergonomics, and a PhD in human factors in air traffic control. Steven is a safety and human factors specialist and European safety culture program leader at EUROCONTROL where he works in countries throughout Europe, and an adjunct senior lecturer at the University of New South Wales, School of Aviation, Sydney, Australia.

Matthew Trigg is the director of Human Applications. Matthew has worked to share ergonomics ideas with nonergonomists, particularly with safety practitioners. He believes that simple ergonomics interventions make work better. He is delighted that, halfway through his working life, there appears to be no end in sight for baffling design. His batting average hovers around the high single digits.

Patrick Waterson, CPsychol, AFBPS, FIEHF, is a reader in human factor and complex systems at Loughborough University, Leicestershire, United Kingdom. His research is broadly concerned with the application of the systems approach and human factors methods to safety across a range of domains including health care, rail, retail, and construction. He has recently edited *Patient Safety Culture: Theory, Methods and Application* (Ashgate), which was published in 2014.

John Wilkinson is a chartered fellow of the IEHF, CMIOSH, and MBPsS. He currently works as an independent consultant, writer, and researcher. John worked for the UK Regulator from 1989 to 2011, was a founding member of its Human and Organizational Factors team from 1999, and led it from 2003 to 2011. The team developed the widely used HF Web Pages, and provided training and support for on- and offshore major hazard inspectors/specialists. John has inspected, investigated, presented, written, and consulted widely on HF/E. He was an expert reviewer for the US CSB reports on Texas City and Macondo. He was a principal consultant with the Keil Centre, Edinburgh, from 2011 to 16.

Claire Williams is a chartered ergonomist and human factors specialist (C.ErgHF) providing advice and training in risk management to a wide range of industrial and government organizations. She holds a BSc in biological sciences, an MSc in ergonomics, and a PhD in ergonomics expertise. She is a senior human factors/ergonomics consultant at Human Applications and a visiting research fellow in human factors and behavior change at Derby University, Derby, United Kingdom.

Anne Williamson, BSc, PhD, is a professor of aviation safety at the School of Aviation, University of New South Wales, in Sydney, Australia. Her research is in the area of human factors, primarily focusing on two related areas—the effects of fatigue and the role of human error in injury and safety. Ann established the NSW Injury Risk Management Research Centre and was Deputy Director. Ann has published extensively in the scientific literature and also been an invited speaker at a wide range of national and international conferences and an invited member of a number of government committees on road and workplace safety.

Roel van Winsen is a safety strategy advisor for a large European Electricity Transmission System Operator (TSO). He also works as a mentor and lecturer in the human factors and system safety master's program at Lund University's Centre for Risk Assessment and Management, Sweden. He holds a master's degree in cognitive psychology from the University of Leiden, the Netherlands, a master's degree in human factors and system safety from Lund University, Sweden, and a PhD in human factors and safety science from Griffith University, Australia. He has trained as an air traffic controller for two years at ATC The Netherlands.

Part I

Reflections on the Profession

Part I of the book has six chapters, which present views on aspects of the state of the profession and discipline of human factors and ergonomics from various viewpoints. The key questions focus on the progress of the discipline and profession in achieving its aims, including "where did we come from?," "where are we now?," and "where might we go?" The questions are examined from an applied point of view, considering the contribution of HF/E in work and life.

In Chapter 1, we consider the evolving identity and nature of HF/E as a discipline and profession. The diversity of HF/E is proposed as an outstanding feature, and several factors that help characterize HF/E in practice are outlined. Chapter 2 (Margo Fraser and David Caple) goes on to consider further the nature of HF/E, both in developing and developed countries. Issues of professional competency and the research–practice relationship are opened up for exploration.

The history of HF/E helps to understand where we are now, and why. Chapter 3 (Patrick Waterson) reflects on the development of HF/E over time and proposes some lessons for HF/E practice from the past and present, with some pointers to the future based on what we can learn from the past.

Chapter 4 (Erik Hollnagel) goes on to consider how HF/E began as a solution to a practical problem, but that over time there has been a proliferation of theories, methods, and solutions that are intellectually attractive but with limited practical effects. Erik argues that human factors as a practical solution should be based on a small number of simple principles with a strong empirical foundation.

Chapter 5 (Roel van Winsen and Sidney Dekker) points out some practical and ethical implications of "human error" (and its subcategories) as an explanation for why sociotechnological systems sometimes fail. Roel and Sidney argue that as a human factors community we need to engage in ethical discussions and take responsibility for the effects of the practices that we promote.

Some of these issues are played out in the media. Taking an outside-in view, the media perspective on "human factors" and "ergonomics" is considered in Chapter 6 (Ron Gantt and Steven Shorrock). This chapter offers some proposals on this neglected area that may help minimize the negative consequences of media coverage, and take advantage of the rare positive effects.

1 Introduction
Human Factors and Ergonomics in Practice

Steven Shorrock and Claire Williams

CONTENTS

THE DIVERSITY OF HUMAN FACTORS AND ERGONOMICS IN PRACTICE

Human factors and ergonomics (HF/E), as a professional activity, has now been introduced to almost all economic sectors. In the primary sector, HF/E helps to improve human involvement in mining, oil and gas extraction, agriculture, and forestry. In the secondary sector, HF/E is embedded in manufacturing and construction to produce finished products. In the tertiary (service) sector, hospitals and health-care organizations, telecommunication, wholesale and distribution organizations, and governments all employ or contract HF/E services. In the quaternary (knowledge-based) sector, HF/E practitioners are employed in information and technology, media, education, research and development (R&D), and consultancy organizations.

While some professionals work in one or a small number of industries, HF/E practitioners work in most industries. From toothbrushes to trains, smartphone apps to flight deck displays, farms to production lines, and warehouses to nuclear power control rooms, the idea of designing to optimize well-being and performance is just as relevant. In practice, this means optimizing several goals related to the effectiveness of a purposeful activity (such as efficiency, productivity, maintainability) and particular human values (such as safety, security, comfort, acceptance, job satisfaction, and joy). For particular applications, some goals generally have higher priority than others, but these goals can also (and frequently do) conflict and compete. This means that HF/E must retain a holistic view, but trade-offs and compromises are nearly always required.

HF/E specialists—practitioners and researchers—are from various academic backgrounds (e.g., psychology, engineering, design, biological science, as well as ergonomics) and increasingly come from a wide variety of professional backgrounds and industries (e.g., health care, aviation). They work with all types of people at all levels: consumers and service users, frontline and support staff, supervisors and senior management, regulators, and policy makers.

In fact, one word that might best characterize HF/E in practice is *diversity*. This diversity is evident from the HF/E literature, for instance considering the contents page of a typical book of conference proceedings (e.g., Sharples and Shorrock, 2014) and the many HF/E books and journals, including the leading journals *Ergonomics*, *Applied Ergonomics*, and *Human Factors*.

However, the literature only hints at the diversity of *practice*. How HF/E works on a day-to-day basis in real environments—not just as the application of theory and method—is not often written about. It is difficult to integrate insights and reflections from practice into regular textbooks for several reasons that we will discuss later. As a result, there is a lack of reports from practitioners on practice in industry, along with a risk of a gap between literature (what we might call "HF/E-as-imagined" or "HF/E-as-prescribed") and practice (what we might call "HF/E-as-done") (of course, there are exceptions such as Broberg and Hermund, 2004; Whysall et al., 2004; Dul and Neumann, 2009).

This book aims to locate HF/E in the various settings of application, capture some of the realities of practice, and celebrate its diversity.

BUT WHAT IS HF/E ANYWAY?

Before we go on, it is worth clarifying what we mean by "human factors" and "ergonomics," and why both? These terms are understood in many different ways, especially in industry and among the general public. The definition of the International Ergonomics Association—the umbrella association for national HF/E societies and associations—is as follows:

> Ergonomics (or human factors) is the scientific discipline concerned with the understanding of interactions among humans and other elements of a system, and the profession that applies theory, principles, data and methods to design in order to optimize human well-being and overall system performance.

(IEA, 2016)

This definition makes several points—explicit and implicit—that are worth exploring in a little more detail. It refers to some abstract terms such as *interactions, system,* and *elements.* For our purposes, *interactions* are kinds of action between two or more elements of a system that have an effect upon one another. The *elements* may be human, technical, informational, social, political, economic, organizational (Wilson, 2014), and physical. A *system* can be defined as "a set of elements or parts that is coherently organized and interconnected in a pattern or structure that produces a characteristic set of behaviors, often classified as its 'function' or 'purpose'" (Meadows, 2009, p. 188). A system can be conceived on several dimensions: from simple to complex, hard to soft, closed to open, conceptual to practical, static to dynamic, deterministic to probabilistic, and linear to nonlinear.

The definition also talks about *human well-being* and *overall system performance*. Some argue that this joint "and" purpose characterizes the holistic nature of HF/E (e.g., see Wilson, 2014). There are difficulties in defining well-being, and this "optimization" is far from straightforward. It is, however, dependent on prior *understanding*, of human work and of the system within which work takes place.

The definition uses the form "Ergonomics (or human factors)," seemingly treating the two as equivalent—one being a synonym for the other (and you may notice that we use a slash between "HF" and "E" throughout this book). The HF/E discipline does indeed treat them as equivalent. This is reflected in many books on HF/E, even where one term or the other is used. Despite this, there is a growing tendency in industry and by some in the profession to see human factors and ergonomics as somewhat different. This is similar to how many people think "counseling" is different to "psychotherapy." Like "human factors" and "ergonomics" (see Chapter 3), the terms "counseling" and "psychotherapy" have different origins, but the British Association of Counselling and Psychotherapy (BACP) states: "BACP sees no evidence of any difference between the functions of counseling and psychotherapy" (BACP, 2010). Another example of confusion, but where there is a substantive difference, is between "dietetics" and "nutritionism," the former requiring a degree qualification, and is regulated and governed by a code of ethics in the United Kingdom. We will come back to this later and it comes up again in other chapters, but for the purpose of this book, we treat human factors and ergonomics collectively as "HF/E" to denote equivalence and also to avoid confusion with the more specific "human factors engineering." In some cases, authors may use one term or the other, depending on the industry (for instance some industries, standards, and organizations only use one term).

The IEA definition highlights explicitly that HF/E is both a discipline ("a branch of knowledge, typically one studied in higher education," oxforddictionaries.com) and a profession ("a paid occupation, especially one that involves prolonged training and a formal qualification," oxforddictionaries.com). Specifically, the IEA calls HF/E a scientific discipline. This is a matter for discussion, which we will come to later in the Introduction and which Patrick Waterson deals with extensively in Chapter 3. The nature of the profession of HF/E is not often discussed in books, which focus overwhelmingly on discipline aspects. As such, professional issues such as ethics, roles, contexts, competency, communication, and so on, receive less attention than theoretical and methodological issues. But such issues are becoming ever more pertinent, as HF/E becomes embraced by far more than "professional" HF/E specialists, as we will see later. There are many other definitions. The late John Wilson defined HF/E as follows in 2000:

> The theoretical and fundamental understanding of human behaviour and performance in purposeful interacting sociotechnical systems, and the application of that understanding to the design of interactions in the context of real systems. (p. 560)

He later defined "systems ergonomics and human factors" as follows (extract):

> Understanding the interactions between people and all other elements within a system, and design in light of this understanding.

(Wilson, 2014, p. 12)

Wilson's more specific emphasis on *human behavior and performance* and *sociotechnical systems* in his earlier definition will be familiar to most who have encountered HF/E, though the emphasis on *human behavior and performance* is a common lay interpretation at the expense of the wider system and design.

From Wilson's definitions, we might see HF/E as a *design discipline*, since the purpose of HF/E is achieved via design; it is design-driven (Dul et al., 2012). Wilson (2000) argued that HF/E blends craft, science, and engineering, but as the purpose of HF/E (human well-being and overall system performance) is achieved via design, it seems sensible to view HF/E as a design discipline. But remember that we are designing *interactions* (among human, physical/technical, informational, social, political, economic, and organizational elements of a system), not just "stuff"—physical artifacts. The interactions may be experienced physically, cognitively, and emotionally from different perspectives: micro (e.g., interactions with a telephone panel in a control room), meso (e.g., communication and coordination between team members), and macro (e.g., organizational communication). HF/E in practice may move between these various perspectives, but still adopting a holistic and systems perspective, with purpose in mind. Wilson's early definition also emphasizes that the domain of application is "real systems." This is the emphasis of this book: practice in the context of real systems.

THE REALITIES OF HF/E IN PRACTICE

Next we offer a few reflections on what we perceive to be some of the realities of HF/E in practice, which hopefully set the scene for the other chapters in this book.

HF/E IN PRACTICE EXISTS IN A FAST-CHANGING AND MESSY WORLD

The context of economic activity around the world is changing fast and, in many respects, might be described as "messy." This messiness is associated with the varying interrelated features of organizations, economies, and societies that create uncertainty, unpredictability, flux, complications, and "systems of problems" (Ackoff, 1974). Problems, possible solutions, and opportunities are often ambiguous and intractable, and can be viewed and approached in various (possibly contradictory) ways. Different people and groups have different values and perspectives concerning them, and there is great resistance to change among some and great appetite for change among others. The consequences of intervention are unclear and interventions often create new problems. Those in charge of solving problems (e.g., politicians, policy makers, and managers) may be ill-equipped to do so, lacking knowledge or power or both. This messiness necessarily affects HF/E practice. We can examine the sources of mess in various ways, none of which is really adequate or can be comprehensive. But for this introductory chapter, we will consider a few aspects of the messy world, starting at a macro level.

At an economic level, markets have changed quickly over the recent years. These changes affect organizations in a variety of ways and also affect the demand for HF/E services. Over the past 20 years, the world's manufacturing base has become geographically more diverse, with production chains expanding to include some of the least developed countries in Asia (e.g., Cambodia) and a number of European countries following European Union (EU) expansion (e.g., Poland, Hungary), for example (World Trade Organization [WTO], 2015). The so-called BRICS nations (Brazil, Russia, India, China, and South Africa) have more than doubled their world export share, from 8% in 2000 to 19% in 2014 (WTO, 2015). China has risen to become the world's top exporter. From the early 2000s, there has also been significant growth in the transport sector. This growth has brought challenges for HF/E in manufacturing, transportation, raw materials, and energy sectors.

In the midst of this growth, there have been crises and complications. The years from 2000 have seen the collapse of the dotcom bubble (1999–2001), the global financial crisis (2007–2008), the European sovereign debt crisis (from late 2009), the Russian financial crisis (2008–2009, 2014–), the Chinese stock market crash (2015), continuing fears of another financial crisis in 2016, and the energy crisis (2000s) and ongoing energy price volatility. In fewer than 20 years, crude oil prices per barrel (West Texas Intermediate) went from a low of $16.28 (December 1998) to a high of $144.78 in (June 2008) and down again to $32.60 (January 2016).

The situation has changed rapidly for particular sectors, countries, and regions. The global economic crisis severely affected transport and finance; in 2009, world transport exports plunged by 22% (WTO, 2015). Conflict in the Middle East, North Africa, and Ukraine, for instance, affected industrial activity, whereas improved situations in some countries (e.g., Iran) triggered a boost in exports.

Such economic volatility can affect HF/E in various ways. HF/E activity could be curtailed in some sectors; new opportunities may open in other sectors (e.g., renewable energy). In response to economic conditions, changes can come about from major national projects that may be announced, cancelled, or put into an uncertain state (e.g., new nuclear power plants, runways, and railways).

The world can change even more quickly and unpredictably, with implications for HF/E, in response to sudden unwanted events, such as natural disasters, accidents, technological incidents, and terrorist attacks. The Brussels airport terrorist attack (March 2016) raised question over security screening in airports, with security checks moving temporarily to airport entrances in some airports following the attack. The German wings tragedy (March 2015) raised questions over cockpit security and reinforced cockpit doors. The Malaysia Airlines flight MH17 shot down in eastern Ukraine (July 2014) raised questions about flight routings, which were changed in response. Air France Flight 447 from Rio de Janeiro to Paris (June 2009) raised questions about automation and pilot training and experience. Many more frequent nonfatal incidents occur, which have short- and long-term HF/E implications. For instance, Belgian airspace was closed and the skies cleared after a power surge (May 2015), and interference on radio frequencies in Scottish airspace led to many flight delays (October 2015). Only a few years earlier, controlled airspace was closed in several European countries following the eruptions of the Eyjafjallajökull volcano in Iceland (April 2010). These are just a few examples of events in one sector that raise questions for HF/E, questions that may need answers in a short timescale. There are of course many other events in other sectors with implications for HF/E, including the Fukushima Daiichi power plant disaster (March 2011), and the Mid Staffs hospital scandal (Francis report published in February 2013).

Politics, laws, and regulations (e.g., health and safety, major hazard safety, public safety, accessibility) interact with economic and social constraints and opportunities, partly by influencing the focus of attention (e.g., compliance needs) of industry. In occupational health and safety, the phrases "ergonomics" and an "ergonomic approach" have been enshrined in guidance to the UK legislation on display screen equipment and manual handling, keeping HF/E on the agenda for these issues way beyond the expectations of most practitioners. Changes in political parties and policies, and statements by political figures, can bring about changes in attitudes to certain human values (e.g., "health and safety"), which may affect corporate decision-making and public attitudes. Political uncertainty, for instance concerning "Brexit" or Scottish independence, or uncertainty over major infrastructure decisions (e.g., airport runways and high-speed rail) may also affect investment behavior and demand for HF/E services.

Of course there are many technological advancements, the most visible in everyday life being improved internet access, digital technology, and associated products and services. This is bringing about changes in our everyday relationship with, and dependence upon, technology, as well as changes in work (e.g., teleworking and remote communication), health care (e.g., electronic records), and citizen behavior (e.g., access to government services, quantified self, online shopping), all of which has wider implications (e.g., increased small parcel distribution). Other developments exist in most industrial sectors. Some are specific to certain sectors (e.g., agriculture, transport, energy), others cross sectors and seep into everyday life, or will in the future (e.g., artificial intelligence, robotics, 3D printing, internet of things, augmented reality, drones, exoskeletons). But generally, there is more technology, more technology performing functions previously performed by humans, and disruptive technology fundamentally changing human interaction within systems and disrupting markets.

There are also various slower changes that affect us. These include aging populations and rising obesity in some countries (see Dul et al., 2012), mass migration and more multicultural communities, and population growth, with impacts on agriculture (see Chapter 27), transport, energy, and other sectors.

Overall, the changes that we are observing can be seen as opportunities for HF/E (e.g., to promote user-centered design principles; to attract new students) and threats to HF/E (e.g., diluting principles; misapplication of methods; neglect of theory;

competition). Whatever happens, HF/E must adapt. It will need to become even more agile, more integrative, and more participative. We will have to make compromises, while sticking to the core features of what defines us as a discipline and profession.

HF/E IN PRACTICE IS EMBEDDED PRIMARILY IN ORGANIZATIONS

Most practitioners of HF/E are employed in organizations: consultancies, primary producers, manufacturers, service providers, and organizations that interface with industry [e.g., government (including regulators), intergovernmental organizations and nongovernment organizations, and universities]. Throughout this book, the authors describe the context of HF/E in practice and the embedded nature of HF/E in organizations and industry (especially Chapter 8 and Part III).

The organizational context is a great source of mess that affects possibilities for HF/E integration. Resources are far from optimal, and there are many constraints and influences that make any plans to "optimize human well-being and overall system performance" seem like a pipe dream. HF/E in practice is constrained by many factors, which interact to affect possibilities for practitioners. Staffing levels and competencies are often overstretched, and sometimes inadequate to meet demand. Many countries publish lists of skill shortages [e.g., nurses, dentists, and aircraft technicians in Australia (2015)], and organizations may struggle with staffing due to rates of pay, work locations and conditions, the nature of demand, and training constraints, for example. Since HF/E usually requires participation of system actors (Dul et al., 2012), the conditions for HF/E interventions are variable. People may or may not be available to work with HF/E professionals. Participation (e.g., in workshops or simulations) can take months to arrange and can change at short notice. Even where staff are available, competencies may be variable, patchy, or absent. For instance, some industries are experiencing a shortage of competencies in older programming languages, and are bringing retired software engineers back as contractors.

Similarly, there is a shortage of qualified HF/E professionals, and many countries lack HF/E education (see Chapter 2). For instance, of more than 40 organizations that provide air navigation services in Europe, only a few have one or more qualified HF/E professionals (e.g., eligible for certification as a European Ergonomist with the Centre for Registration of European Ergonomists). In some organizations, this has been partially offset by other practitioners, sometimes operating under the labels of "ergonomist" (e.g., physiotherapists), "human factors specialist" (e.g., psychologists), "human factors trainer" (e.g., pilots with crew resource management training), and "UX specialist," "usability specialist." "interaction designer" and a plethora of other terms (many varieties of software and computing specialists). This has, of course, created new problems, as previously predicted (see Corlett, 2000, also Chapter 24). Related to the availability and competencies of staff, it is also worth mentioning that employment patterns are changing with more contractors and short-term contract staff fulfilling roles previously undertaken by "permanent" staff. This has a range of implications, and the same pattern can be seen with HF/E.

Equipment in organizations is also both a constraint and an opportunity. HF/E practitioners in many industries find themselves in environments that are a mix of old, legacy technology, installed over time, and newer technology, both commercial-off-the-shelf (COTS) and bespoke. COTS software may be amenable to local adaptation (e.g., colors) but not fundamental change (e.g., scrolling behavior), which may well affect the software deployed across several units and organizations. Equipment frequently varies in operation, sometimes in fundamental ways even within the same working environment (e.g., hospitals). Some is close to unusable, or in various states of disrepair. Tools may be missing or not fit for purpose, forcing workers to adapt their method of work ("violate the rules") to compensate. Spare parts may be lacking; it is not unusual to hear of organizations using eBay to obtain certain parts. Innovation cycles are getting shorter, and equipment is replaced more frequently than before. Consequently, workers (especially technical/maintenance staff) spend much more time in training to understand the technology.

Procedures and work processes are often over- or under-specified, sometimes unworkable. An airport tower control unit may have 1,000 or more pages of operational procedures, plus tens of temporary procedures that make everyday work a challenge (this can be even worse on the railways). Procedures are often designed from afar—policies almost always—and may not reflect reality. Working relationships may be strained. The organization may be in a continual state of reorganization and change. There are new threats to human values (e.g., security).

Still other constraints exist in our organizational structures, goals, finances, processes, measures, and incentive systems. It makes a significant difference whether an HF/E practitioner is located in engineering, operations, safety, occupational health, or other divisions, for instance (see Kirwan, 2000; Wilson, 2014). Organizations are reorganized, HF/E teams are split and scattered, and key decision makers change, along with attitudes to HF/E. Company systems and standards (e.g., safety management systems, internal standards) are developed, sometimes in response to regulations and national or international standards, which can cement how HF/E is (or is not) integrated into organizational functioning. Integrating new concepts and approaches faces many barriers and can take years (see Shorrock, 2013). Subcultures (e.g., along geographical and professional lines) can also vary markedly within organizations, as well as the diversity of staff members' national cultural backgrounds and primary languages. Company incentive systems, punishments, performance targets, and measures can encourage unwanted behaviors. All of these factors have implications for HF/E practitioners, often working against HF/E solutions.

People have to work around the mess, through the mess, despite the mess. This is also the context for HF/E practitioners. HF/E practitioners need to work in an agile and resilient way, to adjust, adapt, and learn in response to demands, opportunities, and constraints. Like the people we work with, HF/E practitioners must manage trade-offs (e.g., between thoroughness and efficiency [Hollnagel, 2009]; between tasks and relationships), balance conflicting goals (e.g., safety, security, health, productivity, efficiency, usability, pleasure, satisfaction), with limited time and resources and under various constraints (e.g., regulatory, organizational, economic, political). Yet HF/E interventions often come with emergent properties—both positive and negative—which must also be detected, understood, and handled. In our experience of working with HF/E practitioners in industry, their success is often less about the latest knowledge of HF/E theory and tools, and more about the ability to reflect, compromise, make trade-offs, and adapt to the conditions of the work environment, organization, and the industry.

This messiness can make it hard to write about HF/E in practice, because the "process" of "doing HF/E" inasmuch as there is one, is not linear and standardized; it is a blend of craft, engineering, and applied science, and varies in order to adapt to conditions. The actual practice of HF/E is less clean and tidy than might be portrayed in the "success stories" that we sometimes read. The organizational context can also mean that much of HF/E in practice cannot be written about publically without stripping away much of the context, due to its sensitivity. These sensitivities relate to technology (e.g., weapons, consumer products), processes (e.g., military, commercial), data (e.g., safety, security), and politics. Sign-off may be at the CEO level, which can immediately discourage writing. And publication is not usually part of the organizational reward structure for practitioners. Some of these issues mean that some HF/E practitioners are put off writing about practice, and even put off using social media for professional purposes.

HF/E in Practice Requires an Effective Research–Practice Relationship

The issues that most affect HF/E in practice are, then, woven into context. Singleton (1994) argued that knowledge and experience in industry are usually essential in HF/E, since a particular task may not be fully comprehensible without a context. But HF/E practitioners draw on an understanding of interactions (between people and other elements within a system) that is both fundamental and applied, from research and practice. On this, Singleton also argued that research should inform practice through ecologically valid and usable methodology (see Chapter 10), and practice should ideally be based on research evidence, and raise more questions for research (see Chapter 9). Similarly, Sind-Prunier (1996) argued that, to be relevant and useful, research should be responsive to the needs of practitioners.

But from his survey of experienced Human Factors and Ergonomics Society (HFES) members, Meister (1999) concluded that "HFE research is not useful to what should be two of its primary consumers: the practitioner and designer working in system development" (p. 264). He bluntly stated that researchers and practitioners "see little value in the products of each other's activities" (p. 223). Green and Jordan (1999) wrote "academics regard industrial approaches as sloppy and lacking in rigor and validity, whilst industrialists regard academic practice as over complex and impractical" (p. 113).

This is addressed elsewhere in this book, but it is important that we do not allow a research–practice gap to widen in HF/E. Issues of research relevance, access to research, research format, and time to read, understand, and apply research (Chung and Shorrock, 2011) must be dealt with by researchers and practitioners in tandem. This is important for the relevance and survival of the profession (Meister, 1999; Sind-Prunier, 1996) and its goals of improving well-being and system performance.

We need an effective research–practice relationship in HF/E. But we also need a relationship that acknowledges the realities of practice. The complexity and messiness of application domains means that theory and method cannot be applied in a straightforward way; practitioners need "instrumental" knowledge, not just "explanatory" knowledge (Meister, 1992). Schön (1983) critiqued "technical rationality," a positivist epistemology of practice that views professional activity as instrumental problem-solving made rigorous by the application of the scientific theory and technique. According to Schön, this is "the view of professional knowledge which has most powerfully shaped both our thinking about the professions and the institutional relations of research, education, and practice" (p. 21). He proposed the idea of "knowing-in-action" as a characteristic mode of ordinary practical knowledge: "a kind of knowing [that] is inherent in intelligent action." Skilled practice reveals a kind of knowing that does not stem from prior intellectual consideration. He also proposed "reflecting-in-action": not only do we think about doing but we think about doing something while doing it. We might call this "thinking on our feet." Schön suggested that:

> When someone reflects-in-action, he becomes a researcher in the practice context. He is not dependent on the categories of established theory and technique, but constructs a new theory of the unique case.... Thus reflection-in-action can proceed, even in situations of uncertainty or uniqueness, because it is not bound by the dichotomies of Technical Rationality. (pp. 68–69)

Usher et al. (1997) characterized technical rationality as "the dominant paradigm which has failed to resolve the dilemma of rigor versus relevance confronting professionals" (p. 143). Going back to the discussion of messiness, Schön remarked "the scope of technical expertise is limited by situations of uncertainty, instability, uniqueness, and conflict. When research-based theories and techniques are inapplicable, the professional cannot legitimately claim to be expert but only to be especially well prepared to reflect-in-action" (p. 345). These considerations are particularly important to us in our thinking about HF/E in practice, the research–practice relationship, and the roles we take on as we practice (see Chapter 7).

HF/E in Practice Is Dependent on Practitioner Characteristics and Relationships

Many factors that have a great bearing on our success in practice are rarely written about (Williams and Haslam, 2011). Such interacting factors may include, for instance, personal characteristics (e.g., integrity, credibility, empathy, congruence, respect), knowledge (e.g., of industry, technology, regulations, laws), and skills (e.g., in marketing, sense-making, design, storytelling, communicating, relating). When it comes to professional practice, character and the constraints of the practice environment may have much more bearing than background education and knowledge (Piegorsch et al., 2006). While many of these factors are generic to many professions (Shanteau, 1992), they are operationalized in particular ways in the context of HF/E, which is rarely discussed.

Just as important is the network of relationships that brings about the desired outcomes of HF/E. HF/E has one foot in a "system world" of tools, methods, processes, standards, and regulations, but it has another foot in a "relational world" (or "community world") of people in various roles in organizations and society. The ability to relate effectively to these people, drawing on relationship skills such as listening and empathy (Shorrock and Murphy, 2007), is vital. HF/E is dependent on a network of relationships This reflects consulting more generally. Peter Block observed that "No matter how research-based or technical the project, it will always reach a point at which the success of the work will hinge on the quality of the relationship we have with out client. This relationship is the conduit through which our expertise passes." (2000, p. 374). The most advanced methods and latest theory will come to nothing without a network of relationships.

HF/E Is Changing

As society and industry is changing, so is HF/E and our identity, especially in the eyes of industry. HF/E is becoming more popular. In the UK National Health Service, there is now significant participation in human factors, in good part due to the work of the Clinical Human Factors Group (see Foreword by Martin Bromiley and Chapter 13). This is evident to the outsider via social media, particularly Twitter. The same can be seen in WebOps, again in no small part due to one of the authors in this text: John Allspaw (Chapter 25). Frontline workers know that HF/E is relevant, and they often see "human factors" as what they *do*. Going back to the IEA definition, it is about the *interactions among humans and other elements of a system*, and at a work-design level, adapting these via knowledge-in-action and reflection-in-action, within the system of constraints that exists. Or referring to Wilson's definition, it is about *human behaviour and performance in purposeful interacting sociotechnical systems*. In simple terms, for many people in industry, 'human factors' is about human work, and 'design' may involve the design of artefacts and processes, or it may be constrained to adjustment of routines, and so on.

There are relatively few HF/E professionals in many environments, so others are often taking the reins. As Corlett (2010) described, this could have bad outcomes: "if there is no control over those who wear the label of "ergonomist" then not only will improvements in practice be a long time in coming but there will be projects put forward as ergonomic improvements which will eventually be discredited as a result of bad practice" (p. 682). But we can collaborate and invite others in. After all, dietary advice is not the sole domain of dieticians, and talking about problems is not the preserve of counselors. While "deprofessionalization" (or counter-professionalism) may come at the expense of quality, overprotection of a profession may come at the expense of participation, and success. We can work together to ensure that dangers and threats are mitigated, but that opportunities are taken, which will likely increase the desire for HF/E (Williams and Haslam, 2006).

So a middle ground is required. Part of this middle ground may be that certain roles, typically involving a wide and deep-level content and method expertise (see Chapter 7), may require highly qualified and experienced HF/E practitioners (e.g., chartered, registered, or certified). Other roles may require a different sort of practitioner, perhaps familiar with certain aspects of HF/E, but not a specialist as such [e.g., a "technical member" of the CIEHF (CIEHF, 2016)]. These roles may involve applying certain aspects of content and method expertise; using process facilitation; advocating or evangelizing HF/E principles; stimulating thinking or experimentation; or acting as an independent monitor. This middle ground requires collaboration among those with expertise in theory, method, and aspects of context (HF/E practitioners) and those with deep expertise in their jobs, working environments, and industry (field experts [Shorrock et al., 2014]; system actors and experts [Dul et al., 2012]).

The middle ground requires that we invite in those at the edge. In the context of communities, McKnight and Block (2010) put it this way:

> The challenge is to keep expanding the limits of our hospitality. Our willingness to welcome strangers. This welcome is the sign of a community confident in itself. It has nothing to fear from the outsider. The outsider has gifts, insights, and experiences to share for our benefit.... The beautiful, remarkable sign of a secure community is that it has a welcome at the edge.

THE PURPOSE OF THIS BOOK

The previous discussion brings us to the purpose of this book, which is to convey some of the perspectives and experiences of practitioners on the real practice of HF/E in a variety of industrial sectors, organizational settings, and working contexts. The book blends literature on the nature of practice with reflections from experience, and offers insights into the achievement (and nonachievement) of the core goals of HF/E: improved system performance and human well-being.

To achieve this purpose the approach of this book can be characterized as follows:

- **Contextual and systemic:** The book takes a holistic approach to HF/E, emphasizing the context of real HF/E practice and the systemic nature of the discipline and profession.
- **Diverse and eclectic:** The book offers multiple perspectives from practitioners from different industries and settings and industry-focused researchers. The diversity of HF/E in practice means that the chapters have an eclectic feel; we avoided stipulating a certain style or format, or extensive referencing, since this is not the reality of practice. But as a whole, the authors give complementary insights into HF/E in practice.
- **Experiential and reflective:** The book recounts real experiences and reflections of practitioners and other HF/E stakeholders via reflection and narrative, and this is prioritized over citation and referencing. Authors also consider some of the lessons learned and not learned by the profession, and some of the wider implications for the profession.
- **Genuine:** The authors are honest about HF/E-as-done, without assuming ideal conditions. The authors discuss some of the factors that influence practice and outcomes in "messy" and constrained environments, including the compromises and trade-offs that are necessary in practice.
- **Useful:** The book aims to help improve professional practice, via practical wisdom from experienced practitioners (e.g., "advice I'd give myself if I were starting over").

This book speaks directly to the realities of HF/E in practice. Like books on practice and professional issues for other disciplines and professions (e.g., Bransetter, 2012; Brookfield, 1995; Kottler, 2010; Schön, 1991), it offers reflections on the world of practice and "life as a practitioner."

YOU, THE READERS

There should be something of interest to anyone with an interest in the discipline and profession of HF/E, including:

- **Current HF/E practitioners** to help reflect on, challenge, and conceptualize their own practice and the practice of others, and how it may be sustained, changed, or improved
- **Future HF/E practitioners** to gain an understanding of HF/E-as-done, and put their training into a wider context
- **Allied practitioners, HF/E advocates and ambassadors** who utilize certain HF/E concepts and methods, or champion the aims of HF/E, to put their practice into a broader and more systemic framework, perhaps progressing to fuller integration into their work
- **Researchers** to gain an understanding of the nature of HF/E practice in a variety of industrial domains and organizational and working contexts, to help focus and direct their research, and to help identify some of the practical implications
- **Policy makers and regulators** to understand some of the practical and systemic HF/E issues affecting policy and regulation
- **Clients of HF/E services and products** to become more active and instrumental in the success of HF/E in practice

THE AUTHORS AND EDITORS

This edited book features chapters and commentary by many experienced HF/E practitioners situated in a diverse range of contexts, including consultancies, manufacturers, service providers, universities, and other organizations. The chapters are diverse in content, approach, tone, and style of writing, reflecting the individual authors, their roles and experiences, their working contexts, and preferred approaches to writing. The authors have worked in many countries on six continents, though most are from Anglophone backgrounds and work mainly in Anglophone countries, with a few exceptions.

We editors are both practitioners with experience in a variety of industries. We have links to academia (we both still research and teach) and previously held full-time roles in universities.

A blog that accompanies the book is at www.hfeinpractice.wordpress.com. Here you will find practitioner summaries for each chapter and some reflections from early readers, and you are invited to reflect on the general themes within each chapter.

STRUCTURE OF THE BOOK

The book is structured in four parts. Part I presents views on aspects of the state of the profession and discipline of HF/E. Part II considers some of the fundamental issues for practitioners, as well as others associated with HF/E. Part III gives some perspectives on HF/E in practice in a range of industries, from health care to agriculture. Part IV considers issues associated with communicating about HF/E, at all levels and in various forms.

A CONCLUSION FROM ALPHONSE CHAPANIS

To conclude this chapter, we will leave you with a quote from Alphonse Chapanis (1917–2002), one of the founders of our discipline. It gives a brief summary of life as a practitioner, one of the things we hope you get from the rest of this book.

> Human Factors has always been challenging, frustrating at times, rewarding at others, but never dull. I can honestly say in retrospect that I have had a full life—an exciting life—and that I have enjoyed telling people about human factors, educating students and others to take over where I have had to leave off, and grappling with the problems trying to make our material world safer, more comfortable, and easier to cope with.

(Chapanis, 1999, p. 234)

REFERENCES

Ackoff, R.L. 1974. *Redesigning the Future*. London: Wiley.

Bransetter, R. 2012. *The School Psychologist's Survival Guide*. San Francisco, CA: Jossey-Bass.

British Association for Counselling and Psychotherapy (BACP). 2010. FAQs. Accessed March 29, 2016. Available from: http://www.bacp.co.uk/student/faq.php.

Broberg, O. and Hermund, I. 2004. The OHS consultant as a 'political reflective navigator' in technological change processes. *International Journal of Industrial Ergonomics*. 33, 315–326.

Brookfield, S.D. 1995. *Becoming a Critically Reflective Teacher*. San Francisco, CA: Jossey-Bass.

Chapanis, A. 1999. *The Chapanis Chronicles: 50 Years of Human Factors, Research, Education, and Design*. Santa Barbara, CA: Aegean.

Chung, A.Z.Q. and Shorrock S.T. 2011. The research-practice relationship in ergonomics and human factors—surveying and bridging the gap. *Ergonomics*. 54(5), 413–429.

Corlett, N. 2000. Ergonomics and ethics in a changing society. *Applied Ergonomics*. 31, 679–683.

Dul, J. Bruder, R. Buckle, P. Carayon, P. Falzon, P. Marras, W.S. et al. 2012. A strategy for human factors/ergonomics: Developing the discipline and profession, *Ergonomics*. 55(4), 377–395.

Dul, J. and Neumann, W.P. 2009. Ergonomics contributions to company strategies. *Applied Ergonomics*. 40(4), 745–752.

Green, B. and Jordan, P.W. 1999. The future of ergonomics. In: Hanson, M.A., Lovesey E.J. and Robertson, S.A. (eds.). *Contemporary Ergonomics 1999*. London: Taylor & Francis, pp. 110–114.

Hollnagel, E. 2009. *The ETTO Principle: Efficiency-Thoroughness Trade-Off*. Aldershot, UK: Ashgate.

International Ergonomics Association (IEA). 2016. Definition and domains of ergonomics. Accessed March 28, 2016. Available from: http://www.iea.cc/whats/.

Kirwan, B. 2000. Soft systems, hard lessons. *Applied Ergonomics.* 31(**6**), 663–678.

Kottler, J.A. 2010. *On Being a Therapist.* San Francisco, CA: Jossey-Bass.

McKight, J. and Block, P. 2010. *The Abundant Community: Awakening the Power of Families and Neighborhoods.* Oakland, CA: Berrett-Koehler Publishers.

Meadows, D. 2009. *Thinking in Systems.* London: Earthscan.

Meister, D. 1992. Some comments on the future of ergonomics. *International Journal of Industrial Ergonomics.* 10(**3**), 257–260.

Meister, D. 1999. *The History of Human Factors and Ergonomics.* Mahwah, NJ: Lawrence Erlbaum Associates.

Piegorsch, K.M. Watkins, K.W. Piegorsch, W.W. Reininger, B. Corwin, S.J. and Valois, R.F. 2006. Ergonomic decision-making: A conceptual framework for experienced practitioners from backgrounds in industrial engineering and physical therapy. *Applied Ergonomics.* 37, 587–598.

Schön, D.A. 1991. *The Reflective Practitioner: How Professionals Think in Action.* New York: Basic Books.

Schön, D.A. (1983). *The Reflective Practitioner.* New York: Basic Books

Shanteau, J. 1992. Competence in experts—The role of task characteristics. *Organizational Behavior and Human Decision Processes.* 53, 252–266.

Sharples, S. and Shorrock, S. 2014. *Contemporary Ergonomics 2014.* London: Taylor and Francis.

Shorrock, S. 2013. Why do we resist new thinking about safety and systems? *Humanistic Systems.* Accessed March 30, 2016. Available from: http://humanisticsystems. com/2013/04/12/why-do-we-resist-new-thinking-about-safety-and-systems/.

Shorrock, S. Leonhardt, J. Licu, T. and Peters, C. 2014. *Systems Thinking for Safety* (A white paper). Brussels, Belgium: EUROCONTROL.

Shorrock, S.T. and Murphy, D.J. 2007. The role of empathy in ergonomics consulting. In: Bust, P.D. (ed.). *Contemporary Ergonomics 2007.* London: Taylor and Francis, pp. 107–112.

Sind-Prunier, P. 1996. Bridging the research/practice gap: Human factors practitioners' opportunity for input to define research for the rest of the decade. In: *Human Factors and Ergonomics Society 40th Annual Meeting.* Philadelphia, Pennsylvania, 2–6 September, pp. 865–867.

Singleton, W.T. 1994. From research to practice. *Ergonomics in Design.* 2(**3**), 30–34.

Usher, R. Bryant, I. and Johnson, R. 1997. *Adult Education and the Postmodern Challenge.* London: Routledge.

Whysall, Z.J. Haslam, R.A. and Haslam, C. 2004. Processes, barriers, and outcomes described by ergonomics consultants in preventing work-related musculoskeletal disorders. *Applied Ergonomics.* 35, 343–351.

Williams, C. and Haslam, R. 2006. Ergonomics by non-Ergonomists—danger, threat or opportunity? In: Pikaar, R.N. Koningsveld, E.A.P. and Settels, P.J.M. (eds.). *Proceedings of IEA2006 Congress, International Ergonomics Association Triennial Congress.* Maastricht, The Netherlands: Elsevier, 10–14 July, ISSN: 0003 6870 [CD-ROM].

Williams, C. and Haslam, R. 2011. Exploring the knowledge, skills, abilities and other factors of ergonomics advisors. *Theoretical Issues in Ergonomics Science.* 12(**2**), 129–148.

Wilson, J.R. 2000. Fundamentals of ergonomics in theory and practice. *Appl Ergon.* 31(**6**), 557–567.

Wilson, J.R. 2014. Fundamentals of systems ergonomics/human factors. *Appl Ergon.* 45(**1**), 5–13.

World Trade Organization (WTO). *International Trade Statistics 2015.* Geneva: WTO.

2 The Nature of the Human Factors/Ergonomics Profession Today
An IEA Perspective

Margo Fraser and David Caple

CONTENTS

PRACTITIONER SUMMARY

Human factors/ergonomics (HF/E) delineates itself from related professions via a design-driven systems approach with the dual outcomes of optimizing human well-being and system performance. However, use of the terminology "human factors" and "ergonomics" varies, and the variations can create confusion. While HF/E practice is typically unregulated, wide variations in the breadth and depth of practice have resulted in a number of certification bodies that use the competencies set by the International Ergonomics Association (IEA) to certify practitioners to provide a measure of confidence in the HF/E services they receive. However, it can be challenging for students to find academic programs that lend themselves to attaining the required competencies. There is a particular requirement for the HF/E profession to support the development of practice in the developing countries through tools, education programs, and support through organizations such as the IEA, the International Labour Organization (ILO), and the World Health Organization (WHO). Increased emphasis on demonstrating the value of HF/E as a means to improve performance in addition to well-being will be needed by the joint efforts of researchers and practitioners to move HF/E from a reactive endeavor to a necessity in upfront design.

INTRODUCTION

Founded in 1961, the IEA represents the ergonomics and human factors societies from around the world, with a mission to advance ergonomics science and practice. It provides means for societies to collaborate and learn from each other and provides access to the international stage to allow work with related international bodies such as the WHO, ILO, International Standards Organization (ISO), and the International Commission on Occupational Health (ICOH). The IEA has gradually grown over time and currently there are 48 Federated Societies, two Affiliated Societies, and four Networks under its umbrella.

From our work with the IEA, it is clear that the nature of the HF/E profession today varies within and among countries. Diversity in contexts and cultures at the country, organization, and individual levels dictates the means of application and influences the areas of practice. At the same time, the use of a systems approach with the objectives of optimizing human well-being and overall system performance provides the thread of commonality and serves to delineate the HF/E professional from those who focus on only one of these objectives. The central link between those working across the world as ergonomics practitioners is their interest in the person within the context of their physical, cognitive, and psychosocial setting. Some practitioners are more skilled at the physical capacity studies and methodologies, often due to their background in the health sciences or design. Others are more focused on the cognitive or psychosocial aspects, but all require the ability to integrate the three core areas of ergonomics research into their practice. As educational opportunities in HF/E reduce, as in many other professions, there are challenges for the international ergonomics community to work together to support a vibrant and contemporary body of knowledge and practice.

HUMAN FACTORS AND ERGONOMICS: IS THERE A DIFFERENCE? DOES IT MATTER?

When discussing the nature of the HF/E profession today, a natural place to start is the definition of ergonomics. After much work on the part of the IEA to obtain input from Federated Society representatives around the world and representing various facets of HF/E, a consensus definition was published in 2000:

> Ergonomics (or human factors) is the scientific discipline concerned with the understanding of interactions among humans and other elements of a system, and the

profession that applies theory, principles, data and methods to design in order to optimize human well-being and overall system performance.

In spite of this consensus definition, researchers and practitioners alike continue to define ergonomics and human factors in a variety of ways. Some tend to use the terms synonymously, while others define ergonomics as encompassing the physical aspects and human factors encompassing the cognitive elements of human–system interaction, or believe that ergonomics is a subset of human factors or vice versa. However an HF/E professional has come to define HF/E, it seems to be difficult to change their point of view. This can create an issue in practice as organizations that might benefit from integration of HF/E in their work, but have come to understand ergonomics and human factors in a certain way, may not realize the resources available to them depending on how practitioners defines themselves and whether it aligns with the way that it is seen by the organization in question. Therefore, getting in the door to speak to a potential client group can be difficult as the practitioner may already be stereotyped based on how they are titled (ergonomist, human factors specialist, etc.).

In discussion with many professionals in the HF/E field, their views on how they see ergonomics versus human factors are entrenched, in the same way that individuals can be about their political party affiliations and no amount of debate has brought about resolution. This diversity of definitions and sometimes conflicting views can be confusing for consumers and may be diluting our power as a profession. As such, practitioners and the societies representing the profession must become adept at communicating the scope of practice both generally and within the context of the country, organization, business unit, or at the individual level as needed and ensure that, at minimum, they define how they are using the terms (synonymously or not, and if not, the definitions being used).

It should be noted also that in some languages both terms translate as the same word. For example, in French a single term "ergonomie" is typically used.

PROFESSIONAL COMPETENCY

Given the breadth of practice, it is important that those working in the field know their areas of competency. One means seen by societies as a way to advance the profession is through the development of a certification process for HF/E professionals. In some countries, the Federated Society has created a certification body that may reside as an entity under the society or may be a separate body. These bodies are typically unregulated from a government perspective, though the Chartered Institute of Ergonomics and Human Factors in the United Kingdom recently attained chartership. This will provide a measure of protection as only those registered and demonstrating appropriate competency will be able to use the "Chartered Ergonomist and Human Factors Specialist" title. This may set the precedent for other countries to seek a protected status. In the meantime, in many developed countries, anyone can call themselves an HF/E professional with no credentialing and this can create a problem for consumers or clients who need to understand the difference between those who have attained credentials and those who have not.

In an effort to assist with the credentialing process, the IEA has developed minimum standards of competency for societies to use in setting up their certification bodies. Those that meet the IEA criteria are able to attain certification of their process from the IEA. In many developed countries, these bodies have set levels of competency that exceed the IEA requirements.

In addition to demonstrating competency, an important requirement is that the practitioners be held to a code of ethics to help protect service users from poor practice and from those who practice in areas in which they are not competent. A complaint process is also required in case the consumer feels that the practitioner has not been ethical in his or her practice.

It then falls to the certification body and the professionals who have been certified to educate consumers on the difference in competency and quality that can be provided by a certified individual. Having said that, there are not enough certified individuals available to service the demand or potential demand for HF/E. Furthermore, it is unlikely that those who may use HF/E skills in their roles as engineers, occupational health and safety professionals, or health professionals have the time, inclination, or competency, at the level required by the certifying bodies, to become certified. Rather than exclude or ignore these individuals for their contributions to HF/E, our time would likely be better spent at educating them in the practice and helping them to understand at what point a problem is beyond their knowledge and they need to look to an HF/E professional [see Williams and Haslam (2006) for a discussion of the knowledge base of related professions versus those of HF/E practitioners].

EDUCATION TO PRACTICE

HF/E professionals come from a wide variety of backgrounds and often the name of the degree attained is not specifically "Ergonomics" or "Human Factors." Few such specific programs exist and HF/E tends to be a specialty that one can take within a given university faculty or department.

In a university or college setting, students are typically taught scientific research and reporting formats. Experiments, for instance, start with a null hypothesis, which is tested, and either confirmed or denied. Reports follow a standard scientific format with an introduction, methods, results, discussion, and conclusions. Once in practice, it can be shocking to find out that clients do not want a report written in a scientific format. Clients want the bottom line. Saving that for the conclusion part of the report does not work. HF/E practitioners therefore need to be succinct, get to the point in an executive summary, and put details in an appendix. Good communication skills are also essential in dealing with the worker population who may have a large range of education and language skills.

Students also may have access to a range of hardware (e.g., measurement devices), software (e.g., statistical software), and information (e.g., books and journals) that are not available to them in practice, either because of cost and time constraints or because of lack of compatibility with the work environment. In practice, the methods and tools that are likely to be used include surveys, interviews, focus groups, observation, video recording, tape measures and force gauges, and existing data with respect to injuries, incidents, turnover, sickness, and so on. Comparison with normative data that may exist is common; however, a practitioner is often left to provide their best judgment using the closest available data. Gaps in research particularly exist in the cumulative effects of nonrepetitive work such as operations, maintenance, and construction where tasks can vary significantly both within and among days.

Educational programs that expose students directly to businesses though co-operative programs or research partnerships are of great service to the practice as they start exposing these future HF/E professionals to the real-world skill sets they will need.

RESEARCH TO PRACTICE

While tools for practitioner use, such as RULA, REBA, NIOSH Equation, Liberty Mutual Manual Materials Handing Tables, OWAS, MAC, and so on, are helpful [see Neumann (2007) for a catalogue of available tools and methods], there still remains a gap, particularly around methods for easily collecting meaningful data and which have comparative guidelines for work that is not cyclical. Organizations are typically not willing to fund large, lengthy studies. They want to solve an existing problem or implement a design that can be attained on time and remains in budget.

Good work done by practitioners often goes unpublished as a result of constraints set by the organization, lack of rigorous scientific methodology preventing journal or conference acceptance, or simply because practitioners do not believe that the work will be valued or do not have the time or resources to write up or present their work. In a survey of over 500 HF/E professionals by Chung and Shorrock (2011), "carrying out more research and publication" was one of the top five suggestions on how practitioners could improve research application, but several highlighted the lack of outlets to do this. On the other hand, some societies trying to solicit practitioner cases and presentations through awards, contests, or dedicated conference streams have found low participation relative to the practitioner population. The IEA Congress has introduced a stream on practitioner case studies since 2012. More than 100 case studies were received from around the world for the IEA 2015 Triennial Congress, which is encouraging on one hand, but low in relation to the thousands of practitioners worldwide. While practitioners need tools and data from research, learning from the successes and failures of other practitioners is equally as important. Finding ways to encourage frequent sharing of practitioner knowledge and experiences continues to be a challenge.

Qualitative data collection and the ability to conduct interviews with stakeholder groups and individuals are necessities for good practice. However, in many educational programs, quantitative data are important and the use of qualitative methods is not emphasized. Fortunately, the necessity for private sector research dollars has resulted in increased applied research conducted in the field. Publically and privately funded research organizations are often required to interact with practitioners to determine their needs, to assist in setting research directions, and to transfer knowledge to the HF/E profession and end users.

THE HF/E PROFESSION IN DEVELOPING COUNTRIES

The IEA has a goal to support the development of HF/E in developing countries (as per the Human Development Index, HDI). This includes providing financial support to IEA meetings and conferences as well as identifying benefactors typically from large companies with a commitment to support developing countries. An example was a large European agriculture machinery supplier providing financial support for research on women in India working in agriculture. Other examples include a major study on HF/E requirements for a backpack used by coffee pickers in Central America. The IEA has also provided guidelines for the establishment of education programs in developing countries. In order to have an HF/E profession, there needs to be an education system in place that supports the competencies required to become a professional in the area. Lack of academics to provide this education is a major barrier toward cultivating HF/E as a profession in some developing countries. Where some practice is going on within these countries, it has often arisen out of the need to address occupational injuries and illness and therefore has been taken up by medical or safety professionals. Electronic resources that are now available can be very helpful in providing access to educational materials and even attaining a related degree, although language can be a barrier, since many of these resources are in English only. Feedback from societies in developing countries indicates that there are few individuals (in the range of 0%–20% of the members of these societies) who are working specifically as HF/E practitioners. With the growth of online educational programs, there are now more opportunities for people from developing countries to select more courses, webinars, and professional development programs from around the world.

A major difference between developing and developed countries is their high proportion of workers in the informal sector. This is often over 90% of their working population. They are outside the direct control of regulators and industry standard enforcement. Innovative processes to mentor, coach, and support industry leaders are generally more applicable than producing large government agencies.

The ergonomics issues in developing countries that practitioners need to address are similar to developed countries. These include the following:

- Prevention of musculoskeletal disorders
- Aging workforce
- Design to suit the worker populations
- Fatigue management and cognitive demands
- Gender differences in work design

The occupations are also similar though the access to technology is often very different. The main industries that are relevant to developing countries include the following:

- Construction
- Transport
- Agriculture
- Healthcare

- Service industries including office-based work
- Mining

Five significant means to advance HF/E in developing countries are the development of the ILO "Ergonomics Checkpoints" (ILO, 2010, 2012), the joint ICOH and IEA publication "Ergonomics Guidelines for Occupational Health Practice in Industrially Developing Countries" (Kogi et al., 2010), roving ergonomist programs, the formation of IEA Networks, and educational initiatives. These are outlined in the following.

ILO CHECKPOINTS

The ILO Checkpoints is an illustrated, easy-to-use guide to problems and solutions. Its development has been a collaboration between the IEA (content provision) and the ILO (distribution) and it is freely available from the ILO website in a PDF format in a few languages. Recently, an agriculture-specific version was released and a human care work version is nearing completion. An app has been released as well. The Checkpoints publication is available in many languages, and countries are encouraged to use and translate it when needed. China is currently adapting it for use in its main industries with more representative pictures being utilized. The first edition was published by the ILO in 1996 and the second edition in 2010.

IEA AND ICOH JOINT PUBLICATION ON ERGONOMICS GUIDELINES FOR OCCUPATIONAL HEALTH GUIDELINES IN INDUSTRIALLY DEVELOPING COUNTRIES

These guidelines were developed for occupational health professionals working in developing countries to understand the basic principles and processes required to implement a site-based ergonomics program. It focuses on participatory methods for users to identify their local ergonomics risks and to develop simple and cost-effective interventions. It includes a range of simple record sheets for observation studies and a checklist to identify ergonomics risk factors in the workplace. A multifactorial approach to intervention is taken to address the physical, cognitive, and psychosocial risk factors.

ROVING ERGONOMISTS PROGRAM

This program enables ergonomics practitioners and academics visiting developing countries to provide guest lectures and conduct workshops with students and local practitioners to share knowledge and experiences. This program is funded via the roving ergonomist who is possibly travelling to other regions of the world for vacation and interested to include some personal and professional development at the same time. The local IEA Society will often offer local hospitality in recognition of their time.

IEA NETWORKS

IEA Networks are typically groups of Federated Societies and Affiliated societies formed to provide means of collaborating and to assist countries without a society to develop. Four such networks currently exist: Federation of European Ergonomics

Societies (FEES), The Latin American Ergonomics Union (ULAERGO), The South East Asian Network of Ergonomics Societies (SEANES), and in 2014, ErgoAfrica was formed to network the African Societies. With the help of these networks, societies in Malaysia and Peru were recently formed and a society in Algeria is in development.

EDUCATIONAL INITIATIVES

Two examples of educational initiatives to advance the HF/E profession are as follows:

1. Graduate Program in Ergonomics for Latin America: In the summer of 2015, a new ergonomics graduate program will start in Cuba to serve the Spanish-speaking nations. Academic resources from within and outside of Latin America have been retained to provide education covering the competencies required as an HF/E professional. With a waiting list of students, this program should provide a significant boost to the profession in Latin America and is being closely watched by other networks, such as ErgoAfrica, as a potential model for programs in their own regions.
2. In India, HF/E education typically resides within design departments and medicine. A new proposal in the development stage is the creation of a program that operates like some of the MBA (Masters of Business Administration) programs in developed countries where students from a variety of backgrounds, including HF/E, and often having some work experience, will form groups and go out into the common local industries (e.g., agriculture and arts and crafts) to observe the work being done, collect data, and use generative design principles to propose solutions that will help the workers observed use local resources and resources that can be manufactured locally.

THE HF/E PROFESSION IN DEVELOPED COUNTRIES

In contrast to developing countries, the majority of society members from developed countries tend to work directly in the field of HF/E as practitioners with smaller numbers in academia and research. In practice, HF/E professionals may work as consultants or as employees in government departments, medium-sized or large companies to provide HF/E expertise. Many consultants tend to be utilized in a reactive fashion where an unwanted event (injury, incident, accident, or illness) has occurred, to provide recommendations to improve the particular situation. External consultants may not get an opportunity to follow up to determine whether the recommendations were implemented, and if so, whether the problem was resolved. Often they are contracted through human resources, occupational health and safety, or operational safety departments. Since representatives from these departments are often not deeply involved in the upfront design process, particularly at the concept and early development stages, gaining access to groups that could utilize HF/E in a more proactive fashion can be difficult. As ergonomists become more exposed to other related professions, they can develop more opportunities to become involved in more strategic and proactive projects. This is generally referred to as macro-ergonomics. The IEA has a technical committee called "Organizational management and design" that incorporates this broader role of ergonomists.

The range of challenges facing ergonomics as a profession was deemed to be consistent enough among many countries that the IEA established a subcommittee tasked with examining the HF/E profession and making recommendations to move it forward. In 2012, this subcommittee, with Jan Dul as the Chair, produced a white paper on the "Future of Ergonomics," which was published in the journal *Ergonomics* (Dul et al., 2012). Through data collection from a number of societies, it was reported that HF/E has a unique combination of three fundamental characteristics: (1) it takes a systems approach, (2) it is design driven, and (3) it focuses on two closely related outcomes: performance and well-being. However, the profession has not done a good job at marketing these characteristics to stakeholder groups that could benefit. While there are numerous papers showing the outcome of well-being, there are few demonstrating improved performance. Two main strategic directions are recommended:

1. To strengthen the demand for high-quality HF/E by increasing awareness among powerful stakeholders of the value of high-quality HF/E by communicating with stakeholders, by building partnerships, and by educating stakeholders
2. To strengthen the application of high-quality HF/E by promoting the education of HF/E specialists, by ensuring the high-quality standards of HF/E applications and HF/E specialists, and by promoting HF/E research excellence at universities and other organizations

The IEA, Federated Societies, and HF/E practitioners will need to work together in order to determine appropriate tactics to move these strategies forward.

CONCLUSION

The IEA provides an opportunity for representatives from both developing and developed countries to come together and discuss how to advance HF/E within societies and in the world as a whole. Its partnerships with other international bodies will continue to play an important role in the recognition of the value of HF/E.

There are challenges in consistency of how we define ourselves, finding novel ways to capture and share practitioner work, and access to appropriate education and practitioner tools. At the same time, collaborations between practitioners and researchers have been growing in conjunction with institutional mandates for knowledge transfer and the necessity of applied research to gain access to industry funding. As HF/E practitioners, academics, and societies become more adept at promoting the benefits of HF/E to potential users in the language and manner that best suit the targeted group(s) and at all levels of organizations, opportunities in the field can continue to grow.

REFERENCES

Chung, A.Z.Q. and Shorrock, S.T. 2011. The research-practice relationship in ergonomics and human factors – surveying and bridging the gap. *Ergonomics*. 54(**5**), 413–429.

Dul, J. Bruder, R. Buckle, P. Carayon, P. Falzon, P. Marras, W.S. et al. 2012. A strategy for human factors/ergonomics: Developing the discipline and profession. *Ergonomics*. 55(**4**), 377–395.

International Labour Organization. 2010. *Ergonomics Checkpoints*, Second Edition. Geneva: ILO.

International Labour Organization. 2012. *Ergonomics Checkpoints in Agriculture*. Geneva: ILO.

Kogi, K. Mcphee, B. and Scott, P. 2010. *Ergonomics Guidelines for Occupational Health Practice in Industrially Developing Countries*. Zurich, Switzerland: International Ergonomics Association and International Commission on Occupational Health.

Neumann, W.P. (ed.). 2007, October. *HF Design Tool Reference Guide: A Catalogue of Tools and Methods for Workplace Designers*. Accessed March 13, 2016. Available at: http://www.ryerson.ca/hfe/projects/hf_design_tool_reference_guide.html.

Williams, C. and Haslam, R. 2006. Ergonomics advisors – a homogeneous group? In: Bust, P.D. (ed.). *Contemporary Ergonomics 2006*. London: Taylor & Francis, pp. 117–121.

3 Ergonomics and Ergonomists

Lessons for Human Factors and Ergonomics Practice from the Past and Present

Patrick Waterson

CONTENTS

PRACTITIONER SUMMARY

This chapter provides some reflection on the development of human factors and ergonomics (HF/E) over time and examines the various conceptions of HF/E that have emerged in terms of its status as "science" and "practice." The early history and prehistory of HF/E is described, alongside the coverage of developments during the Second World War that ultimately led to the birth of HF/E in 1949. Later sections of the chapter focus on a set of present-day issues for HF/E that have been partly shaped by earlier debates within the history of HF/E. These include how HF/E can be integrated within industry and the wider world of practice, and how HF/E methods should be designed and the trade-offs involved in applying exclusively scientific criteria (e.g., reliability, validity) compared with criteria that emphasize the practical utility and usability of methods. A final section offers some pointers to the future based on what we can learn from the past, as well as making a plea for inclusivity within the discipline of HF/E as a whole.

INTRODUCTION

HF/E is relatively young as compared with other related disciplines such as psychology, physiology, and engineering. Just over 60 years ago within the United Kingdom, Professor Hywel Murrell and a group of colleagues formed what today is known as the Chartered Institute of Ergonomics and Human Factors (CIEHF). Similar developments led to the formation of what is now the Human Factors and Ergonomics Society (HFES) in the United States in the mid-1950s (Meister, 1999). One of the key debates that dominated the early days of HF/E, at least in the United Kingdom, was the relationship between the scientific basis of HF/E and its practical application within industry. On one side were scientists who argued that HF/E provides a unique, unified approach and set of methods for tackling applied, industrial problems. On the other side were those who argued that a pure science of HF/E is hard to justify. Rather HF/E represents a gathering of multidisciplinary professionals, which borrows from these disciplines various constructs, concepts, methods, and tools, to solve a common problem. These arguments and debates continue today and take a number of forms (e.g., the relationship between research and practice in HF/E [Chung and Shorrock, 2011]; the difference between basic and applied sciences [Helton and Kemp, 2011]). Although these debates may appear somewhat abstract and overly academic, they remain important, not least given the efforts of bodies such as CIEHF in the UK and HFES in the USA to raise the profile of the discipline and improve its professional standing.

The aim of this chapter is to stand back and reflect on the development of HF/E over time and to examine these different conceptions of HF/E as science and in practice. Part of this involves examining the early history and prehistory of HF/E and considering how debates concerning HF/E as "science" (ergonomics), "craft," and "practice" (ergonomists) have developed and evolved over time. A final section of the chapter focuses on a set of present-day issues for HF/E that have been partly shaped by these debates.

THE ORIGINS OF HF/E WITHIN THE UNITED KINGDOM

A series of papers that I published with colleagues between 2006 and 2011 provides a detailed history of some of the main HF/E developments in the United Kingdom over the last half-century (Waterson, 2011; Waterson and Eason, 2011; Waterson and Sell, 2006). Other historical material covering the emergence of HF/E in the United States (e.g., Meister, 1999) and other countries (e.g., Kourinka, 2000) is also available.

PREHISTORY

Much of the prehistory of HF/E can be traced back to scientific work carried out in the fifteenth to nineteenth centuries (Monod, 2000). This work includes Le Vauban's investigations of working hours during military campaigns (1682), and Bernadino Ramazzini's (1701) work on disease, which paved the way for the modern study of occupational medicine (Cockayne, 2007). The work of individuals in the eighteenth and nineteenth centuries provided a context for the emergence of a scientific approach to the study of work, during the early part of the twentieth century. Marey (1830–1904), for example, was influential in a number of fields including cardiology, photography, and cinematography. Marey's laboratory experiments on muscle fatigue and movement ultimately led him to develop a theory of the "economy of human work" and laws of time and motion. Some of these ideas were taken up by F.W. Taylor in his work on scientific management (Rabinbach, 1992, p. 117). The publication in 1914 of Jules Amar's "Le moteur humain" (The human motor) represents an important part of the drive toward a more scientific and experimental approach to the study of work. Amar invented a number of devices including the "ergometer," which combined a bicycle with a respirator and a means of recording physical exertion in the form of a graph. Together, the works of Marey and Amar represent some of the precursors of the later studies of fatigue within ergonomics (e.g., Floyd and Welford, 1953, 1954).

NINETEENTH AND TWENTIETH CENTURIES

In the nineteenth century, some most important examples of early work within experimental psychology (Hearnshaw, 1987) were conducted. The work of the German psychologists Gustav Fechner (1801–1887) and Wilhelm Wundt (1832–1920), for example, was instrumental in establishing the fields of psychophysics and vision science (Fechner), as well as the methods of experimentation. The later work of J.B. Watson (1878–1958) on "behaviorism" similarly contributed to the experimental studies of work and influenced prominent figures within ergonomics such as Donald Broadbent (1926–1993). Finally, Hugo Münsterberg's (1863–1916) research helped to move psychology out of a laboratory and toward the study of applied problems within industrial settings. The importance of emphasis on the applied psychological studies within working environments grew in the immediate period preceding the First World War, one of the key influences being the work carried out by F.W. Taylor (1856–1915) and Henry Ford (1863–1947) on scientific management, time and motion study, and the standardization of planning, tools, and working methods (The Science Museum, 2004).

Second World War (1939–1945)

The outbreak of the Second World War in 1939 brought about a huge need to allocate workers and their skills to the most appropriate jobs and tasks needed for the war effort. Within the fields of environmental psychology and physiology, the Industrial Health Board, formed in 1929 from the old Industrial Fatigue Research Board, was charged with studies relating to working hours, rest pauses, and environmental conditions in factories. A number of personnel research committees, one for each of the armed forces, were set up by the Medical Research Council in 1939, in order to investigate and provide solutions to the problem of selecting and training personnel. At the same time, collaboration with the United States began to take place on a firmer basis from 1940 (Hartcup, 2000), and research groups in the United Kingdom and United States were in regular contact and exchanged scientific findings they discovered (Fitts, 1946).

Murrell (1965) described that the main disciplines involved in early ergonomic research were anatomy, physiology, psychology, industrial medicine, industrial hygiene, design engineering, architecture, and illumination engineering. These disciplines were spread around a number of research establishments and war-time committees (e.g., RAF Physiological Laboratory—present-day Institute of Aviation Medicine; MRC Applied Psychology Unit in Cambridge—see Figures 3.1 and 3.2).

One of the most influential of these establishments was the Flying Personnel Research Committee (FPRC). For example, while the various groups associated with physiology and psychology worked on separate problems during the war, it was also clear that, through the committee, these groups established very clear lines of communication and, in some specific instances, collaboration. Matthews (1944,

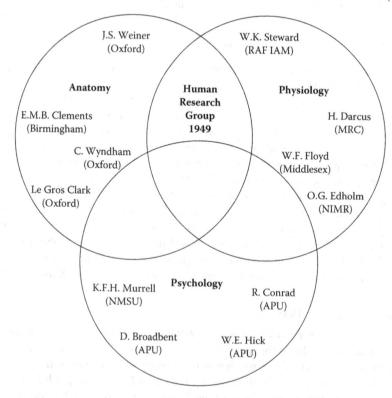

FIGURE 3.1 Disciplinary groups at the 1949 meetings.

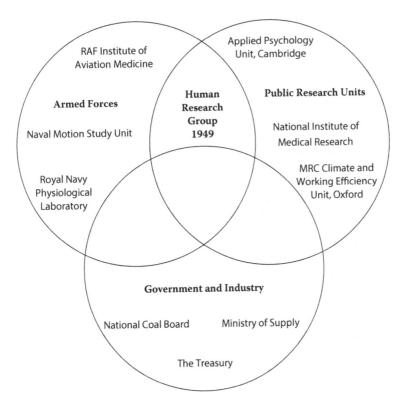

FIGURE 3.2 Organizational groups at the 1949 meetings. (From Waterson, *Ergonomics*, 54(**12**), 1115, 2011.)

1945), for example, describes how problems of stress caused by high altitude and high velocity flying, such as anoxia and resultant pilot blackouts, were solved by new equipment called pressure suits and by making changes to the pilot's posture during flight. These types of solutions often had knock-on effects, such as the need for workspace redesign within the cockpit.

THE BIRTH OF HF/E IN THE UNITED KINGDOM AND EARLY DISAGREEMENTS

In an article looking back on his career, Murrell (1980) noted that as a result of informal discussions with fellow members of the Operational Efficiency Subcommittee of the Naval Motion Study Unit (NMSU), it was clear that there was a desire to meet with the members from other disciplines and to have opportunities to discuss common problems of human work. In July 1949, a dozen people drawn from all three services and from various disciplines met at the Admiralty, Queen Anne Mansions in London (the location of the NMSU) to discuss possibilities. Six months later, in early 1950, the group changed its name to the Ergonomics Research Society (ERS).

Singleton (1982, p. 1) underlined the collaborative, interdisciplinary atmosphere within the ERS at the time:

> The intention was to facilitate the exchange of ideas and expertise between the many disciplines which had made a contribution to the increased effectiveness of human performance during the Second World War.

The first 10 years of the ERS produced a number of important steps forward in bringing these various groups together. In particular, the publication of two books covering HF/E aspects of fatigue and equipment design (Floyd and Welford, 1951, 1953) demonstrates that collaboration similar to that in wartime, continued. There was some evidence of disagreements between groups in the ERS (e.g., physiologists and psychologists). Meister (1999) described similar tensions, what he described as "turf wars," within the American HFES. A report based on the ERS Annual Conference in Oxford in 1959, however, showed the first signs of a split between academics and practitioners. The report mentions, among other things, disagreements that occurred between work study engineers and ergonomists, as well as problems in narrowing down what is actually meant by the term "ergonomics" (Rodger, 1959):

> Has [ergonomics] any distinctive concepts or methods? Is it, perhaps, a convenient gathering place for people belonging to certain technological wings of certain human sciences, and their agents and users in industry?

What emerges from other accounts is the view that HF/E came about not as an explicit attempt to define or form a new scientific discipline. Rather, the intention was much more modest, namely, to facilitate discussion, information exchange, and collaboration between scientists working across a range of specialisms. The aim of much of this work was the improvement of postwartime working environments (e.g., enhanced levels of health and safety, well-being, and productivity). The emergence of the term "human factors engineer" or "ergonomist" was also greeted with some surprise (Murrell, 1970):

> One thing I am sure none of us had envisaged was the development of a professional ergonomist. We were, if you like, society oriented rather than individual oriented; in other words, we felt that ergonomics would provide a forum for the exchange of information between scientists rather than a body of knowledge which would require experts for its application.

To a large extent, this split between the academic and the practitioner groupings within modern-day HF/E continues to be an issue of debate and is reflected in attempts to gain formal recognition and status in the form of the recent drive to achieve Chartership within the United Kingdom (an issue that dates back to the 1960s and was achieved in 2015). Another important change is the worldwide growth of ergonomics (Caple, 2010) and the establishment of bodies such as the International Ergonomics Association (IEA) (see Chapter 2). The context and scope in which the UK CIEHF operates is a global one and much larger compared with the early days of the ERS. One consequence is the volume of information generated by the activities

TABLE 3.1
Changes to the Image and Identity of the UK Chartered Institute of Ergonomics and Human Factors (CIEHF)

Time Period	Images and Identities
Wartime (1939–1945) (Pre-formation of ES)	"Backroom" personnel
Late 1940s–early 1950s (Formation of ES)	"Society of like-minded individuals"
1949	"Founding Fathers" and birth of ergonomics
Early 1950s	A "communicating society" between the disciplines; early schisms; A research body
Early 1960s	An "interface" with the public and industry; supporter of ergonomists working in industry (as well as other groups such as work study practitioners); Still an emphasis on research
Late 1960s–1970s	Links with consumer bodies and widening out of ergonomics concerns; "consultative body"; steps toward becoming a professional body; chartership becomes a priority; schisms between academic and practitioner divisions; breadth of the subject increases; academic and semi-professional body; "learned society"
1980s–1999	Grandparent of the Federation of European Ergonomics Societies professional body; academic and professional body; "learned society"
1999–Present day	Chartered body; academic and professional body; "learned society"; increased interest from other disciplines and interest groups (e.g., within health care, the Clinical Human Factors Group); challenge of maintaining and "owning" a body of knowledge without promoting an image of "possessiveness"

Source: Adapted from Waterson and Sell, *Ergonomics*, 49(**8**), 2006, pp. 791–792.

of the various groups and societies that make up the IEA. Scientific and professional exchange on the scale associated with the postwar ERS is in some respects more difficult to achieve. In short, the identity of HF/E has shifted from what might be termed a loosely coupled group of "backroom wartime personnel" (Waterson and Sell, 2006, p. 791) (Table 3.1) to a much larger organizational entity serving the needs of academics and practitioners. One of the main challenges for the future will be attempting to reconcile some of the differences and problems caused by simultaneously trying to be a "learned society," a professional practitioner organization, and a forum for nonspecialists who nonetheless have an interest in HF/E (e.g., health-care personnel).

MODERN-DAY HF/E: PLUS ÇA CHANGE, PLUS C'EST LA MÊME CHOSE?

Some of the dilemmas and crises of identity that occurred in the past are very much with us today. At conferences and other meetings, one will frequently hear discussions centered on the scientific and practical value of a set of research findings or a description of a new set of tools and methods. Likewise, one of the most frequently mentioned problems, which crops up in discussions with colleagues working in industry, is frustration with the difficulty of applying HF/E methods that have been developed by the academic community. In this section, I want to focus, in particular, on two example issues that reflect these and other types of debates from the past. The first one (human factors integration, HFI) is concerned with the challenges involved in getting HF/E adopted and used within organizations and industry in general. The second issue (the design of methods for HF/E) concerns attempts to improve the utility and usefulness of methods for practitioners, while ensuring that these methods are scientifically valid and reliable. It should be noted that this latter section draws on my own experience in working with HF/E methods over the last decade or so. Some of the examples (e.g., methods for systemic accident analysis) might be said to be on the periphery of traditional HF/E (e.g., methods for training, work design, and evaluation). On the other hand, one of my examples (methods for function allocation) occupies very much what I would refer to as "established territory" within the discipline as a whole.

Human Factors Integration

Over the years, many people have pointed to a range of problems and barriers that underpin the relationship between human factors practitioners and their industrial counterparts. In particular, the work of David Meister and colleagues has addressed these issues in depth. Meister and Farr (1967), for example, found that designers and engineers had little or no interest in human factors, partly since human factors information was perceived as being too inaccessible as compared to charts, graphs, and tables. Later work during the 1970s and early 1980s (Meister, 1982a, 1982b) sought out the views of engineers, research contractors, and government personnel regarding human factors. One of the main conclusions from this research was that individuals were not convinced of the value of human factors and were inadequately trained in the use of HF tools and methods.

Perrow (1983) argued, in an analysis of the influence of organizational context on the work of human factors engineers that the relatively weak position and low profile of HF/E practitioners within the context of the larger organization in which they are employed served to undermine the value and impact of their work. Various other explanations for this type of phenomenon can be given, including the differences in terms of "mindsets" and values that exist between HF/E practitioners and colleagues drawn from other disciplines and backgrounds (e.g., social science compared with engineering) (Cullen, 2007). In addition, many have pointed to the high costs of inputs from human factors into the design process and of HF/E assessments and evaluations (Beevis, 2003; Kerr et al., 2008).

More recently, there has been a focus on the social and organizational processes that act to either facilitate or hinder efforts on the part of HF/E practitioners to influence engineering and other types of design projects. Waterson and Lemalu-Kolose (2010) carried out a study of the work of a human factors team in a large UK company in the defense sector. The findings revealed a number of barriers (e.g., attitudes and perceptions toward HF/E including the value of HF/E and cost considerations), as well as providing insights into the mechanisms and strategies used by the HF/E team in order to improve integration (e.g., attempts to build relationships and establish a working rapport with other groups in the company, as well as other activities aimed at addressing the organizational culture within the company as a whole).

Similar findings have been described by Theberge and Neumann (2010) in interviews with the Canadian HF/E professionals. They found that the process of actively advocating for ergonomics within organizations involved a variety of interactions and collaboration. Some of these practices included "political maneuvering," tailoring data collection and report presentations to clients' concerns, and "goal hooking" in order to make the case for implementing ergonomics in the workplace. Aside from normal, everyday activities such as providing HF/E training, carrying out HF/E evaluations, and using methods, much of the time of the practitioners was taken up in what Theberge and Neumann (2010) refer to as doing "organizational work."

HF/E Methods—Trade-offs and Compromises

The development of practical methods for applying HF/E in the workplace represents a core part of the discipline as a whole and has a long history (Stanton et al., 2013). In the last few years, there have been a number of discussions among researchers and practitioners regarding the reliability, validity, and utility of HF/E methods (e.g., Kanis, 2014). Many of these discussions are related to debates regarding the status of HF/E as a science, engineering discipline, or craft (Moray, 1995) and the nature of the research–practice gap (Chung and Shorrock, 2011). In the last few years, I have been involved in a set of activities related to the design and evaluation of methods for HF/E (Waterson et al., 2014, 2015).

The first of these activities involved looking back and reflecting on the uptake of a method for function allocation that a group of HF/E practitioners at Sheffield University had developed in the early 2000s. Function allocation typically refers to the allocation of tasks and responsibilities between human and machines. In 2002, we published a paper (Waterson et al., 2002) based on some work with the UK Navy, which described seven stages in using the method (from "formation of the overall view of the system" to "proposed allocations"), alongside a set of eight categories of social, organizational, and technical decision criteria that could be used to allocate functions between humans, machines, and teams. Each of the seven stages was accompanied by a set of tools for recording and evaluating function allocations. While a number of criticisms were made about the tools by Navy personnel (e.g., the need for an assessment of cost-effectiveness within "live" projects), it was generally well-received by our clients at the time it was developed. In the years that have followed the publication of the paper, we have monitored the number of times it has been cited (currently 51 times—Google Scholar, March, 2016). Comparing this with other, older attempts to develop methods for function allocation (e.g., Jordan, 1963; Fitts List—de Winter and Dodou, 2011), this rate of citation is modest and in many respects disappointing. Our method clearly has a long way to go in terms of popularity and use (especially given that citation rates are likely to be a poor proxy for actual usage).

One explanation for why the method has proved to be unsuccessful is that it was too complex and unwieldy for HF/E practitioners. They are likely to prefer methods that are more adaptable and tailorable to their work and immediate needs. Ironically, this was one of the key lessons from an earlier piece of work carried out by Clegg et al. (1996) at Sheffield University. A recent discussion with one of the Navy personnel involved in the early development of the method confirmed these suspicions. The method remains very much "on the shelf" and isn't being actively used within the Navy or in other applied contexts.

The second activity involved a program of work on the use of systemic accident analysis (SAA) methods in a number of domains including rail and aviation (Underwood and Waterson, 2013). The SAA methods are intended to cover a range of contributory factors leading up to an accident (e.g., regulatory, organizational, team, individual, and technical factors). Examples of these types of methods include Accimaps (Rasmussen, 1997), Systems-Theoretic Accident Modelling and Processes (STAMP) (Leveson, 2012), and Human Factors Analysis and Classification System (HFACS) (Shappell and Wiegmann, 2000). The findings from this work showed that SAA methods, although widely championed within academic settings, are rarely used in practice. A number of individual, organizational, and industry-related factors impact upon the degree to which accident investigators are aware of the SAA methods and/or decide to adopt and use them. These include perceptions that the methods are too theoretical or conceptual, that they are time consuming and impractical, the existence of a well-established model used within a specific sector or industry, the need to demonstrate accountability or fallibility in investigations, the previous experience of the analyst, and the amount of training required to learn to use the SAA method.

The final activity was carried out with a group of colleagues in the United States and Europe (Waterson and Catchpole, 2016). The focus of the work was on describing and evaluating current sociotechnical methods for workplace safety. Figure 3.3 shows examples of these methods and their development over time.

The motivations for this exercise were to outline a set of requirements that can be used to guide methodology development in the future. Part of the work involved an evaluation of the methods by rating their ability to address a set of theoretical and practical questions (e.g., the degree to which methods capture static/dynamic aspects of tasks and interactions between system levels). The evaluation was based on a set of ratings from each of the group members and a sample of other experts based in academia and industry, active in the field of sociotechnical systems design and safety. The outcomes from the evaluation highlighted a set of gaps as they relate to the coverage and applicability of current methods for sociotechnical systems and safety. As a whole, sociotechnical methods tended to be weakest at representing context and on usability, in terms of the resources required (expertise, analytic time) to apply the methods.

SOME STRATEGIES FOR BREAKING AWAY FROM THE PAST

If we want to build on the success of HF/E within new contexts (e.g., health care and patient safety), we need to stop "reinventing wheels" from our past. By this I partly mean learning lessons from past failures to improve HFI within industry, as well as capitalizing on the unique, interdisciplinary nature of HF/E.

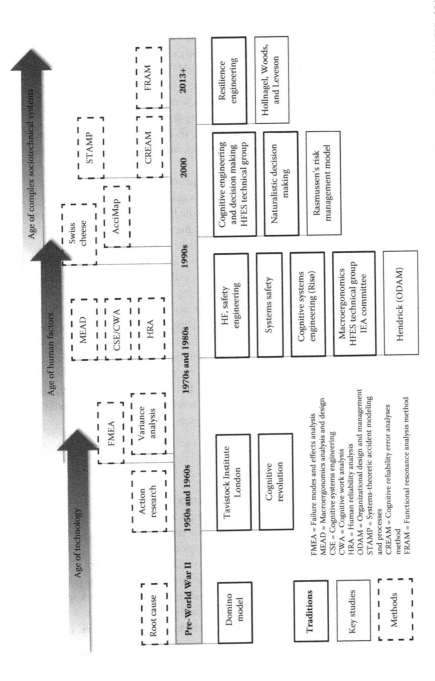

FIGURE 3.3 A timeline of the development of methods for sociotechnical systems and safety. (From Waterson et al., *Ergonomics*, 58(**4**), 565–599, 2015.)

How can this be achieved? Some concrete additional strategies might include the following:

- Acknowledging that HF/E is as much about "practice" as it is about "theory" or "science." Aside from the few studies reviewed earlier on in this chapter, we actually know very little about HF/E and its application outside academia. This is despite the fact that we have many people working in industry and government, many of whom would be willing to share their opinions, given the right forum. Something like this happens within specialist areas (e.g., the journal *Ergonomics in Design* targets practitioners and publishes case studies from their work), but could be extended within other mainstream academic publications.*

- Likewise, I would suggest that the academic community needs to pay more attention to what might be termed the "science of practice." How is HF/E used in industry? How does HF/E relate to the wider business context? How can we relate the practice of HF/E to larger questions concerning strategy in organizations? There are some promising directions out there (e.g., Dul and Neumann, 2009), but this could be taken much further.

- Finally, I think the importance of practice within HF/E needs to be taken into account within our education and training courses. I hear too often the comment from former students that "University lectures were useful when I studied HF/E, but it was a very different reality when I started working as a practitioner." How can we better prepare students for the "workaday world of HF/E" (Moran and Anderson, 1990; see also Chapter 12 by Baird and Williams)? Again, some of these concerns have been raised before by others (see for example Kirwan, 2000, 2012), but they could be taken further and might make for an interesting debate within the wider HF/E community. They are also addressed by other chapters in this book.

SUMMARY AND CONCLUSIONS

A number of patterns of continuity can be discerned across the various time spans of history within HF/E. Even in the stages of prehistory, well before HF/E came into existence, there is evidence of specialist work in areas such as anatomy, physiology, and psychology, which was later taken up and applied within industry. The Second World War appears to have acted as a catalyst breaking down barriers among these specialists and bringing them together. During the immediate postwar period, there seems to have been efforts by many of those involved in wartime work to maintain earlier collaborations and maintain momentum among the specialists. I doubt that at this time there was much enthusiasm for forming a new discipline or science.

---

* It is interesting to note that this used to happen in the past. The journal *Applied Ergonomics*, for example, used to publish industry case studies during its earliest days.

Rather, HF/E was more of an umbrella or portmanteau term that could be used to provide the various specialisms with an identity. In a sense, the term "ergonomics" appears to have been used as a label for the rather unique coming together of the groups of researchers drawn from a variety of disciplines and specialisms. Only much later did discussion begin about HF/E as a distinct body of knowledge or science in its own right.

A second development is the emergence of HF/E as a profession. This appears to have been unexpected among some of the people who founded HF/E after the War. It is summed up in a quote that is attributed to Murrell in the 1950s: "What the world needs is ergonomics not ergonomists" (Waterson and Sell, 2006, p. 782). Despite these reservations, the practice of HF/E has hugely expanded over the last 60 years and has been successful in shaping the course of a number of important areas including, among many others, occupational health and safety. Moreover, work that has examined HF/E within industry (i.e., HF/E in situ), alongside the realities of working as an HF/E practitioner, sheds further light on the importance of regarding HF/E as a discipline firmly grounded within practice (see also Kirwan, 2000, 2012; Chapter 4 by Hollnagel), as opposed to one specifically designed to meet the needs of the scientific or academic community.

To my mind, the argument that HF/E should always adhere to scientific standards such as the use of systematic methodology, replicability of results, reliability, and validly (Kanis, 2014) is misleading. If HF/E is a science, then it is above all an applied science and what this means is that it needs to consider utility and ease of use, alongside other criteria for judging the efficacy of HF/E methods and tools. The science base and the academic community need to be more aware of the realities of HF/E in practice. For example, many of the methods that are designed within universities and other research organizations are difficult to use, expensive, and often take much more time than that available within industry (see Shorrock and Williams, 2016; Chapter 11).

HF/E continues to expand into new areas of application and to absorb knowledge from other disciplines (e.g., sociology, anthropology, political science). I would argue that this is part of its fundamental nature—it is by definition *interdisciplinary* (Figure 3.4). The systems approach within HF/E is perhaps one of the strands that ties together present-day work in HF/E (Wilson, 2014). Arguments about the status of HF/E continue and will continue on for years to come. In many respects, one way of countering the previous quote from Murrell might be to argue that the present-day reality is that there really is no such thing as ergonomics, there are only ergonomists (cf. Gombrich's famous line about the nature of art [Gombrich, 1950]). Rather than drawing lines in the sand and continually arguing about identity and status, HF/E should celebrate what can be achieved by being inclusive (similar arguments have been put forward by the Nobel Laureate Peter Medawar in his discussion of the relation between pure and applied science [Medawar, 1982]). The example of the successes that came about as a result of wartime work shows what can be achieved if we forget for one moment disciplinary boundaries and work as a whole to solve common problems.

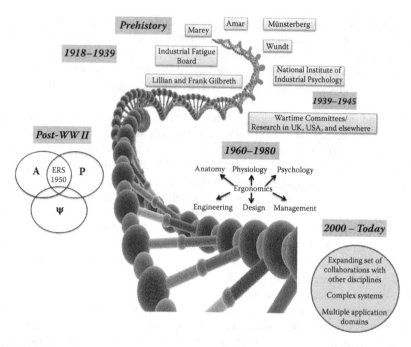

FIGURE 3.4   The "scientific DNA" of ergonomics over time.

ACKNOWLEDGMENTS

I'd like to thank Steven Shorrock, David Caple, and Erik Hollnagel for comments on an earlier draft of this chapter.

REFERENCES

Beevis, D. 2003. Ergonomics–costs and benefits revisited. *Applied Ergonomics*. 34, 491–496.
Caple, D. 2010. The IEA contribution to the transition of ergonomics from research to practice. *Applied Ergonomics*. 41(**6**), 731–737.
Chung, A.Z.Q. and Shorrock, S.T. 2011. The research-practice relationship in ergonomics and human factors – surveying and bridging the gap. *Ergonomics*. 54(**5**), 413–429.
Clegg, C.W., Coleman, P., Hornby, P., MacLaren, R., Robson, J., Carey, N., et al. 1996. Tools to incorporate some psychological and organizational issues during the development of computer-based systems. *Ergonomics*. 39(**3**), 482–511.
Cockayne, E. 2007. *Hubbub: Filth, Noise & Stench in England*. New Haven, CT: Yale University Press.
Cullen, L. 2007. Human factors integration–bridging the gap between system designers and end-users: A case study. *Safety Science*. 45, 6211–629.
Dul, J. and Neumann, W.P. 2009. Ergonomics contributions to company strategies. *Applied Ergonomics*. 40(**4**), 745–52.
Fitts, P.M. 1946. Psychological research on equipment design in the AAF. *American Psychologist*. 2, 93-98.
Floyd, W.F. and Welford, A.T. (eds.). 1953. *Fatigue*. London: H.K. Lewis.

Floyd, W.F. and Welford, A.T. (eds.). 1954. *Human Factors in Equipment Design*. London: H.K. Lewis.

Gombrich, E. 1950. *The Story of Art*. London: Phaidon Press.

Hartcup, G. 2000. *The Effect of Science on the Second World War*. London: Routledge.

Hearnshaw, L.S. 1987. *The Shaping of Modern Psychology*. London: Routledge and Kegan Paul.

Helton, W.S. and Kemp, S. 2011. What basic-applied issue? *Theoretical Issues in Ergonomics Science*. 12(**5**), 397–407.

Jordan, N. 1963. Allocation of function between men and machines in automated systems. *Journal of Applied Psychology*. 47, 161-165.

Kanis, H. 2014. Reliability and validity of findings in ergonomics research. *Theoretical Issues in Ergonomics Science*. 15(1), 1–46.

Kerr, M.P., Knott, D.S., Moss, M.A., Clegg, C.W., and Horton, R.P. 2008. Assessing the value of human factors initiatives. *Applied Ergonomics*. 39, 305–316.

Kirwan, B. 2000. Soft systems, hard lessons. *Applied Ergonomics*. 31(**6**), 663–78.

Kirwan, B. 2012. *Time for Some Human Factors Intelligence*. IEHF Institute Lecture, Blackpool, 2012.

Kourinka, I. 2000 (Ed.) *History of the International Ergonomics Association*. Santa Monica: IEA Press.

Leveson, N. 2012. *Engineering a Safer World*. Cambridge, MA: MIT Press.

Medawar, P. 1982. *Pluto's Republic: Incorporating the Art of the Soluble and Induction and Intuition In Scientific Thought*. Oxford: OUP.

Matthews, B.H.C. 1944. The effects of mechanical stresses on man. *British Medical Journal*, July 28, 114–117.

Matthews, B.H.C. 1945. The effects of high altitude on man. *British Medical Journal*, July 21, 75–78.

Meister, D. 1982a. The role of human factors in systems development. *Applied Ergonomics*. 13(**2**), 281–287.

Meister, D. 1982b. Human factors problems and solutions. *Applied Ergonomics*. 13(**3**), 219–223.

Meister, D. 1999. *The History of Human Factors and Ergonomics*. Boca Raton: CRC Press.

Meister, D. and Farr, D.E. 1967. The utilization of human factors information by designers. *Human Factors*. 9(1), 71-87.

Monod, H. 2000. Proto-ergonomics. In: Kuorinka, I. (ed.). *History of the International Ergonomics Association: The First Quarter of a Century*. Gréalou, France: IEA Press.

Moran, T. and Anderson, R.J. 1990. The workaday world as a paradigm for CSCW design. In: *Proceedings of the 1990 ACM Conference on Computer-Supported Cooperative Work*. New York: ACM, pp. 381–393.

Moray, N. 1995. Ergonomics and the global problems of the twenty-first century. *Ergonomics*. 38, 1691-1707.

Murrell, K.F.H. 1965. *Ergonomics: Man in His Working Environment*. London: Chapman and Hall.

Murrell, K.F.H. 1970. Reflections on the 21st anniversary of the society. *ERS News*, April.

Murrell, K.F.H. 1980. Occupational psychology through autobiography. *Occupational Psychology*. 53, 281–290.

Perrow, C. 1983. The organizational context of human factors engineering. *Administrative Science Quarterly*. 28(4), 521–541.

Rabinbach, A. 1992. *The Human Motor–Energy, Fatigue and the Origins of Modernity*. Berkeley, CA: University of California Press.

Rasmussen, J. 1997. Risk management in a dynamic society: A modelling problem. *Safety Science*. 27, 183-213.

Rodger, A. 1959. Ten years of ergonomics. *Nature*, 184, No. 4688, B.A., 20–22.

The Science Museum. 2004. Making the modern world. Accessed March 27, 2016. Available at: http://www.sciencemuseum.org.uk/visitmuseum/plan_your_visit/exhibitions/making_the_modern_world.

Shappell, S. and Wiegmann, D. 2000. *The Human Factors Analysis and Classification System–HFACS*. DOT/FAA/AM-00/7. Washington, DC: U.S. Department of Transportation, Federal Aviation Administration.

Shorrock, S. and Williams, C. 2016. Human factors & ergonomics methods in practice: Three fundamental constraints. *Theoretical Issues in Ergonomics Science*. 17(5-6), 468-482.

Singleton, W. 1982. *The Body at Work: Biological Ergonomics*. Cambridge: Cambridge University Press.

Stanton, N.A., Salmon, P.M., Rafferty, L.A., Walker, G.H., Baber, C., and Jenkins, D.P. 2013. *Human Factors Methods: A Practical Guide for Engineering and Design*. Farnham, UK: Ashgate.

Theberge, N. and Neumann, W.P. 2010. Doing 'organizational work': Expanding the conception of professional practice in ergonomics. *Applied Ergonomics*. 42(1), 76–84.

Underwood, P. and Waterson, P.E. 2013. Systemic accident analysis: Examining the gap between research and practice. *Accident Analysis and Prevention*. 55, 154–164.

Waterson, P.E. 2011. World War II and other historical influences on the formation of the Ergonomics Research Society. *Ergonomics*. 54(12), 1111–1129.

Waterson, P.E. 2014. Health information technology and sociotechnical systems: a progress report on recent developments within the UK National Health Service (NHS). *Applied Ergonomics*. 2, Part A, 150-161.

Waterson, P.E. and Catchpole, K. 2016. Human factors in healthcare: Welcome progress, but still scratching the surface. *BMJ: Quality and Safety*. 25(7), 480–484. doi:10.1136/bmjqs-2015-005074.

Waterson, P.E. and Sell, R. 2006. Recurrent themes and developments in the history of the Ergonomics Society. *Ergonomics*. 49(8), 743–799.

Waterson, P.E. and Eason, K.D. 2009. '1966 and all that': Trends and developments in UK ergonomics in the 1960's. *Ergonomics*. 52(11), 1323–1341.

Waterson, P.E. and Lemalu-Kolose, S. 2010. Exploring the social and organisational aspects of human factors integration: A framework and case study. *Safety Science*. 48, 482–490.

Waterson, P.E., Older-Gray, M., and Clegg, C.W. 2002. A sociotechnical method for designing work systems. *Human Factors*. 44(3), 376–391.

Waterson, P.E., Clegg, C.W., and Robinson, M. 2014. Trade-offs between reliability, validity and utility in the development of human factors methods. In: Broberg, O. Fallentin, N. Hasle P. Jensen, P.L., Kabel, A., Larsen, M.E., et al. (eds.). *Human Factors in Organizational Design and Management XI*. Santa Monica, CA: IEA Press, pp. 605–609.

Waterson. P.E., Robertson, M.M., Cooke, N.J., Militello, L., Roth, E., and Stanton, N.A. 2015. Defining the methodological challenges and opportunities for an effective science of sociotechnical systems and safety. *Ergonomics*, 58(4), 565–599

Wilson, J.R. 2014. Fundamentals of systems ergonomics/human factors. *Applied Ergonomics*. 45(1), 5–13.

de Winter, J.C.F. and Dodou, D. 2011. Why the Fitts list has persisted throughout the history of function allocation. *Cognition, Technology and Work*. 16(1), 1–11.

4 The Nitty-Gritty of Human Factors

Erik Hollnagel

CONTENTS

PRACTITIONER SUMMARY

Human factors (engineering) started as a solution to a practical problem, namely the challenges to human capabilities that came from uncontrolled technological developments. The practice of human factors was from the beginning justified by various theories about human functioning, specifically the analogy between humans and information processing machines. Through the following decades the dependence on theories increased so that today we are inundated with methods and solutions that are intellectually attractive but with limited practical effects. This chapter provides a condensed survey of this development, and argues that human factors as a practical solution should be based on a small number of simple principles with a strong empirical foundation.

INTRODUCTION

This chapter is about the nitty-gritty, the heart, or the essence of human factors.* But it is about the practical essence of human factors, not the theoretical one.

Human factors started from a recognition that an inadequate or insufficient fit between humans, the technology they used, and the environment or their work conditions could—and in fact did—negatively affect the overall system performance. (Conversely, an adequate fit and compatible working conditions could and should lead to improved performance.) The consequences could be seen in several ways, including diminished productivity, reduced safety (increased number of adverse outcomes), lower well-being, etc. Before the 1950s, the problem was traditionally solved by fitting humans to the requirements of the technology; human factors changed that by showing how the problem could be solved by fitting technology to humans. This was pointed out in the editorial foreword to one of the early reports on human factors:

> In the main, the objective has been to improve efficiency and safety by selecting, training, and otherwise conditioning human beings for adaptation to existing equipment and working conditions. Particularly in the design and operation of machines and tools, the emphasis has been upon the adjustment of man to mechanical equipment, rather than upon the adaptation of the machines and tools to compensate for the physical and mental limitations of man as a working instrument. During the past decade ... there has been growing withdrawal from this classical psychological viewpoint, and ever-increasing acceptance of the principle that "machines should be made for men; not men forcibly adapted to machines." This basic principle of "human engineering" is not new, particularly to industrial psychologists, although it has been consistently neglected, largely because engineers concerned with the development of the machine and scientists concerned with the individual who is to operate the machine have worked in almost complete insulation from one another.

> **(Fitts, 1951, p. iv)**

In order to "make machines for men"—or design technology for humans, as we would say today—it is necessary to know something about humans at work, or at least to make some assumptions about how humans function and behave in typical work situations. This can be done in two different ways, either by taking a theoretical approach based on models of human behavior as individuals and social beings, or by taking a pragmatic approach based on accumulated experience and recognizable patterns. The difference between a theoretical and a practical position has been recognized several times and given various names, for instance, "micro" and "macro" cognition (Hollnagel and Cacciabue, 1995; Klein et al., 2003), or "cognition in the mind" and "cognition in the world" (Hutchins, 1995). This chapter will take the pragmatic approach.

A THEORETICAL APPROACH TO HUMAN FACTORS

When the study of human factors began in the late 1940s, it was natural to look to experimental psychology for a scientific basis. At that time, experimental psychology had for more than half a century excelled in studying a large number of

---

* The common meaning of "nitty-gritty" is: the most important aspects or practical details of a subject or situation. The origin of the term is unknown, but the first usage is from around 1956.

phenomena related to human behavior and performance, although mostly under controlled conditions—*in vitro* rather than *in vivo*. There were thus a plethora of studies about visual and auditory perception and discrimination, remembering and forgetting, problem solving, learning, perceptual motor skills, attention and vigilance, etc. These findings established the scientific foundation of human factors engineering, as can be seen from the organization of the monumental compendium on human factors data that was published in 1988 (Boff and Lincoln, 1988).

As an illustration, consider the well-known limitations of human short-term memory—made famous by the "magical number seven" (Miller, 1956), which quickly came to be used as guidance for human factors design. That there is a limit to the number of items—names, thoughts, syllables, numbers—that we can keep in mind at the same time has been known for ages. That, taken as a fact, is therefore not controversial. The controversy appears when the limitations in our ability to remember things for a short period are explained by referring to some internal mechanism, for example, the capacity of a short-term memory. This confuses the phenomenon—the inability to keep track of more than a limited number of items at the same time—with the characteristics of a hypothetical mental, cognitive, or neurophysiological structure, the short-term memory. (On top of that, there was—and is—a lack of agreement on what the limit more precisely is, as well as how it should be explained, cf. Moray, 1970.) It also substitutes simple descriptions such as "to remember something," with complicated theory-laden accounts such as "to store something in short-term memory."

It is easy to find other cases where assumptions about hypothetical psychological or cognitive "mechanisms" have been used as the basis for artifact design. The most obvious examples are ecological interface design (Vicente and Rasmussen, 1992), situation awareness (Endsley, 1995), workload (Moray, 1982), arousal (Berlyne, 1960; Kahneman, 1973), and limited attention resources (Wickens, 1981). In each case, the concepts—or models—do correspond to very tangible characteristics of human behavior, but in each case the models also turn these characteristics into attributes of the mind in general or of specific types of hypothetical "information processing machinery." These attributes are gradually morphed into a folk model, defined as a set of assumptions about nonobservable constructs that conveniently are endowed with the necessary causal power, but lack any specification of the "mechanism" responsible for such causation (Dekker and Hollnagel, 2004). Since folk models rely on overgeneralizations and *ad hoc* substitutions, they are happily immune against falsification.

When a folk model becomes too powerful, such as the framework of skill-based, rule-based, and knowledge-based performance (SRK) (Rasmussen, 1986) or the levels of situation awareness (Endsley, 1995), its formal expression or the way it is articulated tends to dominate how it is seen and thereby gains a disproportionate influence on how it is used. (This happened to the SRK framework in the 1980s and to the three levels in the situation awareness models in the 1990s.) Looked at more analytically, the articulation is merely a decomposition of wholes into parts that takes place inside a closed system, hence basically a vacuous scholastic exercise. Models should refer to the tangible characteristics of human behavior on their own level, that is, as phenomena in their own right rather than as something that needs to be explained by a convenient substitution. We need, of course, some kind of conceptual framework to be able to decide what is relevant and what is not. This framework should also enable us to design things better, so that we are more likely to achieve the outcomes we want. To the extent that the conceptual framework can be used in practice to "engineer" rather than to "explain" human work, it has construct validity rather than just face validity.

A PRAGMATIC APPROACH TO HUMAN FACTORS

Though the current principles in human factors tend to go from theory to practice, starting from human information processing (cognition in the mind, microcognition) and going to human performance and behavior, a pragmatic approach to human factors does the opposite and goes from individual and collective human performance (cognition in the world or macrocognition) to cognition in the mind or microcognition. The benefit of taking a pragmatic rather than a theoretical approach was recognized many years ago when Neisser (1976) wrote that

> (w)e may have been lavishing too much effort on hypothetical models of the mind and not enough on analyzing the environment that the mind has been shaped to meet.

(Neisser, 1976, p. 8)

A similar concern was raised by Broadbent (1980, p. 113), who argued that "models should as far as possible come at the end of research and not at the beginning." The alternative to starting from a model would be to start from practical problems and concentrate on the dominant phenomena that are revealed in this way. Models should in all cases be minimized to avoid that choices become driven by the model rather than by practical problems. Broadbent emphasized that

> ... one should (not) start with a model of man and then investigate those areas in which the model predicts particular results. I believe one should start from practical problems, which at any one time will point us towards some part of human life.

(Broadbent, 1980, p. 117)

What we need, therefore, are articulated descriptions of the orderliness and regularity that we can find in everyday life and work. The purpose of such descriptions is to capture the recognizable patterns in actual behavior—both for work in a given setting and across work in different settings. It stands to reason that this regularity is determined by the conditions under which work takes place—including the artifacts we have created and that have become part of the environment—as much as it is determined by how the mind works.

Do Humans Actually Process Information?

The discussion about how much of behavior is determined by the "inner" mechanisms and how much by the situation has always been a central issue in academic psychology. One of the most outspoken supporters of human information processing once formulated it thus:

Human beings, viewed as behaving systems, are quite simple. The apparent complexity of our behavior over time is largely a reflection of the complexity of the environment in which we find ourselves.

(Simon, 1996, p. 53)

The purpose of this argument was to support a strong theory of the human mind and of human cognition, specifically that humans were nothing more than information processing systems. Complex human cognition was just a simple reflection of its environment, indeed almost a projection of the environment on the cognitive "mechanisms." But that actually meant, logically, that all we needed to understand was the relation(s) between humans and their (working) environments, which in turn meant the details of how this interaction took place (Card et al., 1983). The argument unfortunately missed the point that the environments were not given, hence not beyond our control. On the contrary, both physical and social environments are actually produced or created by us—by humans. This means that it becomes impossible—in theory as well as in practice—to separate the internal and the external determinants of performance, much to the regret of those practitioners who are looking for "quick and dirty" solutions.

LOOKING FOR PATTERNS

What we need to look for are the perceived patterns and inferred functional principles by which we can account for the regularity of human behavior (Woods and Hollnagel, 2006). This does not mean that we should look for the folk model concepts that "explain" something—or everything. On the contrary, the folk model concepts actually explain nothing since they usually just substitute one concept for another (for some examples, see Dekker and Hollnagel, 2004). We should instead look for articulated principles and concepts that enable us both to make sense or order out of the chaos—the eternal coping with complexity (Hollnagel and Woods, 2005)—and that enable us to predict performance, since prediction is crucial to design, improvement, and prevention. This is possible if we can recognize patterns in how people perform in work environments and use these patterns both to understand what happens (as pragmatic explanations) and to make predictions about what may happen in the future—even in the absence of a strong psychological theory of human behavior.

So rather than use a theory-driven approach, such as the SRK framework, we should use a practice-driven or pragmatic approach. People spontaneously, and perhaps instinctively, organize what they do in terms of goals and subgoals, and think about the means by which they can be achieved. (Since this was pointed out already by Aristotle in the Nicomachean Ethics, Book II.3.8, we may for all intents and purposes consider it as natural behavior.) We have therefore developed a common understanding that human activity (and therefore also the activity of autonomous and semi-autonomous artifacts) is organized in various ways, using terms such as "strategic" and "tactical" (Miller et al., 1960; Schützenberger, 1972). We also know that the realization of plans reflects the current situation, in particular current and anticipated demands. In that sense plans *guide* behavior, although they do not *completely determine* behavior. The fact that plans have to be adjusted as they are carried out emphasizes that we are talking about "cognition in the world" rather than "cognition in the mind" (Hollnagel, 2009).

THE CHALLENGE TO HUMAN FACTORS

The original challenge of human factors—to make machines for men—has not only continued to grow but also transformed in unexpected ways. Practices are never stable, but change continuously for a number of reasons. One reason is the rampant technological developments that lead to more data, more options, more modes and displays, more partially autonomous machine agents, and inevitably to greater operational complexity (Hollnagel and Woods, 2005). A second reason is the ever-growing demands of productivity and performance, the pressure to become "faster, better, cheaper" and thereby make full use of technological innovations. A third reason is what, for the lack of a better term, may be called human nature, which includes the limitations of our brains and bodies as well as our knack for creating the shortcuts and workarounds that make work feasible.

Figure 4.1 shows that these three factors are intertwined by their consequences. An arbitrary starting point is the technology potential that can be used to modify the way things are done, to introduce new functions altogether, and to compensate for the shortcomings of the human that are seen as the causes of unacceptable or substandard performance. Some familiar examples are the use of numerically controlled machines, industrial robots, computer-assisted design, flexible manufacturing, office automation, electronic exchange of data and funds, decision support systems, and the Internet. The growing technology potential is invariably seized upon and exploited to increase system functionality—which in turn leads to higher performance goals or efficiency pressures. This is referred to as the Law of Stretched Systems, originally suggested by Lawrence Hirschhorn:

> Under resource pressure, the benefits of change are taken in increased productivity, pushing the system back to the edge of the performance envelope.

(Woods and Cook, 2002, p. 141)

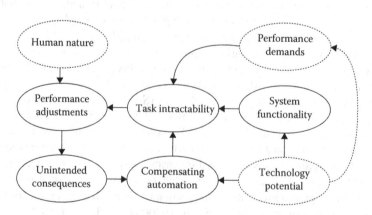

FIGURE 4.1 Circulus vitiosus of human factors.

The increased system functionality that is a consequence of the technology potential will, in combination with the necessary compensating automation and the increased expectations to performance, lead to work situations that are more complicated and to tasks that are increasingly intractable.

The more complicated work situations will require even more adjustments and compensation from the human, leading to more unintended (and unwanted) consequences, which in turn will increase the demand for compensation automation, etc.

Since our technological systems—including, one might add, our social institutions—often seem to be working at the edge of their capacity, technology potential is seen as providing a way both to increase capacity and to compensate for human shortcomings. This has been pointed out ever since Wiener criticized gadget worshippers who "regard(ed) with impatience the limitations of mankind, and in particular the limitation consisting in man's undependability and unpredictability" (Wiener, 1964, p. 53). The present-day version is "solutionism," defined as an intellectual pathology that acknowledges problems based on whether they are "solvable" with a nice and clean technological solution at our disposal (Morozov, 2013). Technology solutions may optimistically be applied in the hope of creating some slack or a capacity buffer in the system, thereby making it less vulnerable to internal or external disturbances. Unfortunately, the result invariably seems to obey the Law of Stretched Systems so that system performance increases to take up the new capacity, hence brings the system to the limit once more. Consider, for instance, the technological developments in cars and in traffic systems. If highway driving today took place at the speed of the 1940s, it would be very safe and very comfortable. But the highways are filled with a much larger number of drivers who all want to go faster, which means that driving has become more complex and more risky for the individual. (More dire examples can be found in power production industries, aviation, surgery, etc.)

WHAT IS HUMAN FACTORS REALLY ABOUT?

When we think about human factors, we think about humans at work—recognizing, of course, that the understanding of what constitutes work itself has undergone considerable change not least since the middle of the last century. Human factors is primarily interested in *how* people typically behave at work—but not necessarily *why* they perform in a certain way unless the *why* has predictive powers. While the scientific study of human factors in the United States dates from the 1940s—with ergonomics beginning in Europe a few years earlier*—the thinking about humans at work is not surprisingly somewhat older and goes back at least to Jastrzebowski (1857).

How to make work as efficient as possible has, of course, always been a goal and a concern. But until a century or so ago, it was not something that warranted specialized study and scrutiny. To understand why that is so, consider work in a different era, for instance during the Renaissance. One example is the printing of books, as shown in Figure 4.2. The drawing shows people from the beginning of the sixteenth

---

* The first volume of *Le Travail humain* was published in 1937.

FIGURE 4.2   Printing press 1520.

century at work using machines (technology). In principle, this situation should qualify for a human factors treatment because it involves work that has been specified, work that is done for pay or profit, and work that uses a machine that (presumably) has been designed by someone else. In other words, a work situation that is an artifact of the sociotechnical culture at the time rather than a natural one.

But just from looking at the work situation, it is easy to see why there was no need of a human factors analysis as we know it today. First of all, the work is simple to understand, even for those of us who have never worked a printing press. Everything is visible and there are no hidden "mechanisms" or functions. (In the modern terminology, we would say that the work is tractable.) No production pressures come from the machines that are used, since they work at the pace of the human who operates them—and who in fact provides the power to use them. (There may, of course, have been other production pressures, but work as it is shown here is paced by humans rather than by machines.) There is furthermore only one machine, the rest of the work being manual. The work activities are only loosely coupled and with little dependence on others, hence no couplings or interdependencies.

This differs starkly from the typical work situation around the middle of the last century, and of course from the typical work situation we can find today. Many contemporary work activities are intractable, meaning that the underlying principles are only partly known. Work may be complex because conditions of work are underspecified and unstable, and there is a high degree of interdependence on external systems. The various parts of work may be tightly coupled, even though the couplings may be difficult to detect and understand.

In order to "make machines for men," as argued by Fitts (1951) it seemed reasonable to think of humans as machines—and in the spirit of the ages, as information processing machines. This is another difference from the Renaissance and from the preindustrial periods in general. Before the middle of the seventeenth century it would be unthinkable to try to describe the human as a machine, for the very simple reason that there were no relevant machines to compare to. The first calculators were invented by Pascal and Leibnitz in 1642 and 1671, respectively, and the famous mechanical writer dolls were not constructed until a century later. The Newcomen steam engine was built in 1712, and La Mettrie published his famous essay on "L'homme Machine" as late as 1748. If there had been a need for a metaphor for the human in relation to work, it would probably not have been as a machine.

But when Fitts wrote the technical note quoted at the start of this chapter, the world was completely different. Although von Neumann (1958) had not yet written his book *The Computer and the Brain* (it was published posthumously), thinking of humans as some kind of computing machinery was commonly known at the time. Indeed, Fitts alluded to it in his famous comparison of humans and machines, using terms such as "information capacity" and "computation" (Fitts, 1951; see also Dekker and Woods, 2002).

DESIGN AND IMAGINATION

Work and work environments (work artifacts) are designed by someone to be used by someone else (usually quite anonymous). Though work environments throughout the history of mankind had evolved gradually and harmoniously, the second industrial revolution—the invention and large-scale use of steam engines—meant that work environments had to be considered and designed by people who were not themselves going to do the work. The situation turned even worse with the introduction of computing technology at the workplace, which became the third industrial revolution.

The first question for human factors is, of course, how to design work environments so that work is safe and efficient. But this leads to a second and equally important question, namely what people typically will do in a work environment that has been specified and designed by others? The design can address many issues. We design tools, or more generally, artifacts. We design the interaction and/or interfaces. We design work—both explicitly and implicitly (Hollnagel, 2003). We design roles and functions in an organization. We train people. We write guidelines, procedures, and instructions. We set down the conditions for organizing work, for planning and preparing, and for monitoring the outcome, for example, in terms of waste (lean), accidents (safety), or quality (statistical process control).

In doing all of this we have to consider or imagine not only how work *should* be done but and how it *will* be done. How it should be done is probably the easiest of the two. Here is it possible to base the design on the purpose of the activity and on the necessary constraints and dependencies, cf., the many advices about task and work design from Taylor (1911) to Wilson and Corlett (2005) or Salvendy (2012).

While there are many guidelines and design principles that represent a theoretical approach to human factors and specific models of human functions, there are relatively few that are based on what people actually do, on the recurrent patterns of behavior. This may be because we often find that these patterns, work-as-done, differ from what should be done or from what we imagined that people would do. When this happens, usually in the context of something having gone wrong, we account for the difference by inferring that what people actually did was wrong—an error, a failure, a mistake—hence that what we thought they should have done was right. We rarely consider that it is our imagination, or idea about work-as-imagined, that is wrong and that work-as-done in some basic sense is right. (To be more specific, work-as-done differs from what we imagine *we* would do in the same situation, but thinking about it from afar and assuming more or less complete knowledge.) Yet we neglect these patterns at our peril.

Listing or naming all, or even the most frequent, patterns that can be found in human performance is beyond the scope of this chapter. But to illustrate the idea, five patterns or principles will be presented and briefly discussed in the following. No claim is made that they are the most important ones, but they are certainly all realistic. It is a sad fact, however, that even though we all know them in a general way—and know them to be true—and even though we easily recognize them in practice, they are in most cases either disregarded or violated by human factors design. The five principles presented here may be an initial step toward a pragmatic approach to human factors, or just an example of the nitty-gritty that must be adhered to.

FIRST PRINCIPLE: TRADE-OFFS AND WORKAROUNDS

To carry out work, individually or collectively, humans must adopt simplifying assumptions. When we reason about how the world works, we tend to reason by analogy, by referring to something that we know well. This is usually deplored because the simplifications may limit the precautions humans take and the range of undesired consequences they envision. These simplifications are often seen as the source of the "human errors" that allow anomalies to accumulate and undesired consequences to grow more serious. In the extreme, it has been proposed that humans and organizations may be defined by what they ignore rather than by what they attend to (Weick, 1998). The simplifications that humans make have also been seen as an expression of the inherent human tendency to do things as easily as possible, for instance in the assumption of Theory X that individuals are inherently lazy (McGregor, 1960).

But the simplifying assumptions may also been interpreted differently.* We rarely, if ever, have the time, the information, the means, or the energy to consider every detail and every aspect of what we are about to do. To get through the day we are therefore forced willy-nilly to make simplifying assumptions and to trade-off thoroughness for efficiency (Hollnagel, 2009). The very design of the work

---

* We may note in passing that the preference for the folk models in human factors described above can be seen as the use of simplifying assumptions on a higher level.

situation and the artifacts we must employ may also require us to make adjustments either as "system tailoring" or "task tailoring" (Cook et al., 1991) or even goal substitution (Koopman and Hoffman, 2003). Indeed, most of these imperfections are due to trade-offs made in the design of artifacts and specification of working procedures. And even if the "objective" work situation did not make trade-offs and workarounds necessary, the social and organizational work situation would (e.g., Homans, 1958).

The trade-offs can be found as heuristics in judgment and decision making (Tversky and Kahneman, 1974), as workarounds in decision making such as satisficing (Simon, 1956) or "muddling through" (Lindblom, 1959), as "unsafe" practices due to company pressure (Choudry and Fang, 2008), and as temporary solutions to pressing problems (Koopman and Hoffman, 2003). They are not only an unavoidable part of everyday work—of work-as-done—but they are also indispensable. Whenever human factors engineering has recognized their existence, the response has often been to try to prevent them or overcome them. Instead we should try to understand them better, and to facilitate them when in the majority of cases they are beneficial rather than harmful.

SECOND PRINCIPLE: THE MINIMAL ACTION RULE

It is nearly a corollary of the first principle that the design of tools, work, and interfaces should strive to allow work to be done without unnecessary effort, whether physical or mental. In other words, simple things should be simple to do. In human factors, this has often been expressed by requiring that systems (tools, artifacts, work environments, etc.) are user-friendly or even intuitive to understand. The emphasis on user-friendliness came to the fore when the use of computers became indispensable for work to be done, both when they were part of the equipment or machinery, as in numerically controlled tools, or when work itself required the direct use of computers such as in human–computer interaction. Today there are few types of work that do not include the use of computing technology in one way or another, from everyday tasks such as buying a train ticket, making a telephone call, or getting money from an ATM to professional duties such as air traffic management or making a CAT scan of a patient.

It is straightforward to formulate the second principle as a set of concrete design rules. One rule could be that the use of tools and equipment should not introduce any unnecessary delays. A second, that it should be obvious how the work should be done, so that people can focus on performing the task rather than on figuring out how the tools or (these days) the interface works. A third, that the possibilities for misunderstandings should be eliminated as far as possible.

The importance of making simple things simple to do is, however, older than the use of computing technology. One example is the research in the United Kingdom in the 1960s to design postcodes that were easy to remember (Conrad, 1967). While this may not be considered work at all, it is certainly an activity where simple things should be simple to do. Another equally mundane example is the size of coins. Here it would seem reasonable if there was a simple relation between the

TABLE 4.1

Size–Value Relations of US Coins

Value	Penny (1 cent)	Nickel (5 cents)	Dime (10 cents)	Quarter (25 cents)	Half dollar (50 cents)
Diameter (mm)	19.05	21.21	17.91	24.26	30.61

size of a coin and its value (Smith et al., 1975). For some currencies, such as the Euro, this is indeed the case. For others, such as the US dollar, it is not (Table 4.1).

More generally, the minimal action rule can be expressed as follows:

- Task descriptions and task sequences should be short. Try to keep the number of user actions in a sequence to a minimum.*
- Break up long task sequences into subsequences.
- Make sure that the user does not have to perform two complex tasks together.
- Action sequences for different tasks should be unique. Try to avoid similar opening actions. Where this is unavoidable draw the users' attention to the point at which the sequences diverge.

A logical consequence of the second principle is that actions that might lead to adverse outcomes should be difficult to do so. Similarly, really dangerous things should be impossible to do, except under very special circumstances. A simple example of that is the use of functional barriers, such as confirmation dialogues when deleting something, or the use of passwords.

The two views, however, sometimes clash. Consider, for instance, opening the front door of your house (or the door of your car). From one perspective, it should be easy to do, so that you can enter without too much effort. Yet on the other hand it should be difficult to do, so that unauthorized persons cannot open the door. Many modern cars have keyless entry, which means that you can open the doors and start the car if the key—or some equivalent gadget—emits a proper electronic signal. One effect of that is that car thieves prefer luxury cars with keyless locking systems (Osborne, 2014).

THIRD PRINCIPLE: FORM SHOULD MATCH FUNCTION AND VICE VERSA

The third principle means that the physical appearance of an artifact, whether a simple tool, a specialized control panel or dashboard, a specialized working environment, or a complicated information display, should match the way it is to be used.

---

* This is possibly a corollary of the Scientific Management principle, cf. Taylor (1911).

This applies to physical qualities such as size, shape, grip, color, etc., as well as to how it functions and how it is controlled.

A simple example is the hammer. Here there are different hammers for different uses, as any do-it-yourself catalogue will show. A hammer for a blacksmith is large with a blunt head and peen, and weighs 2 kg or more. A hammer for an upholsterer is small with both head and peen tapered and a weight of around 200 g. Both are well-suited to a specific use, but will be inefficient or even impossible for other applications.

Another common example is the steering wheel in a car. This used simply to be a steering wheel, which needed to have a certain size corresponding to the type of car—at least before power steering became common. That is why a steering wheel of a bus or a truck was (and still is) larger than that of a private car. Initially, the function of a steering wheel was just to turn the front wheels, to change direction. But as cars have become more sophisticated, other functions have been put on the steering wheel, for instance to control the radio (or the infotainment system), to control the telephone (hands-free, of course), to use the cruise control, to select displays, etc. (My current car has no less than 17 controls on the steering wheel.) The extreme case is a Formula 1 car, where literally the whole instrumentation panel (displays and controls) have migrated to the steering wheel. This is, of course, for the very good reason that the driver neither has the time to look elsewhere, nor to move his hands into a different position. In this way form matches function.

A more elaborate example is the improvement of control rooms following the accident at the nuclear power plant on Three Mile Island (TMI) in 1979. Starting in 1974, the nuclear industry carried out a comprehensive evaluation of control room design, which concluded that design was driven by engineering rather than human factors concerns. The evaluation study found that major control functions were not prominently identified, that it was difficult to relate these functions to individual plant components, and that operators could not get the needed information at a glance but had had to read individual labels painstakingly (Hanes et al., 1982). The sorry state of affairs is illustrated by the control panel layout shown in Figure 4.3a. In spite of these findings, little was done until the accident at TMI made major improvements necessary. A simple solution to some of the problems was to apply the principle of "paint, tape, and label," where lines, labels, and shading were used to group major control functions and make functional relationships easier to identify (cf. Figure 4.3b).

Many other examples could easily be given, for instance that instructions and devices do not always match, perhaps because devices may be modified after the instructions have been written. A final illustration is provided by the need to find icons for the ever-growing number of functions (or apps). How many would know, for instance, that the three icons in Figure 4.4 are supposed to mean "reorder," "group work," and "highlight remove," respectively?

The bottom line is that the artifacts that we use, and in many cases must use, should be designed to fit the activity they are intended for. This applies both to the physical qualities (size, shape, color, weight, etc.), and to the perceptual and semantic qualities (visual distinctiveness, arrangement, Gestalt, ambiguity, etc.). This very fundamental requirement is, unfortunately, often disregarded.

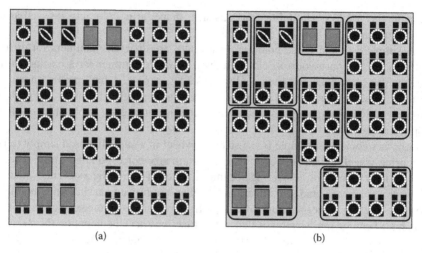

FIGURE 4.3   A control panel (a) before and (b) after the application of paint, tape, and label.

FIGURE 4.4   Three icons, but what do they mean?

FOURTH PRINCIPLE: WHAT YOU LOOK FOR IS WHAT YOU SEE

It is a common misconception that the human mind is like a camera in the sense that it passively registers what the visual field contains. In other words, that we see what is there. In consequence of this misconception, designers may believe that any information that is put on a display or an interface will be noticed. (This is, of course, subject to the condition that the design does not get into conflict with well-established principles such as the Weber–Fechner law or the signal–noise ratio.) Nothing could, however, be further from the truth. Perception is not an objective projection of what is "out there" onto the mind, but rather a narrow selection of features that fit into the current situation. In other words, instead of assuming a "what-is-there-is-what-you-see" principle, we should acknowledge a "what-you-look-for-is-what-you-see" principle.

This simple fact is vividly illustrated by the phenomenon known as change blindness (e.g., O'Regan and Nöe, 2001), which refers to situations where people fail to see large changes that normally would be obvious. A number of studies have shown that under certain circumstances, very large changes can be made in a picture without observers noticing them. All that is required is that the changes are arranged to occur simultaneously with some kind of extraneous, brief disruption in visual continuity, such as an eye saccade, a shift of the picture, a brief flicker, a "mudsplash," or an eye blink, or a film cut in a motion picture sequence. [The latter may be related to the concept of visual momentum as described by (Woods 1984).] Change blindness suggests that humans rely on a combination of external and internal representation of the "world," and that this must be taken into account when designing artifacts for work, in particular displays. Therefore, assuring that "all the information is there" is no guarantee that it will be noticed or used.

Change blindness refers to the fact that we sometimes fail to notice a change, even when it is obvious—at least once it has been explained to us. Similarly, because we usually focus our attention on one part of the scene in front of us, we thereby fail to notice what else may be present or may happen. This phenomenon is called inattentional blindness. Probably the most spectacular illustration of that was the experiment where people were asked to watch two teams of three players passing basketballs. Approximately half the subjects failed to notice a gorilla that walked across the screen (Simons and Chabris, 1999). The robustness of this phenomenon was later demonstrated by Drew et al. (2013), this time with radiologists in a lung-nodule detection task. In practice, the same phenomenon can be found in the "looked but failed to see" phenomenon in traffic (Brown, 2002).

A companion to the "what-you-look-for-is-what-you-see" principle is the "what-you-see-is-what-you-do" principle. Clearly, just as we cannot see something that we do not look for (and we do not even see everything that we look for), then we cannot do something that we have not seen as being useful. The technical label for seeing how something can be used is *affordance*. The term was introduced by Gibson (1977) and explained in greater detail in Gibson (1979). Affordances are defined as the "action possibilities" latent in the environment or "what things can be used for." In Gibson's definition, affordances are independent of the individual's ability to recognize them, but in practice it is more interesting to study what people think that something can be used for—or what the intended use is. This is the essence of design—that the designer's ideas about how something should be used are easy to see and understand. Thus a chair is seen as something one can sit on, while a ball is seen as something that can be thrown. (If a ball is so large that it looks like something to sit on, as in Pilates training, it is also so large that it no longer looks like something that can easily be thrown.) But while a chair is designed to be sat on, other artifacts may be seen as having the same affordance even though they do not.

In most cases where a design works well it is because of long traditions rather than strong theories. Consider, for instance, the teapot and teacup shown in Figure 4.5. The handles on a tea set, on the cup as well as on the pot, provide an obvious

FIGURE 4.5   Affordances.

affordance for grasping and holding the object by hand. The rim of the cup also provides the affordance of drinking, while the rim of the pot does not, not even when the lid is off. In this case there is therefore a good agreement between the intended use and the perceived use—the affordances.

An example of the opposite is provided, for instance, by a door with a large handle and the label "push." Here the sign contradicts the handle, which clearly is there to grip something (the door) and pull it. An even worse, but unfortunately common, example is provided by artifacts where the controls are out of sight. The simplest instance of that is the controls of modern office chairs, which usually are underneath the seat, hence hidden from sight. Regulating an office chair is therefore often a case of trial-and-error, even if it is one you use daily.

FIFTH PRINCIPLE: SHOW WHAT IS GOING ON

A final principle, at least for the present, addresses the actual work situation rather than the preparations for it or the design of it. Work is characterized by being an ongoing activity or an ongoing process that produces something. (This is so regardless of whether the purpose is to assemble a car engine in a factory, get money from an ATM, or write a book chapter.) To manage or control an ongoing activity, it is necessary to know what is happening, or in technical terms to get feedback. Whenever something is done, it is necessary to find out whether the chosen intervention had the intended consequences. In the absence of that, interventions become opportunistic, if not completely random.

Feedback is originally a technical concept that can be defined, for example, as the furnishing of data concerning the operation or output of a machine to an automatic control device or to the machine itself, so that subsequent or ongoing operations of the machine can be altered or corrected. Feedback featured prominently in the basic paradigm for human performance, namely the Test-Operate-Test-Exit or TOTE (Miller et al., 1960). Here the feedback was the outcome of the testing phase that was used to control the operation, so that any differences between the actual state and the intended state were reduced. From a human factors perspective, feedback can be defined as knowledge of the results of any action, where the knowledge is used to influence or modify the following actions.

One basic purpose of feedback is to show that something is happening. This is usually not a problem when working with predominantly technological systems, but it can be an issue when computing technology is involved. There are probably few among us who have not experienced a situation where nothing seems to happen even though we have pressed the right buttons. Typically, the lag in the response tempts us to repeat the command, which usually only makes the situation worse. Another purpose of feedback is to show how fast something is happening, as in the (in)famous status bar when a computer system loads or retrieves something. (While most people experience this with personal computers, irregularities in feedback are unfortunately also common in workplace uses of information technology—such as hospitals, aircraft, etc.) Here one human factors problem is that the progress is not linear, and that the indicated time remaining sometimes increases rather decreases, contrary to expectations. If that happens, the feedback becomes unreliable, that is, it cannot be used to manage or control the process.

The importance of feedback is clearly illustrated by situations where it is delayed. (It would of course also be illustrated by situations where it was missing; in those cases actions would, however, be "blind," i.e., done without any knowledge of what effect they had, which no real work situations could tolerate.) A dramatic example of that is in the steering of supertankers: "A fully loaded supertanker is a nonlinear system which responds very slowly to changes in the rudder position. Moreover, in general it is unstable, that is, it has a tendency to start turning either to the left or to the right. These properties make a supertanker very hard to handle" (Veldhuyzen et al., 1972). The delay can be so large that the rudder must be returned to the neutral position before the ship has started to change direction. While few of us will have the opportunity to steer a supertanker, we can try to imagine a similar situation for driving a car. What would happen, for instance, if there was a lag of 5 s from you turned the wheel and until the car started to change direction?

A similar example, which most of us have experienced in practice, is the delayed feedback in regulating the temperature of water, for instance when taking a shower. At home, you know what the delay is. But when you stay in a hotel, it may sometimes be difficult to achieve the correct temperature in the shower, if the delay is too large. (It does help, of course, if the temperature control clearly shows how to control the temperature—unlike the one shown in Figure 4.6.) To be effective, the arrival of the feedback must be synchronous with the process, hence happen within the 3 s that seem to constitute the "now" (Fairhall et al., 2014).

Human factors, as a discipline, has traditionally paid considerable attention to the design of the work environment, not least of interfaces to technology and computing machinery. But a well-designed or even beautiful interface is of little, if any, use if the processes behind it, the processes that are part of work, are irregular or unpredictable. Precise and timely feedback is important for all levels of human performance, from sensory-motor tracking to double-loop learning. While this is relatively obvious for the former, it is no less critical for the latter because you may have forgotten what the situation was by the time you get the feedback. Although this usually is treated as an organizational rather than a human factors issue, we should acknowledge that efforts to ensure safe and productive work take place on a long time scale as well as on a short time scale. Branding something as an organizational issue by no means reduces the importance of precise and timely feedback.

FIGURE 4.6   Shower control.

A BIT OF PHILOSOPHY

The very fact that there are five rules and not n is an expression of a metaprinciple, namely that there should be minimal requirements to knowledge that cannot easily be associated to something in the work. When designing a work situation, things should be kept simple and manageable to avoid overload, competing tasks, and overly complex structures.

The five principles have been proposed to help in the design of good working environments. But in doing so we should also consider how people possibly might remember the principles, or at least remember how many principles there are. Information processing psychology might recommend—or be seen as recommending—that there should be seven, in honor of George Miller's seminal paper (Miller, 1956). In that case, the number would be based on an assumption about the inner mechanisms and characteristics of the human mind, to wit that short-term memory has a capacity limitation around seven (although this easily can be disputed).

Relying on the characteristics of human memory corresponds to the notion of "cognition in the mind." In other words, what we do is based on our hypotheses about the inner workings of the human mind. There is, however, an alternative that appropriately enough can be called "cognition in the world"—or "cognition in the wild" (Hutchins, 1995). This position argues that whatever we do is the combined result of the processes and functions of the mind, and the physical world in which we find ourselves, in particular the tools that may be available and be used as part of our work.

Following this metaprinciple, we should be able to use something in the environment to remember the number of principles. The "memory" of the value of n should in other words be "in the world" rather than "in the mind." This provides a good reason for setting $n = 5$, since most people have five fingers. (And those who, for one reason or another, have a different number, still know that five is the normal number.) This does not tell us what the principles are, but it does tell us that there are five, and we therefore also know whether we have remembered them all—or even whether we have "remembered" one too many.

In this way, the number of principles illustrates the fundamental fact that everything we do takes place in a context, makes use of that context, and is affected—positively or negatively—by that context. We, as humans, have been extremely effective in shaping our environment, and in particular to design artifacts that can be used as part of work. Human factors must never forget that it is about work-as-done and not about work-as-imagined.

REFERENCES

Berlyne, D.E. 1960. *Conflict, Arousal, and Curiosity*. New York, NY: McGraw-Hill Book Company.

Boff, K.R. and Lincoln, J.E. (eds.). 1988. *Engineering Data Compendium: Human Perception and Performance*. Dayton, OH: Wright-Patterson Air Force Base.

Broadbent, D.E. 1980. The minimization of models. In: Chapman, A.J. and Jones, D.M. (eds.). *Models of Man*. Leicester: The British Psychological Society.

Brown, I.D. 2002. A review of the 'looked but failed to see' accident causation factor. In: *Behavioural Research in Road Safety: Eleventh Seminar*. London, UK: Department for Transport, Local Government & the Regions, pp. 116–124.

Card, S.K., Moran, T.P. and Newell, A. 1983. *The Psychology of Human-Computer Interaction.* Hillsdale, NJ: Lawrence Erlbaum Associates, Inc.

Choudhry, R.M. and Fang, D. 2008. Why operatives engages in unsafe work behavior: Investigating factors on construction sites. *Safety Science.* 46(4), 566–584.

Conrad, S. 1967. Designing postal codes for public use. *Ergonomics.* 10(2), 233–238.

Cook, R.I., Woods, D.D., McColligan, E., and Howie, M. 1991. *Cognitive Consequences of Clumsy Automation on High Workload, High Consequence Human Performance.* In NASA, Lyndon B. Johnson Space Center, Fourth Annual Workshop on Space Operations Applications and Research (SOAR 90), pp. 543–546 (SEE N 91-20702 12–59).

Dekker, S.W.A. and Hollnagel, E. 2004. Human factors and folk models. *Cognition, Technology & Work.* 6(2), 79–86.

Dekker, S.W.A. and Woods, D.D. 2002. MABA-MABA or abracadabra? Progress on human–automation co-ordination. *Cognition, Technology & Work.* 4, 240–244.

Drew, T., Võ, M.L.-H-., and Wolfe, J.M. 2013. The invisible gorilla strikes again: sustained inattentional blindness in expert observers. *Psychological Science.* 24, 1848–1858.

Endsley, M.R. 1995. Toward a theory of situation awareness in dynamic systems. *Human Factors.* 37, 32–64.

Fairhall, S.L., Albi, A., and Melcher, D. 2014. Temporal integration windows for naturalistic visual sequences. *Plos One.* 9(7): e102248.

Fitts, P.M. (ed.). 1951. *Human Engineering for an Effective Air-Navigation and Traffic-Control System.* Washington, DC: National Research Council, Division of Anthropology and Psychology.

Gibson, J.J. 1977. The theory of affordances. In: Shaw, R. and Bransford, J. (eds.). *Perceiving, Acting, and Knowing.* New York, NY: Wiley.

Gibson, J.J. 1979. *The Ecological Approach to Visual Perception.* Hillsdale, NJ: Lawrence Erlbaum.

Hanes, L.F., O'Brien, J.F., and DiSalvo, R. 1982. Nuclear power: Control-room design: Lessons from TMI: Improved instrumentation plus computer-based decision aids under development should help minimize the potential consequences of accidents. *IEEE Spectrum.* 19(6), 46–53.

Hollnagel, E. (ed.). 2003. *Handbook of Cognitive Task Design.* Mahwah, NJ: USA: Lawrence Erlbaum Associates.

Hollnagel, E. 2009. *The ETTO Principle: Efficiency-Thoroughness Trade-Off. Why Things That Go Right Sometimes Go Wrong.* Aldershot: Ashgate.

Hollnagel, E. and Cacciabue, P.C. 1995. Simulation of cognition: Applications. In: Hoc, J.M., Cacciabue, P.C., and Hollnagel, E. *Expertise and Technology: Cognition and Human-Computer Cooperation.* Hillsdale, NJ: Lawrence Erlbaum Associates.

Hollnagel, E. and Woods, D.D. 2005. *Joint Cognitive Systems: Foundations of Cognitive Systems Engineering.* Boca Raton, FL: CRC Press/Taylor & Francis.

Homans, G.C. 1958. Social behavior as exchange. *American Journal of Sociology.* 63(6), 597–606.

Hutchins, E. 1995. *Cognition in the Wild.* Cambridge, MA: MIT press.

Jastrzebowski, W. 1857. Rys ergonomiji czyli Nauki o Pracy, opartej naprawdach poczerpnietych z Nauki Przyrody (An outline of ergonomics or the science of work based on the truths drawn from the science of nature). *Przyoda i Przemysl.* 29, 227–231.

Kahneman, D. 1973. *Attention and Effort.* Englewood Cliffs, NJ: Prentice-Hall.

Klein, G., Ross, K.G., Moon, B.M., Klein, D.E., Hoffman, R.R., and Hollnagel, E. 2003. Macrocognition. *IEEE Intelligent Systems.* 18(3), 81–85.

Koopman, P. and Hoffman, R.R. 2003. Work-arounds, make-work, and kludges. *IEEE Intelligent Systems.* 18(6), 70–75.

Lindblom, C.E. 1959. The science of "muddling through." *Public Administration Review.* 19(2), 79–88.

McGregor, D. 1960. *The Human Side of Enterprise.* New York, NY: McGraw Hill.

Miller, G.A. 1956. The magical number seven, plus or minus two: Some limits on our capacity for processing information. *Psychological Review.* 63, 81–97.

Miller, G.A., Galanter, E. and Pribram, K.H. 1960. *Plans and the Structure of Behavior.* London: Holt, Rinehart and Winston.

Moray, N. 1970. Where is capacity limited? A survey and a model. In: Sanders, A.F. (ed.). *Attention and Performance I.* Amsterdam: North-Holland Publishing, pp. 84–92.

Moray, N. 1982. Subjective mental workload. *Human Factors.* 24(1), 25–40.

Morozov, E. 2013. The perils of perfection. *The New York Times.* 2 March. Accessed March 12, 2016. Available from: http://www.nytimes.com/2013/03/03/opinion/sunday/the-perils-of-perfection.html.

Neisser, U. 1976. *Cognition and Reality.* San Francisco, CA: W.H. Freeman.

O'Regan, J.K. and Noë, A. 2001. A sensorimotor account of vision and visual consciousness. *Behavioral and Brain Sciences.* 24(5), 939–973.

Osborne, H. 2014. Thieves target luxury Range Rovers with keyless locking systems. *The Guardian.* 27 October. Accessed March 12, 2016. Available at: http://www.theguardian.com/money/2014/oct/27/thieves-range-rover-keyless-locking.

Rasmussen, J. 1986. *Information Processing and Human-Machine Interaction.* Amsterdam: North Holland.

Salvendy, G. (ed.). 2012. *Handbook of Human Factors and Ergonomics.* Fourth Edition. Hoboken, NJ: Wiley.

Schützenberger, M.P. 1972. A tentative classification of goal-seeking behaviours. In: Emery, F.E. (ed.). *Systems Thinking.* Harmondsworth, UK: Penguin Books.

Simon, H.A. 1956. Rational choice and the structure of the environment. *Psychological Review.* 63(2), 129–138.

Simon, H.A. 1996. *The Sciences of The Artificial,* Third Edition. Cambridge, MN: The MIT Press.

Simons, D.J. and Chabris, C.F. 1999. Gorillas in our midst: sustained inattentional blindness for dynamic events. *Perception.* 28, 1059–1074.

Smith, H.V., Fuller, R.G.C. and Forrest, D.W. 1975. Coin value and perceived size: A longitudinal study. *Perceptual and Motor Skills.* 41, 227–232.

Taylor, F.W. 1911. *The Principles of Scientific Management.* New York, NY: Harper.

Tversky, A. and Kahneman, D. 1974. Judgment under uncertainty: heuristics and biases. *Science.* 185(4157), 1124–1131.

Veldhuyzen, W., van Lunteren, A. and Stassen, H. 1972. Modelling the helmsman of a supertanker: Some preliminary experiments. *Proceedings of the Eighth Annual Conference on Manual Control,* 1972, Ann Arbor Michigan May, pp. 17–19.

Vicente, K.J. and Rasmussen, J. 1992. Ecological interface design: Theoretical foundations. *IEEE Transactions on Systems, Man and Cybernetics,* 22(4), 589–606.

von Neumann, J. 1958. *The Computer and the Brain.* Yale University Press.

Weick, K.E. 1998. Foresights of failure: An appreciation of Barry Turner. *Journal of Contingencies and Crisis Management.* 6(2), 72–75.

Wickens, C.D. 1981. *Processing Resources in Attention, Dual Task Performance, and Workload Assessment.* Fort Belvoir, VA: Defense Technical Information Center.

Wiener, N. 1964. *God & Golem, Inc. A Comment on Certain Points Where Cybernetics Impinges on Religion.* Cambridge: The MIT Press.

Wilson, J. and Corlett, N. (ed.). 2005. *Evaluation of Human Work,* Third Edition. Boca Raton, FL: CRC Press.

Woods, D.D. 1984: Visual momentum: A concept to improve the cognitive coupling of person and computer. *International Journal of Man-Machine Studies.* 21, 229–244.

Woods, D.D. and Cook, R.I. 2002. Nine steps to move forward from error. *Cognition, Technology and Work.* 4, 137–144.

Woods, D.D. and Hollnagel, E. 2006. *Joint Cognitive Systems: Patterns in Cognitive Systems Engineering.* Boca Raton, FL: CRC Press.

5 Human Factors and the Ethics of Explaining Failure

Roel van Winsen and Sidney W. A. Dekker

CONTENTS

PRACTITIONER SUMMARY

The idea that human performance is systematically connected to the features of people's tools and tasks effectively constitutes the birth of human factors. However, accidents are often still seen as the result of "human error," either at the sharp operational end or the blunt organizational end. This chapter aims to point out some practical and ethical implications of "human error" (and its subcategories) as an explanation for why sometimes sociotechnological systems fail. In briefly discussing some "costs" of relying on this reductionist approach to explaining and dealing with failure, we argue that as a human factors community we need to engage in (ethical) discussions about, and take responsibility for, the effects of the practices that we promote.

INTRODUCTION

Karl is in jail today and human factors might have had something to do with it. Karl Lilgert was the officer in charge of the large passenger ferry "Queen of the North" when it ran aground and sank. Ninety-nine passengers and crew survived the sinking of the ferry. Unfortunately, two passengers were never found and presumed to have drowned. Despite Karl facing poor weather, substandard navigational equipment, inadequate company policies, and a lack of staff on the bridge, he was essentially blamed for the failure to turn the ferry in time (Keller, 2013). Karl was criminally prosecuted, particularly as the "causes and contributing factors" in the accident investigation report strongly suggested that various distractions contributed to his failure to order the required course change (Transportation Safety Board of Canada, 2008). Additionally, the investigation report concluded

> The working environment on the bridge of the Queen of the North was less than formal, and the accepted principles of navigation safety were not consistently or rigorously applied. Unsafe navigation practices persisted which, in this occurrence, contributed to the loss of situational awareness by the bridge team.

(Transportation Safety Board of Canada, 2008, p. 6)

As the court held that, "[m]aintaining situational awareness at all times and in all circumstances is key to proper navigation" (Supreme Court of British Columbia, 2013), the leap wasn't large: The prosecution was able to successfully argue for the dereliction of his duty. Evidently, situational awareness was not maintained, otherwise the ferry would not have crashed and sunk. Karl was convicted for criminal negligence causing the deaths of the two passengers and sentenced to four years in prison. He is there at the time of writing this.

Human factors has embraced "situation awareness" (SA) as a construct to aid our understanding of human decision making in complex dynamic systems and to help with the design of human–machine interfaces. However, now SA—and the loss of it—functions as a causal and normative construct to explain and judge human performance (at least in the case of Karl). This provides an interesting paradox, particularly as the primary concern of human factors is with the well-being of humans in interaction with technology, or better, systems. According to the International Ergonomics Association (2014), human factors and ergonomics (HF/E) is "the scientific discipline concerned with the understanding of interactions among humans and other elements of a system, and the profession that applies theory, principles, data and methods to design in order to optimize human well-being and overall system performance." As such, one of the discipline's central challenges lies in explaining and preventing accidents in our high-risk sociotechnological systems. In this chapter, we will focus on the ethical implications of the popular notion of "human error"—and its subcategories, such as "loss of SA"—as explanations for why sometimes sociotechnological systems fail.

It is here—in "human error" explanations of accidents—that the tension between systems and individuals creates some paradoxes that need both practical and ethical consideration, as exemplified by the case of Karl. The basic tension is one between the parts and the whole. After outlining the historical imperative of human factors as a systems science, in this chapter, we discuss the tendency to default to "human error," particularly in human factors explanations of accidents. This unstated—and possibly unrecognized—commitment to reductionism, that regards the human as a largely independent but problematic component of our complex sociotechnological systems, has resulted in a number of concepts and practices in human factors whose effects run counter to the discipline's aim of optimizing human well-being. As such, it is pertinent that we, as a human factors community, engage in ethical discussions about the goals, methods, and consequences of our discipline's propositions.

HUMAN FACTORS VERSUS "HUMAN ERROR"— SYSTEMS AND COMPONENTS

Historically, human performance—in terms of efficiency and safety—became of interest with the rise of the industrial society at the turn of the nineteenth century. Increasing numbers of miners, factory workers, and other operators were exposed to new machines, working conditions, and risks. Matching humans to new technology was done through increasingly advanced selection procedures and training programmes, intended to better fit the human to the system. Few efforts were made to adjust the system to fit the human. However, with the rapid development of even more sophisticated technologies during the twentieth century, it became increasingly necessary to adjust the task or tools to match the human, that is, to design equipment that takes into consideration the needs and limits of the human (Meister, 1999).

A seminal study that set the agenda for the scientific discipline of human factors was by the experimental psychologists, Fitts and Jones (1947), who adapted their laboratory techniques to study the applied problem of "pilot error" during WWII. The problem they faced was that pilots of one aircraft type frequently retracted the gear instead of the flaps after landing. This incident hardly ever occurred to pilots of other aircraft types. They noticed that the gear and flap controls could easily be confused: The nearly identical levers were located right next to each other in an obscure part of the cockpit. Conversely, the flap and gear controls in the cockpits of the control group of aircraft were not adjacent. Fitts and Jones realized that the gear-up-after-landing problems—the "pilot errors"—were cockpit-design issues instead. Alphonse Chapanis, another human factors pioneer, fixed a small rubber wheel and a small wedge-shaped end, respectively, to the landing gear lever and the flap lever, and this type of "human error" almost completely disappeared. This insight, that human performance is to be understood as being systematically connected to the features of people's tools and tasks, effectively constituted the birth of "human factors."

Later, particularly in the wake of a number large organizational accidents in the late 1970s and 1980s, Fitts and Jones' principles for the design of tools and interfaces was extended to include the broader influence of organizational environments. Reason (1990), for example, proposed that the performance of those at the sharp (operational) end is shaped by latent conditions and upstream organizational factors at the blunt (managerial) end. Organizational pressures for production and efficiency slowly but systematically shape the decision-making and performance of those in the system [see Dekker's (2011) "drift into failure"]. Human performance and behavior is connected to the operational and organizational environments in which people work. By presuming that human performance can only be understood by seeing it as embedded in the larger system in which it takes place, this kind of human factors builds on a holistic—systemic—worldview that sees the system as an integrated whole rather than a collection of disconnected parts. Properties of the system cannot be understood by analyzing the parts, but emerge from the interactions and relationships among the components. In terms of accident explanation and prevention, this means that rather than seeing accidents as resultant of (a linear succession of) failing components, accidents are seen as emergent phenomena that cannot be understood by identifying faulty components—such as the human in error. This systems orientation is nowadays seen as the core strength and characteristic of HF/E:

> It is its very systems perspective and holistic nature that provides the strength of ergonomics. By extension, and importantly for the argument later in this paper, the breadth of concern in ergonomics, to cover all aspects of people's interaction with their environments and the interconnections between these interactions, is what allows it to define itself as a unique discipline.
>
> **(Wilson, 2000, p. 560)**

That said, accidents are often still seen as the result of risk that was not managed well, either at the sharp operational end or the blunt end (whether management, design, construction, or maintenance). Although in the 1960s, "human error" was said to be the main cause of accidents in about 30% of all accidents, it is now commonly accepted that 70%–90% of accidents can be attributed to "human error" (Hollnagel, 2006). This seems at odds with the human factors principle that human performance is systematically connected to the system in which it takes place. Instead of seeing an accident as the emergent outcome of a complex system, and which is thus destroyed when the phenomenon is reduced to isolated components, accident explanations seem to rely on a reductionist search for the "broken part" that caused the accident. The belief that the behavior of the whole can be understood by analyzing the properties of its parts is central to Descartes' method of analytic reductionism. The hegemony of analytic reductionism in our Western scientific approach also shines through in human factors practices, particularly those concerned with accident explanation and prevention.

One possible explanation for this trend toward "human error" explanations of accidents may lie in the inherent contradiction between human factors' pragmatic orientation and the systems approach it tries to build upon. As a practical discipline, human factors aims to provide (design) recommendations to improve the interaction of humans and systems. While (reductionist) science often isolates its objects from their context—to study them in controllable experimental settings—in order to be able to provide causal explanations, systems approaches rely on a more intuitive and holistic approach that does not yield the same "practical" truths (in terms of cause and effect). Systems explanations point away from single-cause explanations, and seek multiple diverging narratives that enlarge the scope of analysis to understand emergent phenomena. As small changes may have large effects in systems theory, at best, it hopes to find coexistence and correlation—as opposed to causation—in the interactions and relations that make up our sociotechnological systems. Consequently, after fixing the obvious, manifest problems in the system, it becomes increasingly difficult for human factors as a systems science to provide "simple" and practical solutions. Where do we intervene or redesign systems when everything in the system is interconnected and interdependent? When we cannot locate a cause for the accident, how can we fix the system so that things do not go wrong again? Also, very important nowadays, from a systems perspective, it may be difficult (if not impossible) to understand who is responsible for any harm.

In understanding why accident investigation reports often set out to explore multiple diverging narratives but then "suddenly" converge in single localized explanations, Galison (2000, p. 40) suggested

> ... if there is no seed, if the bramble of cause, agency, and procedure does not issue from a fault nucleus, but is rather unstably perched between scales, between the human and non-human, and between protocol and judgment, then the world is a more disordered and dangerous place. These reports, and much of the history we write, struggle, incompletely and unstably, to hold that nightmare at bay.

This is where reductionist explanations are useful: Finding a broken component—human error—provides us with an endpoint of an accident investigation, a root cause. It gives us something to fix.

HUMAN FACTORS AND ETHICS—THE IMPLICATIONS OF SOME "HUMAN ERROR" PRACTICES

Although system thinking sees the world as a complex web of interactions and relationships, "human error" is still regarded as a practical target for intervention. This comes with a price—there might be an ethical cost to our reductionist approach. Ethics is the branch of philosophy that is concerned with what the right thing is to do, what is good for individuals and society. Of course, explanations are always simplifications of reality; otherwise our theories and models would be as complex as the phenomena they aim to understand and would yield no explanatory or predictive value. The question is—and here it becomes an ethical issue—what knowledge or solutions do we want our theories or explanations to produce? In the wake of an accident, what countermeasures do we hope to launch, what do we miss by relying on these explanations, and who wins and who loses? Are our practices serving the well-being of society or the operators that human factors claim to have in its best interest? By unearthing the reductionist assumptions underlying some human factors practices, in this section we ask ethical questions of three ways of dealing with "human error"—error taxonomies, automation, and proceduralization.

ERROR TAXONOMIES AND CATEGORIES

In trying to reconcile the tension between practical solutions and its systems ideas, human factors has proposed a great number of performance and error taxonomies intended to provide researchers or accident investigators with a "deeper insight" into understanding sociotechnological failures. We have developed classifications based on different types of work, various cognitive failures, error types corresponding to performance modes (skill-based slips and lapses, rule-based mistakes, and knowledge-based mistakes), and, more recently, taxonomies that are supposed to support the practical management of safety (such as threat and error management [TEM] and human factors analysis and classification system [HFACS]). In addition to these more extensive classification systems, human factors has also proposed and popularized a number of (single) categories that aim to provide specificity to errors: SA, complacency, automation surprise, breakdown of CRM, etc. All these taxonomies and constructs are developed with the best of intentions—that is, in line with the human factors systems philosophy—as tools to help better understand where in the system the failure is located.

One issue with these taxonomies, however, is their fragmented focus on components of the system, most prominently the human. Where human factors is supposed to be about the interaction of humans and technology or systems, very few human factors theories or classification tools take these interactions as its main subject—cognitive systems engineering is the only evident exception, as it explicitly takes the joint cognitive system as its unit of analysis. Otherwise, these classifications provide slightly more specific labels for component failures that often remove the operational, organizational, and social context in which the behavior took place. This (retrospective) labeling of human performance into more general error types removes context and so hampers our understanding of the situation that the people involved in the accident were facing.

As a result, many of the taxonomies and categories we use do not provide a satisfactory explanation for why the person did what he or she did, for why a decision or action made sense at the time. The "local rationality principle" dictates that people do what is rational given the knowledge, goals, and the resources (e.g., time and tools) that they have available at the moment of the decision or action: "People are doing reasonable things given their point of view and focus of attention; their knowledge of the situation; their objectives and the objectives of the larger organization they work for" (Dekker, 2006, p. 13). However, "human error" classifications and explanations seldom take the (local) perspective of the people who were facing a particular situation, but rather judge their behavior based on all the information and resources we have available in hindsight (i.e., in retrospect). Not only do these "judgments"—based on information that we now, in hindsight, know to be important—provide normative standards for behavior (e.g., to always be "situationally aware"), the language that comes with them often allows for naïve allocations of blame. Also, by disregarding the context and local rationality, the entire explanatory load of the accident—as well as the moral responsibility—is placed on the individual (human) component, which takes away all the necessity and opportunity to learn from the event. The systemic conditions that gave rise to the error are left in the system for the next person to run into. Besides being impractical, is it ethically right to put the next person in the same situation or system?

AUTOMATION

One common method to deal with "human error" in sociotechnological systems is by removing the human component from critical operational processes in the system. The twentieth century has seen a tendency to replace humans by technology, from mechanization to computerization and automation—systems being controlled by autonomous technology rather than humans. Initially the introduction of automation had little to do with the performance variability that humans display—it was driven by the sheer possibility of technological progress and the desire to reduce costs through the reduction of human staffing requirements—but later the focus of innovation turned to technology as a manner to remove "human error." Computers were to replace as many human tasks as possible to remove human operators from the control loop; humans were now tasked to simply monitor the automation.

Even if automation is more reliable than human operators, it does not remove the potential for "human error" but merely relocates it. Even in highly automated systems humans continue to be a central element, as they need to install, monitor, maintain, and eventually improve the system. Bainbridge (1983) suggested that the increased reliance on technology, ironically, also increases the importance of the human operator in various ways. First, the human operator's task has increasingly become that of monitoring the automation, even though human factors research has shown that the properties of human attention are generally ill-suited to (long-term) vigilance tasks. Second, automation may decrease operators' physical workload at the price of increasing their cognitive workload, especially during periods in which cognitive workload is already high. Third, as operators have become detached from their original work, when in the case of an emergency the automation fails, they may have lost most of their skills to perform the actual work but are now supposed to take over (from the automation) during highly critical and difficult moments. Replacing human work by technology does thus not simply reduce "human error," but changes the nature of work and has some unanticipated consequences.

Not only does automation change the nature of work, it also does so in a manner that can be seen as practically and ethically problematic. In addition to the unforeseen effects that come with the introduction of new technology, one important ethical issue that deserves some elaboration here is the question of who gains and who loses by the ubiquitous introduction of automation. The widespread introduction of automation in our modern sociotechnological systems was, for example, accompanied by the promise of an increase in quality of life for all the people working in and with these automated systems. Estimations were made that the length of the (human) workweek would shorten to 30 and later even to 20 hours (Hancock and Drury, 2011). Unfortunately, this is not what has happened over the last few decades: Instead of everybody working less, the changing role for humans in modern automated systems required some individuals to work more while others lost their jobs. Studies show that due to the redistribution of work, "workers at the lower end of the income distribution often have difficulty finding enough work, while people with higher incomes, especially those in professional and managerial jobs, work more hours than ever before" (Evans et al., 2004, pp. 1–2). Especially as most people express the desire for a more standard week of about 35–40 h, the findings that nowadays more people are working long weeks (> 50 h) and short weeks (< 20 h) (Hancock and Drury, 2011) foregrounds the question if automation has lived up to its promise of increasing the quality of life. Is this serving the well-being of the individual operator? Is the broad and ubiquitous introduction of automation in the best interest of society?

The intention to remove the potential for "human error" is of course supportive of a basic aim in human factors, and entirely in line with Fitts and Jones' approach for the human-centered design of interfaces. However, the intent to simply replace the human component by automation does not constitute a systems approach. It builds on the reductionist assumption that assumes the system works fine if it were not for the unreliable human components. This rationale, however, overlooks the central role that humans play in enabling the system to exist and function in the first place. It overlooks the fact that humans, more specifically human improvisation and performance variability, are the only resource that can balance goal conflicts and adapt to the various unexpected states that the larger system throws at them. A systems approach would see the system and automation as constituent of the human—as humans we use tools to augment, supplant, and aid ourselves—as well as the human being constituent of the system. As such, a systems approach concerning automation might consider the (possibly less efficient) option of having the automation monitor the human. In the traditional approach of humans monitoring the automation, the nature of work has changed, but not necessarily for the better.

PROCEDURES

To deal with the problem of incidents and accidents in our systems, reductionism dictates that if all parts function as they are supposed to, so will the whole. As such, probably most indicative of human factors' reductionist approach is its solution to deal with the human performance variability in a similar manner as it would deal with technical components: Make the components more reliable by tightly controlling their behavior. Proceduralization provides operators in complex systems with precise instructions (often presented in the form of checklists) on how to resolve both normal and emergency situations and so increases the reliability and predictability of their performance. This kind of performance standardization through proceduralization is not a new idea, nor unique to human factors. Breaking down complex work into smaller—but more manageable and efficient—parts can be traced all the way back to the work of Frederick Taylor in the early 1900s:

> Perhaps the most prominent single element in modern scientific management is the task idea. The work of every workman is fully planned out by the management ..., describing in detail the task which he is to accomplish, as well as the means to be used in doing the work. ... This task specifies not only what is to be done but how it is to be done and the exact time allowed for doing it.
>
> **(Taylor, 1967, p. 39)**

As a management philosophy—with its principles applied to a broad range of industries and still popular today (e.g., Lean production)—Taylorism assumes that by breaking down tasks and meticulously prescribing work, efficiency and reliability can be implemented from the top down. It provides the foundation for the belief that procedures (often specified by supervisors and middle-management) are the best way to specify tasks and roles (for operators). Most safety-critical domains rely on procedures and compliance to increase the predictability and reliability of people's performance. "Over time proceduralisation has become more than an answer of how to increase safety in modern socio-technical systems, it may have become the answer" (Bergström et al., 2009, p. 76).

As a human factors safety solution, the reliance on procedures and compliance has its practical limits. Every time an unwanted event occurs, a new set of increasingly detailed procedures is introduced to respond specifically to just the latest incident. For example, after the friendly fire on two Black Hawk helicopters over Northern Iraq, "higher headquarters in Europe dispatched a sweeping set of rules in documents several inches thick to 'absolutely guarantee' that whatever caused this tragedy would never happen again" (Snook, 2000, p. 201). Over the years, this has resulted in a suffocating level of procedural over-specification, making it impossible for people in almost any system to comply exactly with all the rules. This can clearly be seen in "work-to-rule" strikes in which people take every single procedure in the rulebook in its full and literal meaning, often resulting in production gridlock. We need to ask ourselves if it is ethically right that operators routinely need to work around or loosely interpret many official procedures—a phenomenon discussed in human factors as the gap between "work-as-imagined" and "work-as-done" (see, for example, Hollnagel, 2014)—to get their work done? Additionally, as the current level of proceduralization has made the system increasingly opaque, we need to ask whether safety is served by making it very difficult for the people inside the system to understand what the important rules of the system are?

Of course, everybody in society profits from efficient production machinery, aided by automation and highly efficient ways to manage work. It has brought the (Western) world great prosperity. However, this manner of organizing work seems to lose track of human factors' aim of increasing human well-being. Just as with replacing human work by automation, breaking complex tasks down into smaller more efficient subtasks, to be performed by different people, makes work less meaningful. Instead of constructing an entire bicycle, factory employees are now assigned the task of fitting just one specific part to the bike frame, and they are supposed to do so thousands of times a day. Task and labor specialization, as well as an increased reliance on prescribed behaviors, do not only reduce the meaningfulness of people's work, but may also reinforce people's alienation from work and the loss of professional identities.

Moreover, tightly prescribing work may contribute to making our workforce dumber and less likely to come up with (creative) solutions to problems. This is particularly problematic as procedures are necessarily incomplete specifications for action—they are always simplifications of the real world and can never foresee all scenarios in a complex system. The abilities to recognize, absorb, and adapt to disruptions that fall outside the system's design base have been identified as key aspects of making individuals, and consequently the organizations that they make up, resilient (Hollnagel et al., 2006). However, can we expect this kind of resilience from operators whom we have instructed not to think for themselves, to be "docile bodies"? As such, our reliance on procedures and automation comes with the necessary sacrifices, both practically and ethically.

So What?

The aforementioned instances—practices in regards to taxonomies, automation, and procedures—only represent a small number of the issues that human factors is concerned with. We could also have argued for the need to rethink our insistence on the (self-)reporting of incidents: Is it right to ask people to report on their own or other people's behaviors, especially if this will not be judged in light of the complexity of the context in which work takes place? We could have raised our ethical concerns with the way notions such as "resilience engineering" are sometimes used: Basically asking people to accept more risk by relying on their individual abilities to cope with danger and surprise, when no systemic changes are "engineered." Other examples abound. The point that we are trying to make is that despite human factors' commitment to a systems approach, some of its practices may still be aimed at the individual—human error—as the practical target for intervention. As a result, some of these practices may run counter to human factors' aim of maximizing human (operator) well-being.

One further consequence of human factors' (implicit) commitment to the individual as the major locus of intervention and explanation for accidents might be the steep increase in the criminal prosecution of human error. Where it used to be customary in the aftermath of an accident to launch an investigation aimed solely at preventing similar accidents in the future, nowadays such investigations are almost always accompanied by judicial inquiries aimed at assessing and allocating liability (Dekker, 2009; Michaelides-Mateou and Mateou, 2010). Recent years have shown a growing concern in the safety field about a trend toward the criminalization of professionals for not living up to their professional duties. Not only will the potential criminalization of professionals result in their hesitance to report future incidents, practitioners in safety-critical jobs might also actively start hiding their mistakes (i.e., the evidence that could incriminate them), and start practicing their work more defensively.

The manner in which the juridical discourse takes and uses human factors constructs—such as "(a loss of) SA"—to ask (and answer) questions about error, culpability, and criminality is an ethical problem that we as a human factors community need to take responsibility for. Instead of pointing the finger at prosecutors uncritically using our constructs in "inappropriate" manners, we need to take a critical look at the practices that we ourselves propose. As long as the individual remains the unit of analysis for human factors, we are asking for ethical problems in our systems; we are not helping other discourses (such as the juridical) to move away from accounts of individuals as the fundamental elements of thought and action. We are not doing anything to point other discourses to vocabularies of complexity and systems, to (postmodern) models of causality and accountability in which systems as a whole break down, not their substituent (human) parts.

By recognizing that some of our practices are giving rise to some undesirable effects, it becomes an ethical choice—for us as a human factors community—to denounce or keep promoting them.

CONCLUSION

In this chapter, we have argued that despite the work of pioneers such as Fitts and Jones (1947), who claimed that the features of people's equipment and environment shape human performance, human factors remains (at least partly) wedded to a reductionist focus on the human component of our systems.

Of course there is nothing wrong with reductionism per se. It becomes problematic—both practically and ethically—when it is taken as the only approach.

As a human factors community, we need to ask whether the explanations, labels, classifications, and interventions we propose are serving our purposes. Given the complexity of sociotechnological systems, adverse outcomes may no longer be satisfactorily explained by reference to individual components and cause–effect relations. A systems perspective might be better equipped to help us understand how the system shaped the behavior of those in it and will be able to make lasting changes to prevent future incidents. As the complexity of our sociotechnological systems has reached a point where many of these systems can no longer be modeled or controlled, it may thus be increasingly irrational to keep relying on a reductionist approach.

If we cling to our commitment to find broken parts, as a modern way of scapegoating, we are asking for ethical trouble with our human factors explanations and practices. Constructs, such as SA, which we once introduced to help operators better understand and deal with complex dynamic environments, are now turned against them as they provide normative standards for behavior. Even if SA is a useful human factors construct, the manner in which other discourses take and use it to ask (and answer) questions about error, blame, and criminality is an ethical problem that we as a human factors community need to take responsibility for. Perhaps, in a small but not unimportant way, we are complicit in Karl Lilgert's sentence, and in him sitting in jail today.

REFERENCES

Bainbridge, L. 1983. Ironies of automation. *Automatica*. 19(**6**), 775–779.
Bergström, J., Dahlström, N., van Winsen, R., Lützhöft, M., Nyce, J., and Dekker, S. 2009. Rule-and role-retreat: An empirical study of procedures and resilience. *Journal of Maritime Research*. 6(**3**), 41–58.
Dekker, S.W.A. 2006. *The Field Guide to Understanding Human Error*. Aldershot, VT: Ashgate.
Dekker, S.W.A. 2009. Prosecuting professional mistake: Secondary victimization and a research agenda for criminology. *International Journal of Criminal Justice Sciences*. 4(**1**), 60.
Dekker, S.W.A. 2011. *Drift into Failure: From Hunting Broken Components to Understanding Complex Systems*. Farnham, VT: Ashgate.
Evans, J.A., Kunda, G., and Barley, S.R. 2004. Beach time, bridge time, and billable hours: The temporal structure of technical contracting. *Administrative Science Quarterly*. 49(**1**), 1–38.
Fitts, P.M. and Jones, R.E. 1947. *Analysis of Factors Contributing to 460 'Pilot Error' Experiences in Operating Aircraft Controls*. Dayton, OH: Aero Medical Laboratory, Air Material Command, Wright-Patterson Air Force Base.
Galison, P. 2000. An accident of history. In: Galison, P. and Roland, A. (eds.). *Atmospheric Flight in the Twentieth Century*. Dordrecht, Netherlands: Kluwer.
Hancock, P.A. and Drury, C.G. 2011. Does human factors/ergonomics contribute to the quality of life? *Theoretical Issues in Ergonomics Science*. 12(**5**), 416–426.
Hollnagel, E. 2006. Accident analysis and "human error." In: Karwowski, W. (ed.). *International Encyclopedia of Ergonomics and Human Factors*. Boca Raton, FL: Taylor & Francis, pp. 1889–1892.
Hollnagel, E. 2014. *Safety-I and Safety-II: The Past and Future of Safety Management*. Farnham, UK: Ashgate.

Hollnagel, E., Woods, D.D., and Leveson, N. 2006. *Resilience Engineering: Concepts and Precepts*. Aldershot, UK: Ashgate.

International Ergonomics Association (IEA). 2014. Definition and domains of ergonomics. Accessed March 13, 2016. Available from: http://www.iea.cc/whats/index.html.

Keller, J. 2013. *Karl Lilgert on Trial for BC Ferry Queen of the North Sinking*. The Canadian Press. Available from: http://www.huffingtonpost.ca/2013/01/17/karl-lilgert-bc-ferry-queen-of-the-north-sinking_n_2493214.html.

Meister, D. 1999. *The History of Human Factors and Ergonomics*. Mahwah, NJ: Lawrence Erlbaum Associates.

Michaelides-Mateou, S. and Mateou, A. 2010. *Flying in the Face of Criminalization: The Safety Implications of Prosecuting Aviation Professionals for Accidents*. Burlington, VT: Ashgate.

Reason, J.T. 1990. *Human Error*. Cambridge, UK: Cambridge University Press.

Snook, S.A. 2000. *Friendly Fire: The Accidental Shootdown of U.S. Black Hawks Over Northern Iraq*. Princeton, N.J: Princeton University Press.

Supreme Court of British Columbia. 2013. *Regina v. Karl-Heinz Arthur Lilgert*. (2013 BCSC 1329). Vancouver, British Columbia.

Taylor, F.W. 1967. *The Principles of Scientific Management*. New York, NY: Norton.

Transportation Safety Board of Canada (TSB). 2008. *Striking and Subsequent Sinking, Passenger and Vehicle Ferry Queen of the North, Gil Island, Wright Sound, British Columbia, 22 March 2006*. (M06W0052).

Wilson, J.R. 2000. Fundamentals of ergonomics in theory and practice. *Applied Ergonomics*. 31(**6**), 557–567.

6 Human Factors and Ergonomics in the Media

Ron Gantt and Steven Shorrock

CONTENTS

PRACTITIONER SUMMARY

Human factors and ergonomics (HF/E) as a discipline and profession addresses issues of profound interest to society. But the discipline and its subject matter tend to be represented in a small number of narrow contexts in the news and entertainment media. "Human factors" tends to be associated with failure and "human error," particularly involving frontline personnel. Narrative strands are woven into powerful stories that present the human as either hero or villain. Some features of a story (usually the "human factor") are made prominent while others—such as system and context influences—are routinely glossed over or ignored. "Ergonomics," meanwhile, tends to be associated with physical injury and office work. The entertainment media often reflect HF/E themes in tales of a fundamental distrust of technological advances. Meanwhile, the actual work of HF/E practitioners in improving system performance and human wellbeing is hardly ever represented. In this chapter, we consider these issues and offer some implications for HF/E practitioners to help minimize the negative consequences of media effects on the perceptions of clients and stakeholders, and take advantage of opportunities.

STORIES OF HEROES AND VILLAINS

On January 15, 2009, a US Airways flight left LaGuardia Airport in New York, heading for Charlotte. About 2 minutes after takeoff, the plane encountered a flock of birds and lost thrust in both of its engines. The 155 passengers and crew members, including a lap-held infant, were near disaster. Fortunately, a hero arose. The captain, "Sully" Sullenberger, was able to ditch the plane in the Hudson River, saving the lives of all on board (National Transportation Safety Board, 2010). The pilot was praised as a "hero."

On July 25, 2013, a passenger train was on express route from Madrid Chamartín railway station to Ferrol, Spain. On a section of the conventional track in Santiago de Compostela, 250 m (820 feet) before the start of a sharp curve, the train was travelling at 195 km/h (121 mph), when the train had to slow to 80 km/h (50 mph). Emergency brakes were applied but the train derailed 4 seconds later at 179 km/h (111 mph). Seventy-nine people died. Media reports over the next few days stated that the driver "ignored three warnings to slow down" (Heckle, August 2, 2013) and "admitted speeding." The media reported an old Facebook post that stated: "It would be amazing to go alongside police and overtake them and trigger off the speed camera" (Govan, May 26, 2013), accompanied by a photo of a train's speedometer at 200 km/h (124 mph). The press painted the driver, Francisco José Garzón Amo, as a villain who triggered an accident in an otherwise safe system. Just under a year later, a headline stated: "Confirmed: Human error caused horror train crash." This echoed the conclusion of the investigation, which nevertheless gave many recommendations aimed at improving the system as a whole and none directed at driver Francisco José Garzón Amo.

These events—the "Miracle on the Hudson" and the train crash of Santiago de Compostela—have several things in common: both were accidents; both have human factors implications; both received significant media attention; and both had the frontline actor as the primary focus of media attention. However, one was cast as a hero, while the other was cast as a villain, or at best, a hazard. Chesley Sullenberger received lavish praise from the press,* was the subject of documentaries, and has his own website, blog, and Twitter account. He published two books—one a New York Times bestseller—and works as a speaker and consultant. Francisco José Garzón Amo, at the time of writing, faces 79 counts of homicide by professional recklessness and an undetermined number of counts of causing injury by professional recklessness.

The outcomes of the events are obviously very different and are critical to how we perceive the stories (perhaps more critical than how the outcomes came about). Some reports—and safety specialists—even raised the possibility that Sullenberger "could have landed": "Tucked inside thousands of pages of testimony and exhibits are hints that, in hindsight, the celebrated pilot could have made it back to La Guardia Airport. Pilots who used simulators to re-create the accident—including suddenly losing both engines after sucking in birds at 2500 feet—repeatedly managed to safely land their virtual airliners at La Guardia" (Pasztor, 2010). This counterfactual—unlike most that emerge following accidents—may well be imbued with hindsight, but it might

---

* Sullenberger has consistently deflected personal praise toward his whole crew (including the cabin crew) and referred to crew resource management (CRM, see Chapter 15).

also prove very unpopular if propagated more widely. It throws mud on an otherwise good story: A packed plane in trouble, a heroic former military pilot, a landing on water, passengers standing on the wings—everyone survived. Had the weather conditions or other aspects of the context been different, the story of Flight 1549, and Sullenberger's role, could have been very different. This latter counterfactual could have fuelled moral outrage and alternative narratives.

In the case of the Santiago de Compostela, the context and counterfactuals are also messy, though they have started to seep into news stories. Had the section of track been covered by the European rail traffic management system (ERTMS), as was the section of track 8 km away, the outcome would likely have been very different. The context of the crash also complicates the story. The driver received a call from the ticket inspector to instruct him to enter an upcoming station at a platform close to the station building to facilitate the exit of a family with children. It is unlikely that this practice was unique to this event. The driver was distracted and suffered a "lapse of concentration" as he approached the curve and did not detect three warning signals. The driver told the court he lost sense of where the train was during the call, and believed he was on a different section of the track. The Facebook post may be unwise social media activity, but this was made a year earlier, and such speeds are normal and fully permitted on the high-speed line sections.

STORIES AND BEING HUMAN

Stories provide a means by which people create meaning in their environments (Rae, 2014). Stories are fundamental to our experience of being human, and it seems that we are wired to think in terms of stories. For example, in one study subjects were shown images of shapes in series, so that the shapes appeared to be moving from one location in the image to another. Participants not only described the movement, but also developed intricate stories in which the shapes were people performing various actions together (Heider and Simmel, 1944).

Stories not only help people create meaning, they also involve choice. Consider the stories told above. In each case, the actual events involved significantly more detail. The storyteller decides which details are important and which are extraneous (Rae, 2014). Although other factors have influence on the choices that storytellers make (Hilgartner and Bosk, 1988), these details are often chosen based upon the message the storyteller is trying to get across. Details are chosen purposefully, implying that the detail is important in telling and making sense of the story. Conversely, details left out are, by implication, less pertinent to an understanding of the story. In both stories identified above, only one person was the focus, whereas in each story many more people were involved directly or indirectly. The implication of these choices suggests a particular perceived importance for each character.

THEORETICAL BACKGROUND ON MEDIA EFFECTS

A rich set of literature exists on the media's effect on public perceptions. Perhaps the two most influential theories regarding media effects are agenda-setting theory and framing (Scheufele and Tewksbury, 2007). Agenda setting refers to what the media reports, whereas framing refers to how the story is presented. Agenda setting was first studied in the seminal "Chapel Hill Study" and has been replicated numerous times since, where coverage of stories in the news media correlates with the perceived importance of those issues in the public who pay attention to that media (Scheufele, 2000). The relationship between news media coverage and public perception is thought to be related to the cognitive effects of priming and the availability heuristic (McCombs, 2005). Not surprisingly, increasing media coverage of news stories seems to make stories easier to retrieve from memory. Those stories that are more cognitively available will be seen as more important than those that are less cognitively available (Scheufele, 2000), regardless of how they feel about the story (e.g., whether they are for or against the issue being discussed). Using the lens of agenda setting, it would be reasonable to predict that public perceptions of HF/E would be influenced by how often specific stories related to HF/E are portrayed in the media.

While the agenda-setting theory is based on the idea of cognitive availability or accessibility, media framing is based primarily on the idea of applicability (Scheufele and Tewksbury, 2007). People use frames to help make sense of their worlds, making the complex less ambiguous (Goffman, 1974). As events unfold, we activate applicable frames that allow for the understanding and expectation of events. For example, when one goes to an expensive restaurant, there is an understanding of what is expected in terms of dress, behavior of the staff, and the type of food served. This, in turn, has an effect on behavior. Most people expect the bill to be brought to the table following the meal, and therefore keep their purse or wallet in their pocket until the end of the meal and wait for the waitstaff to bring the bill, rather than track it down themselves. In the news media, journalists use frames to reduce the complexity of a story for the purpose of efficient delivery. This delivery activates frames in news consumers, helping them to identify salient information, including potential problems and solutions, as well as to make predictions about what will happen next (Scheufele and Tewksbury, 2007). Both agenda setting and framing have been shown to affect public perceptions (McCombs, 2005). Using the lens of framing, how HF/E-related stories are portrayed, what features of a story are made prominent, and what are glossed over or ignored, will affect how the public makes sense of those stories.

Adding to the literature on media effects, Hilgartner and Bosk (1988) proposed a particularly influential model to describe how social problems come to prominence and die out based on media coverage. Although media framing and agenda setting describe how the media influences individuals, Hilgartner and Bosk's so-called "public arenas" model describes how the media chooses social problems to cover and how those social problems are portrayed. "After all, there are many situations in society that could be perceived as social problems but are not so defined.... Why, for instance, does the plight of the indigenous people of South America ... receive less public attention than the plight of laboratory animals used in scientific research?" (Hilgartner and Bosk, 1988, p. 54).

The public arenas model explains the life and death of social problems as an interaction not only between an individual social problem and a given culture, but also interactions with other social problems, competition with other interest groups over problem definition, and structural features of media outlets. "Our model stresses the 'arenas' where social problem definitions evolve, examining the effect of those arenas on both the evolution of social problems and the actors who make claims about them … instead of emphasizing the stages of a social problem's development, we focus on competition: We assume that public attention is a scarce resource, allocated through competition in a system of public arenas" (Hilgartner and Bosk, 1988, p. 55).

Social problems have to compete on at least two levels within the public arenas model. First, the frame of the social problem must be chosen and varying interest groups compete to have their frame be the dominant one in public discourse (Hilgartner and Bosk, 1988). For example, accidents involving nuclear power can be framed as the product of bad actors (i.e., individual workers, managers, or organizations), or as a statement on the limitations of the safety structures designed to prevent those accidents (Downer, 2013).

The second area of competition for social problems is competition among social problems. "Second, a large collection of problems—from teenage pregnancy to occupational health to shortages of organ donors—compete with one another for public attention, as a complex process of selection establishes priorities about which should be regarded as important" (Hilgartner and Bosk, 1988, p. 58). Both the competition among social problems as well as competition for defining a given social problem are heavily influenced by preexisting cultural scripts (Hansen, 1991) or political realities (Lawrence and Birkland, 2004).

Another important element of the public arenas model is the idea that various outlets for social problem discourse have a limited "carrying capacity." Each institution (e.g., newspapers, television news, radio, and congressional/parliamentary hearings) has a finite amount of time and/or space to devote to a social problem. "While it is clear that the number of situations that could potentially be interpreted as social problems is so huge as to be, for practical purposes, virtually infinite, the prime space and prime time for presenting problems publicly are quite limited. It is this discrepancy between the number of potential problems and the size of the public space for addressing them that makes competition among problems so crucial and central to the process of collective definition" (Hilgartner and Bosk, 1988, p. 59). Individuals also have carrying capacities, as the amount of compassion and concern they can muster is finite. The most compassion is reserved for those social problems that are of personal significance, which is, again, socially constructed (Hansen, 1991).

In choosing which social problems will win the competition for the precious carrying capacities of the public audience, the public arenas model posits that a set of standard selection criteria are applied. These include drama; novelty and saturation; culture and politics; and organizational characteristics (Hilgartner and Bosk, 1988). Drama refers to the emotional salience of a given social problem. However, even very dramatic social problems lose salience over time if their presentation is oversaturated without change, encompassing the second selection criteria. Cultural factors play a heavy role in defining the salience of a given issue (Hansen, 1991) and political issues may hold more sway than other issues, given that political elites have special access to media outlets and significant influence (Lawrence and Birkland, 2004). Characteristics of the organization that may present the news tend to have more structural influences. For example, news outlets have news desks devoted to identifying and analyzing news in some areas, but not others, making social problems in the areas with news desks more likely to be selected simply because someone is paying attention and the structure of the organization has legitimized the issue as "newsworthy" (Hilgartner and Bosk, 1988; Hansen, 1991).

The public arenas model has implications for HF/E, as any portrayal of HF/E issues would be subject to the carrying capacity of the media outlet, and must meet the selection criteria to provide a minimum level of drama, all while competing with other issues for public attention.

"HUMAN FACTORS" AND "ERGONOMICS" IN NEWS MEDIA

With an understanding of some of these theories and models, a review of how HF/E-related stories are told in the media should provide clues as to how HF/E-related issues are perceived by the average media consumer. What stories receive the most attention (agenda setting)? How are those stories framed? Based on the media outlet, what amount of coverage is possible (carrying capacity)? How is the story portrayed to provide drama?

We conducted a search of the top three "online news entities" (according to Pew Research Center Analysis) of the top 50 listed on journalism.org: Yahoo-ABC News Network (comprising ABC News and Yahoo News), CNN Network, and NBC News Digital. The top two UK-based media websites were also searched (Daily Mail and BBC). The terms "human factors" and "ergonomics" were used via the Google News advanced search feature for a 5-year period from January 1, 2010 to December 31, 2015. Hundreds (or thousands) of other terms associated with HF/E could be searched of course, but such an exercise would be unmanageable. The search was, therefore, restricted to the exact terms used to represent the discipline, to give a sense of the portrayal of HF/E in the news media.

"HUMAN FACTORS"

When searching using the term "human factors," the news stories had the following as a primary focus (in rank order): (1) accidents and safety; (2) technology (e.g., aviation, driving, military, space, and office and consumer products); (3) health (e.g., risks from sitting or standing). Other stories concerned more general societal and environmental issues such as climate change, animal extinctions, population growth, natural disasters, and public health issues.

The most common search result for "human factors" that directly related to HF/E concerned specific accidents and general stories concerning accidents. These stories comprised well over half of the search returns for this query. The accidents overwhelmingly involved transportation (aviation, road, maritime, rail, and space). A minority concerned health-care and process industries.

In almost all cases, human involvement was framed negatively and the human in the system (usually a member of frontline personnel) was isolated as a faulty component. In most stories, the term "human factors" was used in a general, vague, and negative sense with respect to an investigation. Human involvement was sometimes referred to as the "human factor." In a story concerning a roller coaster accident, where a woman fell to her death, Shoichet (2013) quoted an amusement ride and device safety consultant who said, "You're going to look at the operations of it, and the human factor, the human part of the equation.... Was there an error or omission made by an operator or someone in operations?" "Human factors" (used as a plural) were usually equated with accident causes. In most cases, the term "human factors" was used synonymously with "human error" (see also Shorrock, 2013 and 2015, on "human error" in the news media).

None of the stories concerning specific accidents reviewed included an interview or quote from an HF/E specialist, but some included quotes from other professions or industry personnel. In a story on the Germanwings tragedy, Zilmer (2015)—a clinical psychologist—wrote "When complex engineering systems interact with human factors, it is most often the human that causes the anomaly. Humans are far less reliable than machines." This "weak link" narrative ignores the role that humans play in creating safety day-to-day in degraded conditions, and ignores the complex nature of system interaction and emergence. It may be that HF/E practitioners see these stories as too sensitive, and many practitioners will need formal permission to speak with the media, or may be afraid of being misrepresented. But the implication may be that the systems view of HF/E does not reach journalists, and HF/E as a profession is misrepresented by the casual use of "human factors" and related notions.

Other concepts used by the HF/E community were also cited in some stories, though these rarely provided any illumination. Regarding the fatal crash of Amtrak 188 in 2015 (Hoye, 2015), "National Transportation Safety Board spokesman told CNN that their investigation is looking into Bostian's "situational awareness," as part of the "human factors" component of the investigation. The spokesman points out that it is standard procedure to look at every possible factor."

More general stories concerning safety related to driving, plane evacuation, rail safety, drones, oil drilling, and patient safety. Some of these stories did feature an interview with an HF/E practitioner (and some made reference to an HF/E article), but more often than not those cited were not HF/E practitioners. In an article on "The causes of plane crashes" (Yahoo Travel, August 3, 2015), aviation safety analyst, pilot, and FAA Safety Team representative Kyle Bailey was quoted as follows: "If somebody comes up to me and says, 'Planes are unsafe, blah blah blah,' I'll always say, 'You shouldn't be afraid of the plane, you should be afraid of the human element." The story continued: "And when you add *all* human factors—mistakes by mechanics and air traffic controllers in addition to pilots—Boeing estimates human error in general might be a factor in as many as 80 percent of all airplane accidents."

Most of the news stories implied a reductionist model of failure. Human activity was disconnected from the other parts of the systems and features of the work contexts and environments. This view of the "human factor" as a hazard or liability, and as a decomposable part of the system, is consistent with a "Safety-I" interpretation of the human contribution (Hollnagel et al., 2013).

On the rare occasions where the human is painted in a more positive light, the unsystemic focus remains. Stories related to the Miracle on the Hudson, mentioned previously, seemed to focus on the actions of the pilot, while the actions of the system that contributed to the resilient performance were often overshadowed. The fact that the pilot's name, Sully Sullenberger, is well known, whereas the names of no one else involved in the incident are not is testament to this fact. In this case, the story appears to be one of a failure in the system, saved by the actions of a lone hero.

In some stories, a systems view began to shine through, but these were rare. One example related to the flight that landed at the wrong Missouri airport (Mark, 2014). The author talked about "a growing disconnect between the technology created by our smartest engineers and technicians and the pilots who use it," and referred to issues of pilot confusion, monitoring, boredom, and (automation) "complacency." Unlike many others, Mark pointed out that "labelling this all as an easy mistake or as pilot error is too simplistic" and later "even labelling this a 'human factors' problem is too easy."

On the whole, the term "human factors" is one with largely negative connotations in the media, primarily concerning "human error," the "human factor" or (unwanted) factors of the human (the grammatical awkwardness of "human factors" probably doesn't help in this respect). "Human factors" is associated primarily with accidents— "human factors" lead to accidents and "human factors" as a subject area appears to be primarily the concern of accident investigators. It seems that part of the name of our discipline has been associated with rare, unwanted events. Largely, as reported in the media, it is detached from design. Since the media is not interested in ordinary work, or even "safe operations," the full context of human factors is not represented. The input of people to effective operations is ignored; the assumption seems to be that the system is basically safe as designed, contrary to the reality of complex systems.

"Ergonomics"

Over the six news media websites reviewed, the biggest category of articles citing "ergonomics" concerned consumer technology (including many product reviews), followed by office work. A smaller number of articles also featured stories on work-place health and safety, driving and seating posture. Many other issues had only one or two associated articles (e.g., trains, robots, military, space, medicine, cock-pit displays, and hotels). Most often, "ergonomics" was associated with physical ergonomics (associated with comfort, health and injury prevention) and interaction design (associated with product usability). This may well be influenced by the fact that "ergonomics" is more commercialized; stories featuring the term include PR releases, advertorials, and magazine-type reviews. Stories featuring the term "ergonomics" were much more likely to include a quote from an ergonomist, typically a researcher.

The Reality Gap

While we in the discipline tend to see HF/E as synonymous, the Anglophone media (and probably the public) does not. In the minds of journalists, and perhaps the general public, "ergonomics" is more clearly associated with "design for human use," while "human factors" is primarily associated with accidents (and, to a much lesser extent, accident prevention). Most of the stories cited either "human factors" or "ergonomics," but not both unless the story was citing an association (e.g., CIEHF, HFES), job title, or department.

A significant gap seems to exist, then, between media portrayals of HF/E-related issues and how HF/E as a discipline wishes to portray itself. On the whole, the news media does not seem to understand the focus of the HF/E discipline, and uses the terms "human factors" and "ergonomics" in narrow ways that do not accord with our own.

HF/E IN ENTERTAINMENT MEDIA

The theories and models discussed earlier, particularly agenda setting and framing, would suggest that other forms of media could also affect public perceptions. The storytelling of the entertainment media (i.e., film and television) may provide analogies or frames to make sense of everyday life, given the importance of stories to us in making sense of our worlds (Weick, 1995). Entertainment media has been shown to affect perceptions of the limitations and promise of science and technology (Nisbet et al., 2002).

Entertainment media is subject to similar selection criteria (drama, novelty/ saturation, culture/politics and organizational characteristics) as news media. Entertainment media must utilize drama to keep audiences engaged in the story. Entertainment media also has a carrying capacity in terms of time (usually between 30 and 60 minutes per episode for television, and between 90 and 180 minutes for film). Although entertainment media allows for greater creativity than news media in how stories are told, the types of stories told, and the frames in which those stories are told, are often similar to the news media. After all, both entertainment and news media are often situated in the same cultural environments. Therefore the frames that decision makers in entertainment and news media use to construct their stories should be similar. Choices such as what role characters play or which parts of the environment are important for providing context are made based upon taken-for-granted assumptions that are socially constructed.

Reviewing entertainment media successes in 2015, similar themes as those in news media emerge. A consistent theme in entertainment media is the role of the hero and villain. Because of some inherent moral flaw, the villains create a hazard, often threatening to hurt or kill many people. In some cases, this creation of threat by the villain is not intentional, but rather a form of human error, created by the moral failings of the villain (and therefore no less blameworthy). Luckily, a hero usually emerges to save the day.

A review of the top grossing movies of 2015 (Nash Information Services, 2016) shows that in the top three (*Star Wars Episode 7: The Force Awakens, Jurassic World*, and *Avengers: Age of Ultron*), each contained a stark contrast between heroes and villains. In each case a threat emerges via a villain that society is helpless to deal with. In the case of *Star Wars*, a group of villains, known as "The First Order," reminiscent of fascist regimes of the last century in both their methods and the imagery used to portray them, seek to take control of the galaxy for their own evil purposes. In their lust for power, they are willing to commit genocide and amass a formidable army, including a weapon capable of destroying entire planets. The galaxy is helpless against the power of The First Order. A group arises to oppose them, known as

"The Resistance." With the help of a small handful of heroes, including a former First Order soldier, who experienced a moral awakening that caused him to leave the First Order, the Resistance is able to thwart the plans of the First Order and save themselves and the galaxy.

Recalling the recurrent theme of distrust of technology harking back to *Mary Shelley's Frankenstein, Jurassic World,* and *Avengers: Age of Ultron,* all see technology run amok to create the threat in each story. In the case of *Jurassic World,* a corporation uses biotechnology to create a super dinosaur to be a blockbuster attraction at their dinosaur theme park. Lessons from the past (specifically the previous Jurassic Park movies) pointed to the creation of this new dinosaur being risky. But the corporation was overly confident and blinded by greed. The new dinosaur proves to be too formidable for the corporation's safeguards and it escapes, killing many workers and causing many other dinosaurs to escape. Again, a hero emerges who, through his respect for nature, is able to utilize other dinosaurs to stop the super dinosaur and save the day.

In the case of *Avengers: Age of Ultron,* one of the heroes creates an artificial intelligence, called Ultron, designed to protect Earth. Ultron turns on his creator and sets in motion a plan to rid the world of humanity. The world is helpless against Ultron's superior intelligence. The Avengers, a group of superheroes, come forth to stop Ultron's plan and save the day. The hero who created Ultron has a moment of contrition but ultimately works with his team members to make things right again.

In each case, and on multiple screens throughout the world everyday, a similar story plays out—individuals of poor moral character create havoc and it is up to a small number of heroes to save the day. The job of these heroes is to not only save the innocent, but to punish the villains (or to allow their punishment). Those who cause the threat are not always intentionally doing so. But in every case, these individuals are always blameworthy and always must find a way to make up for their moral failings, often involving some form of punishment.

Missing from this narrative is how individuals interact with the wider system. The fate of many is entirely in the hands of human agency, and usually the agency of a very small number of people. How the system interacts to create these choices through affordances is rarely depicted, except perhaps through cursory mentions of the upbringing of individuals. When human–system interaction is portrayed in entertainment media it is often through the lens of humans having to overcome the system through hard work or simply through having innate talent (or superpowers).

Within the entertainment media, the stories told seem to reflect many of the themes found in the news media, with few exceptions. Stories such as those discussed above are based on the idea that human actors are responsible for outcomes, not systems. The success or failure of a system is dependent on the individual. This is consistent with the way that "human factors" is portrayed in the media.

A large number of films are based on the natural human anxiety regarding technological advances. Stories about technology getting out of our control are common, ranging from the simple disasters, to the post-apocalyptic, to the more existential, relational concerns. As technology innovation cycles become shorter, the use of automation grows, and our lives grow ever more integrated with this technology,

HF/E professionals might usefully consider how this might influence the attitudes of human actors in systems now and in the future. A need to be in control has been highlighted in various studies (e.g., Bekier et al., 2011), and should also be part of discussions in HF/E practice.

IMPLICATIONS FOR HF/E PRACTITIONERS

Unfortunately, the future will be challenging for media portrayals of HF/E and related issues. Using the model suggested by Hilgartner and Bosk (1988), a thorough treatment of issues using an HF/E lens would require a large carrying capacity and provide a low level of drama in most cases. The complexity of human–system interaction is not easily reducible to a media soundbite and the more nuanced positions of systems thinking tend to be less dramatic than a typical hero/villain storyline involving human agents. When HF/E-related stories do rise to the level of requiring media coverage, due to the short time frames available and the need for drama, they will often fit into preexisting, simplistic frames, as found in the discussion above.

The ordinary tales of HF/E practitioners helping to improve system performance and human wellbeing, like other aspects of "normal work," tend to be ignored by the media. Successful outcomes are often uninteresting unless there was a perceived risk of failure. Therefore, other more interesting stories will win the competition for public attention. This perhaps reinforces the need for resonant stories about success, and insights into normal work.[for instance, BBC's "Airport Live" (BBC 2013) and "Skies Above Britain" (BBC 2016a)].

Although the discussion above is related directly to media perceptions and actions, we might infer that the public would mirror the presentation of HF/E issues by the media. The frequency with which various themes are presented should make these themes salient in the minds of the public, according to agenda-setting theory. The frame used by the media should influence the frames used by the public to make meaning from the stories. The frame may influence HF/E-related decision making in organizations.

We wonder how HF/E practitioners, as consumers of media, are affected by media effects. Do we, at some level, also buy into the argument that human error is the cause of $x\%$ of accidents, for example—even if "we know what *we* mean by that"? (This notion can be found in many HF/E papers.) Looking back, we have also at some point in the past also fallen for this narrative, using the same sorts of phrases as found in the media, but meaning something different. Dekker (2007) discussed how Western religious narratives regarding the introduction of sin (i.e., the Garden of Eden) suggest ready-made frames of individual culpability following an accident and human error.

So what is an HF/E practitioner to make of these media perceptions and effects? We think that HF/E practitioners can help to shape the popular narrative. HF/E practitioners have opportunities to anticipate and minimize the negative consequences of media effects on the perceptions of clients and stakeholders, and perhaps take advantage of the rare positive effects.

First, HF/E practitioners could engage more with the media. At the moment, other disciplines (and industry representatives) are engaging instead, and the media are reporting an impression of "human factors" (especially) that we may not wish to portray. The HF/E practitioner should, however, be aware of what to expect in the event that the media wishes to use the practitioner for a story, and adjust accordingly. Following an air crash at a nearby airport, one of the authors (Gantt) was interviewed by a local journalist who wanted to write a story about the reactions of passengers in the planes following the crash. After two interviews totaling an hour and a half, discussing the nuances of disaster and emergency psychology research, the rarity of "panic" in disasters, and the fact that panic seemed like a poor way to describe the behavior of the passengers in this crash, it seemed that a proper treatment of the subject had been given. When the story was released shortly thereafter, only one line of the interview found its way into the story, surrounded by a narrative that seemed to suggest mass panic on the part of the passengers. It seemed that the journalist had a frame and the story told by the author did not fit into that frame.

HF/E practitioners who are solicited for news stories can do their best to portray the discipline as accurately as possible. In emotive stories of accidents, it is probably wise to avoid speculation, or to give a lengthy interview, but rather succinctly summarize the HF/E discipline to avoid being quoted selectively or out-of-context (e.g., sticking to the themes in the International Ergonomics Association [IEA] definition, IEA, 2015)). This would achieve the journalist's aim of getting a timely quote, with minimal risk to the practitioner and discipline. At the moment, HF/E specialists who speak to the media are primarily researchers. Where practitioners cannot speak to the media, they can educate clients and stakeholders about the more expansive focus of HF/E. On more general issues, HF/E practitioners can contribute to feature articles and programmes. An example can be seen in the video of British ergonomist Mark Young on a BBC article on "Five ways ergonomics has shaped your life" (Lane, 2009) , and Neville Stanton's interview for Radio 4 on autonomous cars (BBC, 2016a).

Second, the HF/E practitioner can monitor media reports to identify how HF/E and related issues are presented, and respond where it is helpful to do so. New media tend to focus on "first stories" (Cook et al., 1998), and surface features of events, such as "human error." In advertising and public communications, the law of primacy in persuasion holds that the side of an issue presented first will have greater effectiveness than the side presented subsequently. In HF/E our systems perspective encourages us to concern ourselves with the second story, including systemic and underlying features. It is fair to say that the study of "human error" (at least according to some definitions) has helped the development of the understanding of human performance (e.g., in design and crew resource management). But with ever increasing system complexity and obscure causation, might our focus on "human error" as an anchoring object in our analyses and narratives have unintended consequences, perhaps serving to reinforce a populist view of "human error"? The HF/E practitioner can identify patterns of perceptions—perhaps a first story focus—that may be shared by clients and stakeholders. Knowing these in advance, the practitioner can better prepare to counter these perceptions verbally, in writing, graphically, or via video. This can have surprising impact. Following the "bad apple" media reporting of the train crash in Santiago de Compostela, one of the authors (Shorrock) wrote an article for EUROCONTROL's HindSight magazine called "Human error: The handicap of human factors, safety and justice" (Shorrock, 2013). A shortened version of this article was blogged on www. safetydifferently.com, entitled "The use and abuse of human error." This article was spotted by Tom Peters, one of the world's top management gurus, who tweeted the article to over 100K followers, with a subsequent complementary tweet directed at the blog generally. In the 24 hours that followed, the post received over 1000 views, a spike clearly attributable to the influence of Tom Peters. At the time of writing, the post has been viewed many thousands of times, plus distributed to thousands of controllers and other readers via HindSight magazine and downloaded via www. skybrary.com. The theme was subsequently the theme of a discussion at Ergonomics and Human Factors 2014 entitled "Is human error the handicap of human factors?" which was recorded and written as another post (http://bit.ly/1c5w1Ax) and a magazine article (Shorrock, 2015). What started as an internal reaction to media misreporting resulted in an article shared to thousands of people outside the HF/E discipline. This was not really the intention, but it did illustrate the value of outreach and public engagement (see Chapter 31 by Dom Furniss).

More generally, knowing how the media will tend to portray HF/E, the HF/E practitioner can prepare materials and arguments to address these beliefs (for clients, the public, the media, etc.). This might involve writing to newspapers and journalists to address aspects of their reporting. At the least, the HF/E professional can be aware of the fact that the meanings we attach to labels that we use might not be shared by the media, or the public, and this poses questions for how we communicate.

Third, knowing that people use stories to make sense of their world, the HF/E practitioner can identify examples in the media (news, TV, or film) that may provide useful analogies when explaining concepts. Activating these frames may provide the HF/E practitioner opportunities to explain complex topics in relatable ways.

It would be easy for HF/E practitioners to deride the media for its simplistic and inaccurate portrayal of HF/E and related issues. But we could instead take a systems view of the media, its influencers, and its influences. Prominent figures in related disciplines (e.g., psychology) do seem to influence the media, and HF/E practitioners could have influence too, without necessarily becoming celebrity scientists (a double-edged sword). Organizations exist to influence public debate and media coverage (e.g., Sense About Science), and some HF/E societies have this in their remit. While it is difficult to influence the media's portrayal of HF/E and related issues, it can be done in collaboration with media channels (if done carefully), or through social media. In this way, HF/E practitioners can help to moderate the distorting portrayal of HF/E, and show the true scope of the discipline and profession.

ACKNOWLEDGMENTS

We are thankful to Martin Bromiley, Sidney Dekker, Daniel Hummerdal, Roel van Winsen, and John Wilkinson for comments on an earlier version of this chapter.

REFERENCES

BBC 2013. Airport live. First broadcast 17 June 2013. http://www.bbc.co.uk/programmes/p018t3xg/episodes/guide

BBC 2016a. Autonomous cars, Bees and neonicotinoids, Marden Henge, Royal Society Book Prize. BBC Radio 4. First broadcast 18 August 2016. http://www.bbc.co.uk/programmes/b07nn8l5

BBC 2016b. Skies above Britain. First broadcast 17 August 2016. http://www.bbc.co.uk/programmes/b07pmwyy/episodes/player

Bekier, M., Molesworth, B.R.C., and Williamson, A. 2011. Defining the drivers for accepting decision making automation in air traffic management. *Ergonomics*. 54(**4**), 347–356.

Cook, R.I., Woods, D.D., and Miller, C. 1998. *A tale of two stories: Contrasting views of patient safety*. Chicago, IL: National Patient Safety Foundation at the AMA. Accessed March 3, 2016. Available at: https://c.ymcdn.com/sites/npsf.site-ym.com/resource/collection/ABAB3CA8-4E0A-41C5-A480-6DE8B793536C/A-Tale-of-Two-Stories%281%29.pdf.

Dekker, S.W.A. 2007. Eve and the serpent: A rational choice to err. *Journal of Religion and Health*. 46(**1**), 571–579.

Downer, J. 2013. Disowning Fukushima: Managing the credibility of nuclear reliability assessment in the wake of disaster. *Regulation & Governance*. 8(**3**), 287–309.

Goffman, E. 1974. *Frame Analysis: An Essay on Organizational of Experience*. New York, NY: Harper & Row.

Govan, F. 2013. Spain train crash: Driver had boasted about speeding on Facebook page. *The Telegraph*. May 26, 2013. Accessed March 3, 2016. Available at: http://bit.ly/1RBn2XZ.

Hansen, A. 1991. The media and the social construction of the environment. *Media, Culture and Society*. 13, 443–458.

Heckle, H. 2013 Spain train driver ignored 3 warning signals to slow down before crash: Investigators. *CTV News*. August. Accessed March 3, 2016. Available at: http://bit.ly/1RAWllX.

Heider, F. and Simmel, M. 1944. An experimental study of apparent behavior. *American Journal of Psychology*. 57(**2**), 243–259.

Hilgartner, S. and Bosk, C.L. 1988. The rise and fall of social problems: A public arenas model. *Journal of Sociology*. 94(**1**), 53–78.

Hollnagel, E., Leonhardt, J., Licu, T., and Shorrock, S. (2013). *From Safety-I to Safety-II: A White Paper*. Brussels, Belgium: EUROCONTROL.

Hoye, M. 2015. Amtrak to install inward-facing cameras in train cabs. *CNN*. May 27, 2015. Accessed March 3, 2016. Available at: http://edition.cnn.com/2015/05/26/politics/amtrak-cameras-train-cabs/.

IEA. 2015. *Definition and domain of ergonomics*. Geneva, Switzerland: *IEA*. Accessed March 3. 2016. Available at: http://iea.cc/whats/index.html.

Lane, M. 2009. Five ways ergonomics has shaped your life. *BBC*. Accessed March 3, 2016. Available at: http://news.bbc.co.uk/1/hi/magazine/8363862.stm.

Lawrence, R.G. and Birkland, T.A. 2004. Guns, Hollywood, and school safety: Defining the school-shooting problem across public arenas. *Social Science Quarterly*. 85(**5**), 1193–1207.

Mark, R.P. 2014. Wrong runway landings an urgent wake-up call. *CNN*. January 14, 2014. Accessed March 3, 2016. Available at: http://cnn.it/1FV6qGB.

McCombs, M. 2005. A look at agenda-setting: Past, present and future. *Journalism Studies*. 6(**4**), 543–557.

Nash Information Services, Inc. 2016. Top grossing movies of 2015. *The Numbers*. Accessed March 3, 2016. Available at: http://www.the-numbers.com/market/2015/top-grossing-movies.

National Transportation Safety Board. 2010. Loss of Thrust in Both Engines After Encountering a Flock of Birds and Subsequent Ditching on the Hudson River, US Airways Flight 1549, Airbus A320-214, N106US, Weehawken, New Jersey, January 15, 2009. *Aircraft Accident Report NTSB/AAR-10/03*. Washington, DC: NTSB.

Nisbet, M.C., Scheufele, D.A., Shanahan, J., Moy, P., Brossard, D., and Lewenstein, B.V. 2002. Knowledge, reservations, or promise? A media effects model for public perceptions of science and technology. *Communication Research*. (**5**), 584–608.

Pasztor, A. 2010. 'Hudson miracle' gets closer look. *The Wall Street Journal*. May 4, 2010. Accessed March 3, 2016. Available at: http://on.wsj.com/1eJ9CuT.

Rae, A. 2014. *Tales of Disaster: The Role of Accident Storytelling in Safety Teaching*. London: Springer.

Scheufele, D.A. 2000. Agenda-setting, priming, and framing revisited: Another look at cognitive effects of political communication. *Mass Communication & Society*. 3(**2&3**), 297–316.

Scheufele, D.A. and Tewksbury, D. 2007. Framing, agenda setting, and priming: The evolution of three media effects models. *Journal of Communication*. 57, 9–20.

Shoichet, C.E. 2013. Roller coaster ride became 'nightmare' for Texas woman's family. *CNN*. July 22, 2013. Accessed March 3, 2016. Available at: http://edition.cnn.com/2013/07/21/us/texas-roller-coaster-death/.

Shorrock, S. 2013. Human error: The handicap of human factors, safety and justice. *Hindsight Magazine*. 18, 32–27. Brussels, Belgium: EUROCONTROL.

Shorrock, S. 2015. 'Human error' in the headlines: Press reporting on Virgin Galactic. *Humanistic Systems*. July 30, 2015. Accessed March 3, 2016. Available at: http://humanisticsystems.com/2015/07/30/human-error-in-the-headlines-press-reporting-on-virgin-galactic/.

Weick, K.E. 1995. *Sensemaking in Organizations*. Thousand Oaks, CA: Sage.

Yahoo Travel. 2015. What really causes plane crashes? (It's not what you think). *Yahoo! News*. August 3, 2015. Accessed March 3, 2016. Available at: https://www.yahoo.com/style/what-really-causes-plane-crashes-its-not-what-125605316542.html.

Zilmer, E.A. 2015. What was in the mind of Andreas Lubitz? *CNN*. March 29, 2015. Accessed March 3, 2016. Available at: http://edition.cnn.com/2015/03/27/opinions/zillmer-germanwings-co-pilot/.

Part II

Fundamental Issues
for Practitioners

This part considers some of the fundamental issues for practitioners of human factors and ergonomics, as well as others associated with HF/E. These issues concern practitioner roles and contexts, the research–practice relationship, tools and methods, and practitioner development.

Chapters 7 and 8 consider the roles that HF/E practitioners (and others) may take on, and the types of organizations in which practitioners are situated. Each of these roles and contexts has different implications for practice. Roles for HF/E specialists are explored in Chapter 7 (Claire Williams and Steven Shorrock). Practitioners may take on roles of content expert on the one hand, and process facilitator on the other hand, but practitioners may also shift between other more specific roles, such as "teacher," "detective," and "advocate." Others associated with HF/E aims may also adopt certain roles. Such roles have implications for us and others, and the projects, organizations and wider contexts with which we engage. The organizational contexts within which we are embedded also affect our work profoundly. They include consultancies, producers, manufacturers, service providers, universities and research institutes, governments (including regulators), and intergovernmental organizations. Each of these may have economic, organizational, social, and personal upsides and downsides. These are discussed in Chapter 8 (Steven Shorrock and Claire Williams) via interviews with 17 HF/E practitioners, all contributing authors to this book.

Chapters 9 and 10 go on to examine research–practice issues. Integrating research into practice is critical for maintaining competency; theory, models, concepts, and methods change over time. But there are perceived barriers to the use of research for HF/E practitioners, and a research–practice gap in HF/E is a continuing concern in the discipline and profession. This is explored in Chapter 9 (Amy Chung, Ann Williamson, and Steven Shorrock), via a review of the literature and a survey of over 600 HF/E specialists. The other side of this equation concerns doing practice-oriented research. Chapter 10 (Claire Williams and Paul Salmon) outlines the goals, demands, resources, methods, outputs, trade-offs, compromises, and constraints of practice-oriented research with examples and advice for HF/E practice-oriented researchers.

Practitioners know, however, that there are complications and constraints in using the methods that often result from research, and from practice. Trade-offs and compromises are needed here too. These issues are explored in Chapter 11 (Matthew Trigg and Richard Sciafe).

But how does one develop as an HF/E practitioner? This issue is explored in Chapter 12 (Andrew Baird, Claire Williams, and Alan Ferris), which addresses the blend of knowledge, skills, abilities, and other factors that are needed, from initial training though to ongoing training on the job, and the constraints and challenges to the various parties to this process.

7 Human Factors and Ergonomics Practitioner Roles

Claire Williams and Steven Shorrock

CONTENTS

PRACTITIONER SUMMARY

Human factors and ergonomics (HF/E) practitioners do not all work in the same way. We take on various roles depending on the situation and context, and our own preferred ways of working. While we may favor particular roles over others, we need to be able to reflect on and communicate about our roles, adapt, and consider who else might be better placed to adopt certain roles. Whatever roles we take on, we need to hone process and interpersonal skills alongside content knowledge.

INTRODUCTION

In our roles as HF/E practitioners, what we do as individuals can have effects that have wide consequences for others as well as ourselves (Corlett, 2000). Understanding these effects and consequences is important for professional practice. In this chapter, we consider a variety of roles we may adopt as HF/E practitioners, which are critical to the kinds of effects that we have. Some of these roles we choose up-front, while others we may find ourselves inhabiting in response to specific circumstances with respect to a project, organization, industry, or the HF/E discipline and profession.

In this chapter, we will consider a number of different role taxonomies. We focus on the work of Schein (1978) to link the *roles* of a practitioner (content expert vs. process facilitator) with models of their *functions* (purchaser and doctor–patient, catalyst, and facilitator), and discuss examples of each of these. We then move on to consider roles in which HF/E practitioners may find themselves, irrespective of intent, by considering Steele's intriguing 1975 taxonomy of consultant roles: teacher, student, detective, barbarian, clock, monitor, talisman, advocate, and ritual pig. We will draw out some points for reflection using examples from our own experience, as well as feedback from a conference workshop on multidisciplinary teams with 27 HF/E participants (at the Ergonomics and Human Factors 2015 conference).

A WORD ABOUT WORDS

Throughout this book we have chosen "practitioner" as our descriptor for those who apply HF/E methods and theory in "real" contexts (typically organizations, but also potentially communities, homes, etc.). Much of the literature discussed in this chapter, however, refers to "consultants"—people who provide expert advice, professionally. We are particularly looking at the roles of HF/E practitioners whose expertise is being sought by others, and so in this chapter we will treat "consultant" and "practitioner" as synonymous terms. Some of the roles may apply to others with an interest in HF/E, for instance clients or those who are not qualified and experienced HF/E practitioners but promote HF/E or integrate aspects of HF/E theory and method in their work.

LIPPITT AND LIPPITT'S CONTINUUM

Consultants in many fields have been described using a continuum, for instance with increasing or decreasing degrees of direction by the consultant. One example was proposed by Lippitt and Lippitt (1986) (see Table 7.1). In some situations, consultants may be more directive, giving specific advice on the goals and methods of problem solving, and taking more responsibility for specifying solutions. In other situations, consultants may be less directive, perhaps observing, asking questions, and reflecting back to the client. In the middle, there is a more balanced, collaborative problem-solving relationship. HF/E practitioners will recognize many of these roles, and will likely have adopted several.

TABLE 7.1
Lippitt and Lippitt's (1986) Consultant Roles

Directedness (Continuum)	Role	Description
Nondirective (client focus)	Objective observer/ reflector	Observes behavior, provides feedback, asks questions, and raises questions for reflection
	Process counselor	Observes problem-solving process, interviews, raises issues, and makes suggestions
	Fact finder	Gathers data via interviews, questionnaires, observation, documents, tests; stimulates thinking; and interprets
	Alternative identifier and linker	Identifies alternative solutions, helps to evaluate and determine likely consequences but does not assist in selecting final solution
	Joint problem solver	Consultant and client work to identify and solve the problem, including interpreting the problem, identifying causes, generating and evaluating solutions, choosing a solution, and generating an action plan
	Trainer/educator	Provides instruction and other directed learning opportunities. Assesses training needs, writes learning objectives, and designs learning experiences
	Informational expert	Defines right and wrong approaches to a problem. Consultant plays a directive role until the client is comfortable with the approach
Directive (consultant focus)	Advocate	Consultant uses power and influence to impose their ideas and values about content or process issues, including goals and/or method of problem solving

MULLIGAN AND BARBER'S CONSULTANT AS ARTIST AND CONSULTANT AS SCIENTIST

Sometimes, rather than a continuum, a dichotomy is presented. For example, Mulligan and Barber (2001) define two overarching roles: consultant as artist (yin) and consultant as scientist (yang) (Table 7.2).

The practice of HF/E has variously been described as art, craft, and science (e.g., Spielrein, 1968; de Moraes, 2000; Wilson, 2000), and this debate is ongoing (e.g., Sharples and Buckle, 2015). While the International Ergonomics Association (IEA) definition of HF/E labels it a "scientific discipline," Moray (1994) argued that "Our discipline is an art not a basic science, and one which only makes sense in the full richness of the social setting in which people work" (p. 529). The need in HF/E practice for both scientific skills (yang) and those more aligned to the artist (yin) has been acknowledged and discussed (e.g., Shorrock and Murphy, 2005, 2007; Williams and Haslam, 2006, 2011).

SCHEIN'S CONTENT EXPERT AND PROCESS FACILITATOR

Schein (1978) also outlined two models of consultation: "content expert" (advising or telling others what to do) and "process facilitator" (supporting others to generate solutions). For simplicity, we will discuss these two roles in more detail.

Schein explained that each of the roles can function within two models. The content expert role concerns the task to be performed and problem to be solved, and has a *purchase* model and a *doctor–patient* model. The process facilitator role concerns the way in which the problem is addressed, and has a *catalyst* model and a *facilitator* model.

CONTENT EXPERTISE AND THE PURCHASE MODEL

The content expert may function in the *purchase* model, where a client buys information or expertise from the consultant. For example, a client may ask "simple" questions such as "How high should this shelf be?" or "What size should this escape

TABLE 7.2

Artist versus Scientist

Yin—Consultant as Artist	Yang—Consultant as Scientist
Soft focus	Hard focus
Inner world directed	Outer world directed
Explores through experience	Applies theories
Attuned to feelings and intuitions	Attuned to thoughts and senses
Attends to the relational dance	Attends to boundaries and rules
Concerned with being	Concerned with doing
Expresses and creates	Diagnoses and tabulates

Source: Mulligan, J. and Barber, P., The client-consultant relationship. In: Sadler, P. (ed.), *Management Consultancy: A Handbook for Best Practice*, London, Kogan Page, 2001, pp. 83–102.

hatch be?" or more complex questions like "Can you design this control room?" In principle, they are purchasing expertise already held by the HF/E practitioner, though immediately, corollary questions are likely to be crying out to HF/E practitioners reading this, such as "What's the shelf for?", "Who is using the escape hatch, wearing what?", and "Can I talk to the stakeholders for the control room?" So in spite of the fact that HF/E practitioners will have a good deal of technical knowledge, the simple purchasing of content expertise already held probably rarely describes the situations that most HF/E practitioners face.

CONTENT EXPERTISE AND THE DOCTOR–PATIENT MODEL

In the *doctor–patient model,* a client may come with "symptoms" and the content expert diagnoses problems and proposes remedies. The question may come as "Why are we having so many accidents?" and the HF/E practitioner will undertake a range of activities to "diagnose" and propose solutions to the problem. This description is likely to resonate more with the experience of a typical HF/E consultancy project, though in the diagnosis phase, the practitioner may not take on the content expert role, but instead may draw on the process facilitator roles and models described below.

PROCESS FACILITATOR AND THE CATALYST MODEL

In the *catalyst* model the consultant has no answers per se, but supports the client to find solutions to their problems; the consultant is a catalyst, facilitating arrival at solutions but not generating them. While not depending on content knowledge, clearly the practitioner will be using considerable skill in this approach. In reality, this is an unlikely role and model for the HF/E practitioner in its purest sense, because we bring content knowledge to any project by virtue of being the HF/E specialists on the project. If facilitation was the only thing required, then HF/E competencies may not be required, but others with an applied interest in HF/E may well take on this role. Having said that, the role of "honest broker" and challenger is one regularly frequented by us for our clients. In this sense, HF/E as facilitation is often more directive than process facilitation in its purest sense.

Take, for example, the scenario where one group in an organization wants a particular solution to an HF/E problem (say, a specific piece of equipment for a job) while another group wants something different. We have been involved in providing content expertise to affirm that either product is fit for purpose, but then have been facilitators of the discussion to arrive at a final decision, where we have no HF/E view as to which the client should use beyond our commitment to the participatory process for choosing it.

In one project, operators in a new airport control tower had a new large interactive touchscreen map that was oriented north up, but the operators were looking south at the airfield. In the previous tower, all operators had always faced north. Close to the date of operation of the new tower, some operators had difficulties using a north-up touchscreen map when looking south, since the physical and mental activity flows were the opposite of those observed visually. But not all operators agreed. Some could work with the screen 'north up' while others insisted that they could only use it flipped 180deg south up – corresponding to the outside view. (The display made it possible to flip the screen in either orientation.) While there was HF/E content knowledge on the issues of compatibility, population stereotypes, mental rotation and transfer of training and experience (with north-up maps), for instance, there was no clear answer to which was best for all. An in-situ workshop probed the needs of operators (experienced and inexperienced) and assessed the risks, and participants agreed a trial period where each could choose an orientation – north or south up. After a period of time, all became comfortable using the screen in a north-up orientation, but all also felt listened to and taken seriously.

(Steven Shorrock)

For HF/E practitioners, process facilitation in the catalyst model may resemble "agents of change" (Argyris and Schön, 1978; Caple, 2007) or "skilled helpers" (Egan, 2001; Shorrock and Murphy, 2005), with a particular emphasis on communication and interpersonal and relationship skills.

PROCESS FACILITATOR AND THE FACILITATOR MODEL

In the *facilitator model*, the consultant may well have ideas and possible solutions, but withholds these in favor of helping the client to arriving at their own. Again, in our experience this is unlikely to exist entirely without some HF/E content expertise, but we have worked on projects where the implementation decisions are entirely left to the client, with only a facilitation role for the HF/E practitioner.

After a long project of content expert/doctor-patient work with a food production company, we provided the client with a list of prescribed needs. These included: a more usable near miss reporting system; better break-taking facilities; more time to clean down the line at shift end. We expanded on what we meant by each point but did not describe how to achieve any of them (though we had some of our own ideas). We then facilitated sessions for the client to generate how they would provide for each of these needs, ensuring representatives of all of the stakeholder groups were there.

(Claire Williams)

SOME POINTS FOR REFLECTION FROM SCHEIN'S MODEL

This relatively simple description of consultancy is a useful way of describing possible approaches we may take to a given piece of work. Different projects will require different role/model combinations and it can be a challenge to make clear what role we have (and to remain aware of that throughout a project), and also to agree this with the client so that their needs are met. There is no single right approach. In the 2015 workshop, of the 27 HF/E participants, 15 most commonly found themselves as content expert and 12 as process facilitator. The discussions revealed that most worked in both roles at times. In some participatory ergonomics programmes, for example, the ergonomist may take aspects of both roles, providing scientific information as a "content expert," but empowering and facilitating nonergonomist team members to generate their own solutions with the information (Devereux and Manson, 2008). This affirms that our training needs to include both content and process skills.

STEELE'S ROLES FOR CONSULTANTS

Having considered these more general roles or approaches that we tend to choose up front, we also want to consider some more transient roles that may apply to HF/E practitioners, in response to client needs and circumstances as they unfold during a project or, more generally, in an organization, industry, or in society (e.g., via social media). Some of these roles may also apply to other HF/E stakeholders. Steele's (1975) nine-point taxonomy is a useful way to discuss this. We give some reflections from our own experiences for some of the roles.

TEACHER—IMPARTING KNOWLEDGE BY SEMINARS, WORKSHOPS, OR COURSES

This sits within Schein's "content expert/purchasing model" and is also reflected in Lippitt and Lippitt's roles. It is a common role for many HF/E practitioners. The ability to educate others in HF/E is probably a goal for most HF/E practitioners. Many practitioners have done seminars, workshops and training, and some have formal roles on industry or higher education courses.

> We have been 'teachers' when starting out on an ergonomics change team project in broad industries including pharmaceutical and automotive manufacture and food production; educating the in-house or external teams in HF/E, and have also taken on this role throughout the lifetime of a project where the acquisition of additional HF/E expertise is required by the client.
>
> **(Claire Williams)**

> Part of my role has involved helping to educate members of the judiciary of many European countries on systems thinking and human factors in the context of 'just culture'. Many prosecutors and judges have never heard of human factors or systems thinking, and the key ideas are completely new to them. This educative activity therefore helps to address an important issue that is neglected in HF/E: judicial attitudes to people involved in accidents. Another educational activity involves teaching a course on HF in transportation. Most of the students are frontline personnel such as pilots, and teach me at least as much as I hope to teach them.
>
> **(Steven Shorrock)**

STUDENT—MODELING A LEARNING BEHAVIOR THAT WE WOULD LIKE TO ENCOURAGE IN THE CLIENT.

Often during a project, we will uncover a need beyond our own technical expertise, requiring input either from a colleague in our own organization or one from outside. When this happens, the practitioner may, for instance, move to the "student" side of the table to learn from the technical expert and apply this to the situation.

> We have recently required expertise on the properties and storage requirements of specific substances, where we were advising on layout and flow issues at a plant. We

learnt from experts, alongside our client, and put the new learning into our advice about the plant layout.

(Claire Williams)

It is also important that we can empathize with students who may have little prior exposure to HF/E concepts and methods or their application, and perhaps little prior formal education.

DETECTIVE—TRYING TO DISCOVER HARD AND SOFT DATA TO GAIN AN ACCURATE PICTURE OF THE SYSTEM, ITS PROBLEMS, AND ITS STRENGTHS.

This is a common role for HF/E practitioners, and was cited most commonly as the role the workshop participants took on in their work. It may be akin to the "diagnosis" stage in Schein's doctor–patient analogy, or to the catalyst model of the process facilitation role. The term detective, of course, implies the need to look beyond the superficial and obvious, amassing information from multiple sources.

> In safety culture workshops where there may be a messy or confused picture of a situation, and the HF/E practitioner needs to work with participants to unearth multiple perspectives on a situation, perhaps including a discussion of very sensitive issues that are otherwise rarely discussed.

(Steven Shorrock)

BARBARIAN—VIOLATING COMFORTABLE BUT LIMITING NORMS AND TABOOS THAT ARE PREVENTING THE SYSTEM FROM BEING AS EFFECTIVE AS IT MIGHT BE. (A COUNTERMEASURE AGAINST TUNNEL VISION.)

Undoubtedly, some people find this role easier than others, but we have certainly needed to take on this role to "point out the elephant in the room." While "Barbarian" implies an uncivilized or primitive individual (one therefore not bound by the social norms of a civilized setup) our experiences of "speaking the unspeakable" are that it can be done tactfully and gently, even in its directness, perhaps using humor, analogy, or storytelling.

> Recently when I was working with a senior team on safety culture, they mourned the lack of trust within the business, particularly lack of trust towards them. My response was to ask them what they did that made them trustworthy. I had witnessed several broken promises from them to their staff, even in the time I was working with them. Each broken promise was understandable (feedback from a report not provided because it was too contentious; capital spend on an improvement left undone because of cashflow issues) but to their colleagues, the senior team was not trustworthy. My conversation with them was relatively gentle though its content was uncomfortable. I was the Barbarian.

(Claire Williams)

Acting as the HF/E specialist on a major infrastructure project can put the practitioner into this space. In one example, toward the latter stages of a project, a new facility was reaching the date when it was meant to go live. Observations from the training simulator suggested that the end users were not ready for the transition. I put my observations into a memo to the local management, and presented this to them in person. I was challenged on what I could know so soon into a new project, but I had spent several days observing operators in the simulator. No one else had, and this experience – confirmed by the front-line operators – helped to ensure that the 'go live' date for the project was delayed by several months, following additional training in the simulator and in shadow mode, and the facility opened very successfully.

(Steven Shorrock)

ADVOCATE—ADVOCATING CERTAIN PRINCIPLES PERTAINING TO THE RELATIONSHIP BETWEEN AN INDIVIDUAL AND THE ORGANIZATION.

Steele's advocate is different to the "advocate" proposed by Lippitt and Lippitt. HF/E practitioners (and others aligned with HF/E goals) will often find themselves in this role, essentially advocating for the "user." It may be combined with the *Barbarian* role, perhaps challenging decision makers, especially when there is a clash of values or priorities. In our experience, this role is a fairly common one, for example when running stakeholder workshops, when presenting project findings to different groups, and even when collecting data at the start of the project.

One railway project involved understanding the user needs for a particular railway system, and documenting these in a specification. There was pressure from within the organisation to exclude one group of users from the specifications. This was for political reasons. However, the purpose of the project (spurred by a previous accident) dictated that this user group was safety-critical staff. This was a source of conflict within the organisation, between powerful individuals who wanted certain users excluded from documentation (but could not openly say why), and the project members who believed that these users were safety-critical staff. Ultimately, the user group remained in the documentation and their needs were met in the specifications.

(Steven Shorrock)

CLOCK—STIMULATING THE CLIENT INTO GETTING SOME THINKING OR EXPERIMENTING DONE, BY ACTING AS A REGULAR "TIME SIGNAL."

In our experience, this role is more often the job of a project manager, and so may be a role for HF/E specialists in human factors integration projects, for example, but is not particularly specific to HF/E. The 27 workshop participants cited this as a role that they were least likely to take on.

MONITOR—PROVIDING AN INDEPENDENT VIEW OF HOW THE CLIENT IS PERFORMING IN CONNECTION WITH SOME MUTUALLY AGREED TASK RELEVANT TO THE PROBLEM.

The role of *Monitor* is a useful description of the auditing function that we might provide to our clients. It will therefore tend to be more of an up-front role, specifically chosen for a piece of work, as opposed to many of the other Steele roles, which might be adopted in response to circumstances. Again, the workshop suggested this was one of the less common roles for HF/E consultants, perhaps because of how the role is defined by Steele. However, more generally, it is common for HF/E specialists to have independent assessment and auditing roles, formally (e.g., as a regulator) or informally.

> Following a safety culture survey, clients will tend to draw up an action plan of changes or further activity that they want to perform. This typically concerned changes to the sociotechnical system that are hoped to influence how people think and work. To get an independent view of the effectiveness of their actions, clients will ask an HF/E practitioner to discuss progress with staff.

> **(Steven Shorrock)**

Talisman—providing a sense of security and legitimacy by our presence, thus allowing the client to feel comfortable enough to experiment in new areas or with new ideas.

This is the role that made the workshop participants most uncomfortable. In discussion, a number of stories shed light on when HF/E practitioners were the *Talisman* and why this was problematic. In one example, from the oil and gas sector, an experienced practitioner described how the presence of an HF/E specialist was mandated for the various review stages of the design of a new asset. Instead of this supporting meaningful input, however, he was akin to a "good luck charm"; they wanted the person there so they could tick the HF/E box, and this did empower them to experiment with new ideas. The problem was, their new ideas took no consideration of the HF/E issues and the practitioner soon moved from *Talisman* to *Barbarian*. However, in other circumstances, the HF/E practitioner may add a sense of security while a client experiments with new ways of thinking.

> In a set of workshops, a client wished to experiment with some different ways of thinking about safety. Over the course of a few days, the experimental 'safety lab' led to insightful use of several systems thinking techniques, and adaptation of other thinking techniques (e.g., Edward de Bono's six thinking hats). This was partly led by me, but often led by participants who would take the lead for different activities.

> **(Steven Shorrock)**

Ritual Pig—inadvertently serving as an "outside" threat, creating solidarity and action within the organization by having to deal with you.

This is probably not a role any of us would seek out, though it is one we have occasionally found ourselves in. Commercially, in line with its name, the ritual pig role is likely to be deathly to fulfill commercially, and may preempt the end of working with that client. However, as a regulator, it may be part of the role, especially when it comes to enforcement action (but the purpose would not normally be to "help create solidarity and action within the organization by having to deal with you"). Even in some projects where HF/E objectives conflict with other project objectives, this role may surface to some extent.

A subtle variant of the ritual pig might be called the *omega wolf*. In wolf packs, the omega acts as both a scapegoat for the other wolves and as a kind of social glue to prevent fights or all-out war by taking on the frustration of the other pack members. It may sound little better than a ritual pig, but in this case, the omega wolf is not killed. It exercises a role skillfully and actively in order to achieve a purpose.

There can be some interesting dynamics and role transitions in some of the roles proposed by Steele. Consider the following example. An HF/E practitioner is brought into a project as a *Detective*, but the role transitions to *Talisman* and subsequently *Barbarian*, as problematic features of the project emerge. If this is not welcome to the project team, and there is an inadequate relationship, there is a risk that the practitioner (perhaps seen as part of an out-group) is thrown into the role of *Ritual Pig* by the project team (the in-group).

> This is an experience we have had recently in an organisation that already had taken against the individual who had hired us. Even though the work we were doing fulfilled exactly the brief he had given us, his colleagues used the feedback opportunity that we provided at the end of our work, to first coalesce in their rejection of us (because we had been hired by him), and then to continue their solidarity in firing him.

(Claire Williams)

ADDITIONAL ROLES FOR HF/E PRACTITIONERS

Related to the role of *Advocate* above is another role, which our experience suggests is important: the *Evangelist* role (Hicks, 1991). Our definition would be: "Enthusiastically espousing a message with the goal of converting others to our way of thinking." HF/E has at its heart particular humanistic and systemic values, and these values may need to be promoted (e.g., to counter contrasting values) (see Chapter 6 by Gantt and Shorrock on the media). HF/E practitioners have taken on this role at every level of an organization, from the shop floor to the boardroom, with project-specific messages, but also more widely to other professions, governmental bodies, and society as a whole, through professional and industry bodies and via social media. This has opened up opportunities for emerging groups in specific sectors, such as the Clinical Human Factors Group (CHFG), which has excelled in this role.

Evangelism may involve evoking positive emotional reactions (as well as some negative ones), and hence may be outside of some practitioners' comfort zones. As Hicks notes, this role requires credibility and carries some risk; some may be turned off by the message or even feel criticized.

We believe that another important role is that of the *Connector*. In a community context, McKnight and Block (2010) describe how some people naturally seem to have a capacity for making connections in a community, but each of us can be encouraged to discover our own connecting possibility. According to McKnight and Block, these connectors typically are gift-centered, seeing the "half-full" in everyone; are well-connected themselves; are trusted and create trusting relationships; believe in the people in their community; and get joy from connecting, convening, and inviting people to come together. Connectors are seeking to join people, not lead them. This role would seem to be especially important in the contexts in which HF/E practitioners find themselves, where we may communicate with CEOs and senior managers, middle managers, frontline staff, end users, regulators, policy makers, and many other professions, often in multiple industries, within and between organizations, in person and also via social media.

CONCLUSIONS AND REFLECTIONS

In this chapter, we have presented a few ways of describing the roles of HF/E practitioners. The experiences of the authors and those attending the workshop from which some of these findings are drawn suggest there are many roles. It is important to reflect on the roles that we adopt, along with the implications of the roles for us and others, in the context of a project, organization, and wider context, including the cultural and political environment (e.g., see Broberg and Hermund, 2004). We may shift roles as we go through a process or project, and in different situations. Role transitions that are not reflective and reflexive may be a source of difficulty. For example, we can be driven to the role of *Barbarian* by frustration rather than by project needs, and in doing so may lose our right to a dialogue by the nature of our delivery. We may slip from facilitator to content expert without awareness. There may also be helpful and unhelpful role combinations, roles that we have not really considered before, and roles that may be better performed by others who have a stake in the goals of HF/E. Our roles also have ethical implications. In the more directive roles, or content expert model, we must be particularly careful. We may be having significant influence over client decisions, which places ethical responsibility on us as practitioners, in terms of our competencies and role. Considering the broad range of outcomes (e.g., efficiency, safety, comfort) and stakeholders (e.g., staff and management), our interventions have the potential to do harm as well as good. We may also be taking on some responsibly for client decisions, even co-creating a dependency relationship. Alternatively, the client may not be committed to acting on our advice. It is important to communicate and agree roles with other stakeholders, so that immediate and longer term needs are met in an effective, sustainable, and ethical way.

In summary, to be successful as HF/E practitioners, we need to be able to reflect on the roles of ourselves and others, to adapt, and to communicate, but also to be aware that forces outside of us may affect the roles we take on, and the extent of our success.

REFERENCES

Argyris, C. and Schön, D. 1978. *Organisational Learning: A Theory of Action Perspective.* New York, NY: Addison-Wesley.

Broberg, O. and Hermund, I. 2004. The OHS consultant as a 'political reflective navigator' in technological change process. *International Journal of Industrial Ergonomics.* 33, 315–326.

Caple, D.C. 2007. Ergonomics–Future directions. *Journal of Human Ergology*. 36, 31–36.

Corlett, N. 2000. Ergonomics and ethics in a changing society. *Applied Ergonomics*. 31, 679–683.

de Moraes, A. 2000. Theoretical aspects of ergonomics: Art, science or technology–Substantive or operative. *Proceedings of the Human Factors and Ergonomics Society Annual Meeting*. 44(**33**), 264–267. doi: 10.1177/154193120004403350.

Devereux, J. and Manson, R. 2008. A follow-up study using participatory ergonomics to reduce WMSD's. Contemporary ergonomics, 2008, 406.

Egan, G. 2001. *The Skilled Helper: A Problem-Management and Opportunity-Development Approach to Helping*. Belmont, CA: Wadsworth.

Hicks, M.J. 1991. *Problem Solving in Business and Management: Hard and Soft Creative Approaches*. London, UK: Chapman and Hall.

Lippitt, G. and Lippitt, R. 1986. *The Consulting Process in Action*. Second Edition. La Jolla, CA: University Associates.

McKnight, J. and Block, P. 2010. *Abundant Community*. San Francisco, CA: Berrett-Koehler.

Moray, N. 1994. 'De maximis non curat lex' or how context reduces science to art in the practice of human factors. In: *Human Factors and Ergonomics Society 38th Annual Meeting*. Stouffer Nashville, Nashville, TN, 24–28 October, pp. 526–530. Santa Monica, CA: HFES.

Mulligan, J. and Barber, P. 2001. The client-consultant relationship. In: Sadler, P. (ed.). *Management Consultancy: A Handbook for Best Practice*. London: Kogan Page, pp. 83–102.

Schein, E.H. 1978. The role of the consultant: Content expert or process facilitator? *The Personnel and Guidance Journal*. 56(**6**), 339–343.

Sharples, S. and Buckle P. 2015. Ergonomics/human factors–Art, craft or science? A workshop and debate inspired by the thinking of Professor John Wilson. In: Sharples, S., Shorrock, S., and Waterson, P. (eds.). *Contemporary Ergonomics and Human Factors 2015*. London: Taylor & Francis, pp. 132–132.

Shorrock, S.T. and Murphy, D.J. 2005. The ergonomist as a skilled helper. In: Bust, P. and McCabe, P.T. (eds.). *Contemporary Ergonomics 2005*. London: Taylor & Francis, pp. 168–172.

Shorrock, S.T. and Murphy, D.J. 2007. The role of empathy in ergonomics consulting. In: Bust, P. (eds.). *Contemporary Ergonomics 2007*. London: Taylor & Francis, pp. 107–112.

Spielrein, R.E. 1968. Ergonomics: An art or a science. *Australian Occupational Therapy Journal*. 15(**2**), 19–21.

Steele, F. 1975. *Consulting for Organizational Change*. Amhurst, MA: University of Massachusetts.

Williams, C. and Haslam, R. 2006. Core competences in ergonomics–Do we have them? In: Pikaar, R.N., Koningsveld, E.A.P., and Settels, P.J.M. (eds.). *Proceedings of 2006 IEA Congress, International Ergonomics Association Triennial Congress*, Maastricht, the Netherlands, ISSN: 0003 6870. [CD ROM]. Amsterdam, the Netherlands: Elsevier.

Williams, C. and Haslam, R. 2011. Exploring the knowledge, skills, abilities and other factors of ergonomics advisors. *Theoretical Issues in Ergonomics Science*. 12(**2**), 129–148.

Wilson, J.R. 2000. Fundamentals of ergonomics in theory and practice. *Applied Ergonomics*. 31(**6**), 557–567.

8 Organizational Contexts for Human Factors and Ergonomics in Practice

Steven Shorrock and Claire Williams

CONTENTS

PRACTITIONER SUMMARY

The way that human factors and ergonomics (HF/E) fits within an organization and interacts with other organizations has a strong influence on the practice of HF/E. In this chapter, we consider four broad contexts of HF/E practice via interviews with 17 of the practitioners who have contributed to this book. We consider four organizational contexts: External consultancy; working in producers, manufacturers, and service providers; working in universities and research institutes; and working in governments (including regulators) and intergovernmental organizations. The practitioners reflect on the economic, organizational, social, and personal aspects of HF/E practice in these four contexts.

INTRODUCTION

HF/E in practice is mostly situated in organizations, and how HF/E fits into and interacts with organizations affects our opportunities and possibilities for practice. Wilson (2014) referred to this "embedding" as a *"feature of a systems approach"* and *"the way ergonomics fits within the organisational system and is embedded within practice"* (Wilson, 1994). The various organizational contexts affect our possibilities in HF/E, both facilitating and constraining our practice. The real issues are not so much about HF/E theory and methods, but rather concern our social, organizational, and economic interactions.

In this chapter, we consider four broad contexts that seem to fit a majority of the following HF/E practitioners:

1. Consultancy (sole-trader or small partnerships, small consultancies, and larger multidisciplinary consultancies)
2. Producers (e.g., oil and gas, mining), manufacturers and service providers
3. Universities and research institutes
4. Governments (including regulators) and intergovernmental organizations

When planning this chapter, we felt that that the multiple perspectives of practitioners might give the best reflection of what it is like to be an HF/E practitioner in these contexts, including the key upsides and downsides. We therefore interviewed a selection of authors of the chapters in this book, and asked them to reflect independently on their experiences of working in these contexts as HF/E practitioners. We then took a narrative approach to writing, creating a tapestry of practitioner experiences. In the following sections, we integrate reflections from 17 of the authors, who collectively have many decades of HF/E experience embedded in each of the four organizational contexts (all are based in United Kingdom, United States, Australia, New Zealand, or Canada, but some have worked in many other countries).

- David Antle
- Fiona Bird
- David Caple
- Ken Catchpole

- Ben Cook
- Ben O'Flanagan
- Dom Furniss
- Ron Gantt
- Nigel Heaton
- Daniel Hummerdal
- Dan Jenkins
- Dave Moore
- Dave O'Neill
- Richard Scaife
- Sarah Sharples
- Pat Waterson
- John Wilkinson

Many of the issues in this chapter overlap with those raised in the chapters in Part 3 of this book, on HF/E in different industries. Some also overlap with the Chapter 7 on HF/E roles (Williams and Shorrock).

HF/E PRACTICE IN CONSULTANCIES

Perhaps the most common context for HF/E practitioners is the consultancy environment, where services are sold to a variety of clients, often in a wide range of sectors. The practitioners cited below have experience in a range of consultancy environments, from one- or two-person consultancies, through small consultancies with tens of staff, to large organizations with hundreds or thousands of staff, offering HF/E consultancy alongside other services. The following are the main themes that emerged from their reflections.

DIVERSITY AND VARIETY

The consultancy context has a number of upsides, some of which lie in the diversity or variety of work, clients, sectors, and countries. *"I love the variety and pace of the work. I literally work on everything from trains to toothbrushes on a daily basis,"* said Dan Jenkins, a consultant in a medium-sized design company in the United Kingdom. David Caple has been managing a small consultancy in Australia with five to 10 staff for over 30 years. He also reflected on the variety of consulting: *"We typically have around 30 projects running at the same time and each has a story to tell. After 32 years we have been behind so many large fences to understand work in most industry sectors and to implement a wide application of ergonomics knowledge."*

David Antle, an HF/E consultant in Canada, noted that the consulting environment allows for a great deal of diversity when applying knowledge and skills in HF/E: *"Without being tied to one workplace or industrial sector, a consulting firm can take on work in construction and trade services, public utilities, manufacturing, food processing, educational, and office/professional contexts – to name just a few."* He noted a potential downside for those who need or prefer routine: Consultants need to be comfortable with unpredictable changes to schedules—particularly when compared to working for one company or industrial sector.

John Wilkinson, Principal Consultant in a small consultancy in the United Kingdom, said that consultancies *"need some diversity to insure against market shocks and industry business cycles. HFE can also drift in and out of fashion in some industries so the risk needs to be spread."* He also brought up the issue of travel. With diversity comes travel, and this can be very heavy for consultants.

AN EXPANDED ROLE

In small consultancies, responsibilities have to be shared, so the practitioner's role may expand further. John Wilkinson remarked, *"I share the business responsibilities, build up my own client and expert network, provide input to marketing, sell directly, contribute to innovation, and design, propose and deliver my own work, and have direct input to the business strategy and planning."*

For sole traders, accounting, quality, administration, contracting, insurance, marketing, etc., may take significant time, or need to be outsourced. Dan Jenkins reflected on his former experience as a sole trader (in a Limited Company): *"I really enjoyed the diversity of doing everything, learning about marketing, invoicing, the importance of a well structured set of terms and conditions, accounting, etc. For me, it was great fun trying to build a small brand and spread a coherent message."* Dave Moore, formerly Director of a small ergonomics consultancy in New Zealand, similarly remarked, *"Running your own business is an immensely valuable experience for HF/E practitioners. Most of us see the value of 'having a go' when it comes to trying specific tasks hands-on as part of our data collection. The same applies to building understanding about the thrills and spills of being a micro/small/medium business owner."* This can, however, be an extra burden. *"Support for administrative work, marketing and finances, technical support, drafting and design, and other speciality skills may not be available to you all of the time,"* said David Antle. *"This means you, as the practitioner, must take on ancillary tasks which are not typically associated with your profession."*

FINANCES AND WORKLOAD MANAGEMENT

Nigel Heaton co-founded Human Applications, a small consultancy that has grown from a micro-concern to an SME. He discussed the challenges of cash flow and workload management: *"We learned about cash flow and how cash is king. On a number of occasions we were owed astonishingly large numbers, were making considerable sums and yet never had any cash in the bank. We also learned that if you are too busy working you are not winning your next contract and this can lead to incredible peaks and troughs ... and without income we have no business."*

Richard Scaife made a similar remark: *"I could work hard to meet my project delivery targets one month only to find that I had spent too little time developing business so I had little to do the next month. I had to work hard to get that balance right so that my workload was more constant over the year."* He learned that there is a need to *"multi-task between several concurrent smaller projects, plus run the business and consider a range of commercial issues at the same time. This makes time management and delegation a critical skill that needs to be learned quickly."*

Ben Cook, reflecting on his past experience as a purchaser of consultancy services, noted a downside of consultancies where financial targets are at the forefront: *"The pressure for individual consultants working within such firms tends to place them under conditions of long working hours and high stress to deliver and meet targets. This often results in dissatisfaction with the organisation, balanced in the short to medium term by the reward of increased financial remuneration. The longer-term result is often burnout, which is counter-intuitive for human factors practitioners, and misaligned with the behaviours and values required of their own client: a double standard."*

For sole traders, this problem can be even more concentrated, since financial security may be limited to the last contract and the practitioner's war chest. Dan Jenkins recalled: *"My goal at the start was one that I hear from a lot of others – go after only the most interesting projects targeting a 60% utilisation. A trap that is very easy to fall into is to take on every opportunity that comes along based on two fears. First, you never know when the next project is going to come along, and second, you can't say no to an existing client. The reality then can be that you end up doing jobs more based on when they come in and less based on interest. I also found myself working 60 hour weeks to deliver everything on time."*

Solo and two-person consultancies do, however, have a particular financial advantage: low overhead (office space, human resources, payroll, accounting, administrative staff, etc.), which can affect operating costs dramatically. Dave O'Neill mentioned the ability to vary fees, and another upside: *"I managed my projects without involving a hierarchy and without having to cover the associated overheads. There were some new opportunities, being financially independent."*

Ben Cook recently left his government role in Australia to start a small consultancy. He observed that remaining small allows for greater capacity to vary client fees accordingly *"to maximise the quality of the process itself, e.g., allow for additional days to ensure the quality of the work is not compromised."* Ben discussed the importance of not allowing cash flow insecurities to affect the freedom to say "no" to client requests. *"A recent request to review and update fatigue risk management practices for a client that conducts emergency medical services helicopter operations began with a fairly honest and frank conversation. If the organisation just wanted to achieve regulatory compliance then we were not the organisation for them, but if they genuinely wanted to identify and manage fatigue risks because they believed it would enhance long-term well-being of employees, then we were interested."* Ben noted that tokenism and rubber-stamping is frustrating and professionally demoralising for HF/E practitioners who may find themselves in a "talisman" role (see Chapter 7 on HF/E roles). This may result in false reassurance for the client. Related to this, John Wilkinson remarked, *"Some clients simply want to satisfy internal or external (regulatory) demands as quickly as possible and do not grasp the opportunities that can go with this."*

OVERCOMING ISOLATION

Isolation can be expressed in various ways for consultants. The solo consultant may be especially vulnerable. Dan Jenkins recalled, *"It can be incredibly lonely. Most of the time there is lots of collaboration with clients, but there are times when it's just not appropriate to discuss certain things that you might discuss with a co-worker."*

The family unit of the small consultancy can provide a rich environment for teamwork, but consultants still often work as individuals and lack of access to non-HF/E specialists (e.g., for insight into operational issues) may also be an issue. John Wilkinson, based in a small consultancy, reflected that, *"Having come from a large regulatory bureaucracy I miss some of the wider collegiate opportunities offered there, including wider interchange with my own colleagues, other specialist disciplines and teams, researchers, work in the European Community and interchange with regulators elsewhere."*

The issue of isolation can be overcome, at least partly, via a social network, and especially via social media. Social media, especially Twitter, LinkedIn, and blogs, can be a sounding board and a source of ideas, from multiple perspectives. This issue is discussed in much more depth in Chapter 31 (Dom Furniss).

LIMITED CAPACITY AND OPPORTUNITY

Consultancies are typically seen as highly responsive to client needs, with less bureaucracy than larger organizations and with standard products, which can be deployed quickly. But for small consultancies (especially sole traders), a downside is limited capacity. Dave O'Neill reflected that *"Some opportunities were lost – customers who didn't want to sub-contract to a chap 'working from his dining room table' and without the back up of library, laboratory, workshop and other facilities."* He made the decision not to employ anybody else but stay solo and subcontract as needed and get specialist assistance as required.

Dave Moore, thinking back to his consulting days, recommended setting up alliances in order to help counter the capacity problem: *"For me HF/E is a team game and my advice would be for people setting up on their own to form alliances with mates who have complementary weaknesses as a first priority. I would argue that Solo Ergonomics is generally a contradiction in terms."*

This was the action taken by Ben Cook, who also mentioned the need to limit the size and scope of projects to the capacity of the consultancy: *"I've established a number of strategic partnerships with other small consultancies, which allows my own business to operate with a larger footprint and no additional overheads. This requires a high trust relationship and complementary HF/E skill sets and/or aligned business needs."* Ben gave the example of a partner organization that specializes in applied risk management for civil and military maritime operating environments. Since they had no HF/E specialists, a memorandum of understanding and strategic alliance was established. Ben stated that this allows complementary business, and mutual recommendations. It also means a small consultancy can act as sole provider of services in different countries and act as a combined entity for contract submissions. He also noted that approved supplier list submissions can take a long time to complete and must be focused on clients that have a clear alignment with the HF/E consultancy skills and experience.

Ron Gantt reflected on his work in a safety consultancy in the United States. He talked about collaboration for large contracts: *"It makes sense to identify other small consultancies or practitioners that you can leverage for short-term large projects. Identifying that network in advance will give you more flexibility should the proverbial 'big fish' contract come your way."* He mentioned that in the United States, there are requirements that a certain percentage of public contracts be awarded to a small business, so large companies have to contract with small consultancies. *"Developing relationships with larger consultancies to be their go-to small business for such contracts can get you in the door for some projects you may not otherwise be able to get,"* said Ron.

In many countries, larger firms serve as a shell organization or preferred supplier, through which smaller consultancies with specific expertise have to operate, often performing nearly all of the work for a fraction of the fee charged to the client. While this can feel unfair, it is sometimes the only way for small companies to access certain contracts.

GETTING AND KEEPING CLIENTS

Getting and keeping clients is an obvious priority for consultancies. Small consultancies often have a loyal client base and the relationships between the practitioner and the client are crucial. Working for a large, well-known consultancy, on the other hand, provides an easily recognizable brand, which can be helpful in establishing trust and confidence. The solo consultant, meanwhile, is fully dependent on her or his reputation and relationships, though able to rely on the reputation of previous employers and clients, for some time at least. Lack of capacity to do multiperson work (e.g., workshops) and to do large amounts of work at the same time (e.g., interviews) can extend the time needed for projects.

Interviewees emphasized the importance of relationships. Ben O'Flanagan, based in Australia, said that, *"If you're in a small consultancy it's about building a market to generate income, and this often comes down to the quality of personal relationships with key stakeholders in client organizations."* Similarly, David Caple recalled that *"Over half our work involves clients and industry sectors we have worked for the entire 32 years. Sustainable relationships are key to a successful consultancy."*

Dave O'Neill commented on the importance of repeat business: *"By and large, I continued working for the same clients with many of my former work colleagues. Just about all my work was through my contacts or recommendations from colleagues or clients."* He added, *"It definitely helps being personable and also having a strong track record. I find using the 'participative approach' in projects for/with clients facilitates relationship-building (as well as delivering good results)."*

Association with a larger entity can be another way to attract clients. Nigel Heaton recalled that having the contacts to work in large organizations was key: *"We were hugely privileged to start the company* [Human Applications] *with close links with Loughborough University. Our clients were the BBC, Unilever, the MOD and London Underground Limited. This meant that we instantly became the sort of company who worked for big companies, so we had considerable credibility in tenders."* The need for marketing, however, became clear soon after starting the business: *"At a University the phone rings. People actively sought us out. We thought that this was normal. Five years into working for a small consultancy we discovered how abnormal this is in the real world."*

Nigel Heaton also reflected on previous experience in a large multidisciplinary consultancy: *"We were part of a team of 20+ people: software and hardware engineers, AI gurus and techies of all flavours. HF/E was viewed as a bit of a joke.... One of the first jobs that the two of us in the ergonomics team had was to produce a brochure to justify our existence and to 'sell' our services."* He noted that the HF/E team was expected to market and sell the benefits of ergonomics, while other teams did not need to do this. At that time, technology-centred design processes were generating poorly conceived, unusable systems, with a few notable exceptions such as the Xerox Star and subsequently Apple. *"It led to a renaissance in HF/E. Suddenly we were required expertise in a multitude of programmes. We learned to sell ergonomics and then how to work as part of a multi-disciplinary team, initially as a bolt-on extra that no-one really wanted, then as a built-in service that people wanted but were not quite sure what they were getting (or indeed what they needed)."*

KNOWLEDGE AND INTELLECTUAL PROPERTY

Another challenge for small consultancies relates to intellectual property. Richard Scaife saw this as *"the lifeblood of a small consultancy"* and reflected that *"As I developed into a business owner, it became very much part of my thinking to protect our intellectual property not just in publications and delivery of project work, but also in proposals that we submitted for work. Obtaining registered trademarks for products and developing license agreements were all new areas for me that initially I struggled to feel comfortable with, but with time managed to integrate into my daily work."*

Another approach is to give away products, attract wider use, validation, and recognition, and sell expertise in applying the products. Dave O'Neill remarked, *"Personally, I do not expect intellectual property rights. But I have submitted a patent application in collaboration with a client."* Similarly, John Wilkinson said, *"if the right client with the right needs and attitude comes along then such products can be developed jointly and reasonably profitably in collaboration with a client."*

Ron Gantt reflected on the issue of loss of competency as a business, when a staff member or associate leaves the business: *"There's a lot of room for cascading failures here. If you have a key service provider leave, there is a chance they take big clients with them. Or you can have a service provider in a key area leave, making it so you cannot conduct a particular service, which causes you to lose a client or potential client. In either case these can be huge hits for a small company, sometimes enough to put them out of business."*

INDEPENDENCE: A MIXED BLESSING

Independence is both an upside and a downside for the external consultant. David Caple noted some of the benefits: *"Often the clients have ergonomics knowledge and expertise within their staff. However, the independent consultant can bring fresh eyes and ears to the matters of concern and take all parties into their trust to identify the core issues and develop a way forwards."* On a related note, David Antle suggested that in many workplaces, tools are defined and developed at a corporate level while *"as a consultant, you have far greater control over your tools and approaches."*

Dave Moore noted the constraints on consultants from not being embedded in the organization: *"A defining feature of HF/E according to Sanders and McCormick is dedication to a systems approach. The reality for HF/E practitioners who live off small contracts is that the scope of the systems they can consider is usually defined by the client. Another defining characteristic of an HF/E approach is that it is as multi- and cross-disciplinary as it needs to be. Again, the client will have the final word on the breadth of professional help. For small contracts this will invariably be narrower than the HF/E practitioner would recommend."* Similarly, Fiona Bird recalled from her UK-based consulting experience, *"In consultancy I was able to experience a range of industries but these were generally discrete pieces of work and once delivered it was difficult to see the implementation or understand successes and identify areas for improvement."*

Ben Cook discussed the need for consultants to stay involved through the life cycle, which can be difficult where the focus is more on report delivery than outcome: *"The longer-term lessons regarding the success (or otherwise) of the outcomes and recommendations are often not gauged by the consultants themselves. This can regularly result in a high-cost but rather shallow business process, particularly if the consultant is not able to be involved with the complete life-cycle of the project at hand."*

Dave O'Neill reflected that he once lost a client when his analysis did not conclude with what the client wanted: *"I was aware of this as the project progressed but did not compromise."* There is a risk, though, that the need for cash flow may compromise independence in some cases.

EDITOR REFLECTIONS

For me, consultancy requires the most agile practice of all contexts; the mix of excitement and terror at the constantly changing project landscape is hard to match whilst the vulnerability and demand of being a sole trader or small consultancy can be exhausting. Advice from practitioners about building networks and alliances with others is wisdom worthy of application.

(Claire Williams)

I have worked in a number of environments as a consultant (e.g., aviation, rail, border security, chemical manufacturing, other process industries), each with very different technologies, staff competencies, risks, maturity of experience, and levels of HF/E integration. The upsides of consulting included the diverse range of contexts, and an ability to innovate and respond quickly to client needs. A challenge was to maintain sufficient knowledge of fast-changing industry sectors, and a lack of integration into the fabric of organisations, making it difficult to stay with changes in the longer term, and adjust HF/E input in light of on-going feedback, post-implementation.

(Steven Shorrock)

HF/E PRACTICE IN PRODUCERS, MANUFACTURERS, AND SERVICE PROVIDERS

A second context for HF/E practitioners is producers (e.g., mining, oil and gas extraction), manufacturers (e.g., chemical or train manufacturers), and service providers (e.g., air traffic control service providers or train operating companies). In such contexts, the HF/E practitioner often acts as an internal consultant or advisor, perhaps occasionally doing consulting work for external clients. The practitioner may be the only HF/E specialist or one of a (usually small) team. The practitioners interviewed work in a range of organizations, including transport, healthcare, building, and manufacturing. The following are the main themes that emerged.

END-TO-END INVOLVEMENT

Dan Jenkins started work in a large automotive environment. He suggested *"the real attraction of these roles is the opportunity to see a project through from start to finish."* Similarly, Fiona Bird, also considering a manufacturing context, noted, *"it is rewarding to see the work of the team embodied in a product that is in service and to receive feedback from the customers and users so that we can evolve our approach and understanding of user needs."* This is an aspect of working in-house that is less often experienced in other contexts.

DEEP IMMERSION

Ken Catchpole highlighted the deep contextual immersion that is only usually possible in-house. Reflecting on his work in a hospital, he argued, *"in order to understand and address patient safety, you need to stay close to the 'coalface'. It's been an amazing way to understand clinical work, language, and the process of making change in a complex organisation."* Ken stated that, had his work been a consultancy or academic role, his experiences would probably have been "project based"—more confining and less immersive. *"I've had incredibly enlightening ad-hoc conversations with my clinical colleagues that have provided all sorts of insights into the challenges we face in applying better HF/E, and the benefits thereof. In working with clinicians, I have been able to contribute to clear improvements in working lives and patient outcomes. I have had clinicians, initially very resistant to HF/E, come to me later to express how much benefit they now see in this. It's very rewarding."*

Fiona Bird made a similar point: *"In the multidisciplinary environment I was able to expand my domain knowledge and evolve my approach to HF so that I was cognisant of the requirements of other disciplines."*

The downside of deep immersion is that HF/E practitioners are usually tied to a particular sector (though in some cases may work across many, e.g., ergonomic furniture and equipment suppliers). Another downside is that deep immersion, and the deeper relationships and shared goals, may threaten independence and role clarity.

INTEGRATION

In this context, HF/E practitioners can benefit from greater integration into the business. Deeper immersion can bring the HF/E practitioner closer to a range of stakeholders. But being a sole practitioner in a large organization has its downsides. Daniel Hummerdal, reflecting on his experience in a large organization (>13,000 employees), noted a *"risk of patchy contribution – that the perspective stays as little more than an add-on and is not integrated into systems and the way work is setup and managed."*

Ken Catchpole similarly noted a downside of being a sole HF/E practitioner in a clinical context: *"Clinicians may not recognise the legitimacy of the HF discipline as a philosophy and a science, nor understand the value that I could bring. Being removed from my professional colleagues can be isolating and it can be difficult to get support or keep up with the latest theoretical developments."* This isolation is a problem shared with sole traders, although within a wider organizational context.

FINDING A NICHE, BUILDING AWARENESS, AND ADDING VALUE

It can be significantly easier to access clients and end-users who are in-house, often with less travel. However, internal silos and relationship issues may affect this. Most HF/E specialists will relate to the difficulties in building understanding of HF/E. *"If you're in a large organisation, it's about generating awareness so that senior managers will invest in HF/E sufficiently early in the process and that it is built into standards, etc.,"* said Ben O'Flanagan.

Where the HF/E perspective is new or challenging to the organization, there are particular acceptance problems. *"The challenge can be internal politics and the pace of change – it can be easy to get bogged down by a 'we don't do it like this' culture",* remarked Dan Jenkins. Daniel Hummerdal, meanwhile, recalled that in an organisation heavily informed by mechanical engineering, *"the potential value addition of the perspective is likely to be recognised by most people as meaningful and something that has been missing. However, the same gap – between the HF/E perspective and the traditional way of doing things – can also make it challenging and frustrating that the perspective is not automatically included in considerations. Furthermore, as the organisation is not aware of what the perspective can bring, the role may come with wide latitude for interpretation of what can be done, should be done, and how it is done."* He went on to say that this is a great opportunity for any HF/E practitioner who enjoys working under uncertain conditions, but perhaps not if a clear role and accountability is preferred.

Nigel Heaton reflected on his time in Europe's largest collection of HF/E experts at BT's Research Laboratories in Martlesham, United Kingdom. At the time, BT was exploring areas such as public access systems (e.g., Prestel), speech recognition and speech synthesis, and exploring how faster, wider bandwidth would deliver more usable services. Nigel noted that HF/E gave a competitive edge: *"In this environment, HF/E was seen as a way of being better and taking on competitors. The aim was for people to use services more, to make things more accessible and usable."*

Daniel Hummerdal also noted that *"with successful local applications the perspective can be increasingly recognised as value-adding, and can gradually come to inform more and more of the design of systems, procedures, and practices.... One benefit of working in a large business is that the HF/E role can access a wide range of opportunities to contribute. However, as businesses tend to be focused on a small niche, the work context can become repetitive."*

Daniel also reflected on the importance of one's place in an organization, which can help or hinder the integration of the perspective: *"As a central (corporate) resource the access to opportunities that have cross-organisational impact is likely to be higher, relative to having a role which is closer to a specific service delivery. On the other hand, a more centralised role may limit the access to the sharp-end of the operations, which brings its own limitations and creates a need to develop relations with 'gate-keepers' in projects."*

WHO PAYS?

Another aspect impacting the degree to which an HF/E practitioner is called upon is financial. For Ken Catchpole, *"probably the biggest challenge is funding. The regulatory requirement for HF/E in healthcare is solely in device design, so funding must come either direct from the (resource constrained) hospital, or through research funding. Neither offer job security."*

Daniel Hummerdal commented that *"Some organisations have internal billing for internal services, in which case the service is likely to be less requested by projects where such a cost had not been considered in the budget. In contrast, if the HF/E role is available free of charge it becomes a welcome addition to any project, but in particular the more resource constrained ones."* He suggested that the risk in such a context is that conflicting priorities arise.

Overall in this context, however, commercial issues are less likely to be in focus in the same way as external consultancy. Reflecting on his time at BT, Nigel Heaton said, *"I realised with hindsight that we were running like a University Department, sheltered from the harsh realities of the commercial world."*

EDITOR REFLECTIONS

Working in a service provider, especially when deeply involved in major infrastructure projects, can be deeply satisfying or deeply frustrating, depending on many team, organisational and regulatory factors. In this context, however, the importance of relationships, between oneself and other stakeholders, and between teams, became particularly clear to me. The working alliance influences one's credibility as a practitioner, the leverage possible, client commitment, and the satisfaction of those involved. The quality of relationships is therefore something that those involved tend to remember fondly.

(Steven Shorrock)

HF/E PRACTICE IN UNIVERSITIES

The third environment in which HF/E practitioners tend to be situated in is universities and research institutes. Academics perform occasional independent consulting jobs, or more coordinated contract work as a team, though this kind of work is necessarily secondary to research and teaching.

BRIDGING THE RESEARCH–PRACTICE GAP

For Pat Waterson, a senior lecturer in HF/E in the United Kingdom, *"One of the frustrations of working in an academic environment and being involved in HF/E at the same time is the difficulty of finding out what practitioners actually do on a day-to-day basis. Much of my work is about developing methods, tools and techniques with the ambition of not just publishing papers about these, but also (and this is crucial) getting them actually used."* Pat reflected on a recent conference that featured primarily industry speakers: *"I think I found out more about aviation in a few hours than I would have done by reading some of the main books covering the domain."* Pat emphasized the need for real engagement with practitioners, including opening up our conferences and avenues for publication. Chapter 10 of this book (Williams and Salmon) discusses practice-oriented research in more detail.

Some research on the research–practice relationship in HF/E has found a need for greater collaboration (see Chapter 9 by Chung, Williamson, and Shorrock). Dave Moore suggested that *"great collaborations can be formed between academics providing the theoretical knowledge, and the practitioners providing the expected professional speed of response and reporting style."* He recommended getting involved in consultancy, research, and teaching.

Dom Furniss, a UK-based researcher, stated that, *"Academics value peer review and citations, but there is more and more encouragement to engage with industry and demonstrate impact. I think most would want to disseminate the value of their research and lessons learnt widely. NDA's [Non-Disclosure Agreements] may prevent this."*

Offset against practitioners' privileged access to organizations is researchers' privileged access to publications. Access to conferences, books, journals, and other library materials is significantly easier for academics. David Antle, reflecting on his time in academia, argued, *"the major benefit is the access to current peer-reviewed evidence to support your work. On-going research in your home institute, and through access to journals and academic resources, are incredibly helpful."* This is, however, changing. Sarah Sharples, Head of a large HF/E Research Group in a UK University noted, *"all journals submitted as evidence of our excellence now have to be open access."*

GETTING INVOLVED

A key issue in universities and research institutes is the reward system, which can fluctuate between rewarding research or teaching as the main priorities, but direct work with industry is usually lower priority. David Antle noted that academic institutes *"give little to no credit for applied consulting/research that rely on small samples or case studies.... This means you would be unable to apply your consulting time/projects towards metrics for bonuses, promotion or academic merit."* He noted that this meant that involvement in consulting must be limited to allow contribution to core activities of research and teaching.

This is changing to some extent. In the United Kingdom, "impact" has come into the spotlight, and "services rendered work" can help to improve academic ratings, as well as provide useful income. For Sarah Sharples, this is now considered part of her group's impact work: *"It can be as important to us that we have used our methods or skills externally, as it is that we have generated new research data ... the need for research data is much less – we have often worked on the basis of 'emergent theories' so our work in consultancy actually often helps our interpretation of work that we do elsewhere."*

Dom Furniss suggested that formal partnerships through government or inter-governmental schemes can be a good source of funding and motivation for HF/E practitioners to work with organizations. However, he warned that *"This needs good communication and negotiation to manage competing goals between the two, e.g., there might be tension between pursuing what is theoretically interesting and what it likely to deliver practical value to the company."* For academics, there may be a requirement to obtain (publishable) research data, which can compete with the client's needs (this idea is explored in more detail in Chapter 10 by Williams and Salmon).

The location within a university is also relevant. Nigel Heaton worked at the Human Sciences and Advanced Technology (HUSAT) research institute in Loughborough, which had won a contract to create a commercial Human Computer Interaction service, alongside the Loughborough University of Technology Computer Human Interface group (LUTCHI). Its goal was to exploit commercially the research and ideas of HUSAT and LUTCHI, selling these ideas to the UK Government and Industry (e.g., to help introduce the so-called six-pack regulations in the late 80s early 90s). *"My role was to sell the HF/E services and deliver fast and practical advice to a wide range of organisations. We worked on schemes to provide HF/E design advice to SMEs in the UK; we worked for the Republic of Ireland's Design Council, delivering the same service. We became adept at translating client needs into well-scoped projects that under promised and over delivered, making a few horrendous mistakes of the poorly-scoped, over-promised and under-delivered variety along the way. I discovered the crucial need to figure out the stakeholders early and engage them very actively."*

Organizational access may be challenging, but academics—like consultants—often have access to a wide variety of sectors, depending on the credentials of the practitioner. Sarah Sharples noted that *"we often have other routes into organizations through colleagues from other disciplines, which can be useful,"* but that *"it is very much about the individual's reputation for us as well."*

FUNDING

Funding for applied HF/E work is also an issue in university contexts. David Antle noted that *"there are often perceptions among industry that if you are working out of the university/college (with the school being a publically funded institute), that the service should come at minimal or no cost."*

Dave Moore, currently a lecturer, brought to light another issue: *"Salaried University lecturers that dabble in consultancy are in a privileged position, and one from which they can easily undermine the full time consultants out there who need to charge for every hour and don't have access to free kit and support. Many in academia are oblivious to their impact on the struggling local private sector."*

In my experience, in comparison to working in a small consultancy, endeavouring to consult from within a university is incredibly frustrating and the opposite of 'agile'. This is a shame, because there is a real desire from within academia to use the resources and knowledge for real world applications, but the systems to allow for it are often not there.

(Claire Williams)

Except for small pieces of work, contracts can take a lot of time to scope and set up. Being removed from the contractual and financial departments means that client relationships can be harmed when it takes months to get contracts up and running. My experience was that being in a university offered initial client contact and access to books and journals, but otherwise offered few benefits.

(Steven Shorrock)

HF/E PRACTICE IN GOVERNMENTS AND INTERGOVERNMENTAL ORGANIZATIONS

Governments (including their various departments, agencies, and regulators) and intergovernmental organizations (e.g., for transport or energy) are another employer of HF/E specialists (as well as a buyer of HF/E services). HF/E practitioners situated in these environments face a different range of upsides and downsides.

THE POLITICS AND PRIORITIES

One issue that HF/E specialists in government and intergovernmental organizations face is the politics. This can be both positive (having and using influence to further HF/E aims) and negative (silos and power politics). Nigel Heaton worked for some time in the UK central government (treasury), and reflected that *"no matter what your expertise, everyone was sheep dipped in management, the workings of central Government and how the democratic process runs in the UK.... Learning basic management skills has served me well, as has an understanding of how politics (with a small 'p') work."* This highlights that politics is not necessarily divisive (though it can be); politics is a social lever and means of influence.

On 'big P' politics, John Wilkinson remarked, *"Regulators have to respond to political and other demands, and also to the economic cycle. Strategic priorities are not always a good fit with on-going HFE needs."* He noted that support for HF/E can become more difficult as the strategic focus moves on.

BUREAUCRACY?

A typical perception about government and intergovernmental work is stifling bureaucracy. John Wilkinson reflected on the opportunities of his previous regulatory role, and didn't recognize this picture: *"My initial experience was the opposite, we were empowered to get on with designing and delivering HF/E for the field and industry, and we were allowed to get on with this unhindered."*

Client Base and Funding

Dave Moore reflected on 12 years in an HF/E team in a Crown Research Institute in New Zealand (state-owned companies that carry out scientific research). *"These roles are often seen as attractive to new entrants because of the higher perceived job security in comparison to private consultancy work. Personally I found it to be the opposite. A small consultancy with an established client base is a resilient force that can adapt should a client be lost. An HF/E practitioner running mostly government-funded projects in reality has only one customer. And this client turns the tap on and off in response to political forces, not the technical, regulatory, or operational ones that we as service providers understand and, with experience, predict."*

Access and Influence

John Wilkinson recalled a particular benefit to HF/E specialists in this context: *"Such a role offers unparalleled access to organisations, sites and processes."* On the other hand, for regulators, access may be constrained by lack of openness. Client relations may be more formal and defensive for government departments and regulators. John remarked that: *"People choose what they want to say to regulators and the exchanges are necessarily skewed a little by your professional and legal role. I think it is healthy for there to be good permeability between industries and the regulator, but the regulator role requires more than just technical, process or management expertise. The regulator can start to believe that 'work-as-imagined' (what the ideal organisation does to work safely) should always match 'work-as-done' (the 'real world' of business). The right position lies somewhere in-between to allow for innovation, business improvement, real-world challenges, regulatory learning and so on."* John noted that two possible positions each come with problems: *"Improvement is unlikely if the regulator simply accepts the status quo and focuses on the technical and engineering aspects without addressing the organisational issues. Equally if expectations are too unrealistic then industry can simply end up working around the real intent of the legal requirements."*

Editor Reflections

My experience of working in an inter-governmental organisation has provided access to, and contacts within, organisations in tens of countries in one sector (aviation), spanning from frontline workers up to Board level in many organisations. It has also meant working with other inter-government organisations (rail, maritime, aviation), and with professional associations (e.g., for air traffic controllers and pilots) and even with the judiciary of many different countries. The macro-level access and strategic influence is rare in other contexts that I have worked (in-house or external consulting). A downside of deep and broad integration is being tied to a particular sector.

(Steven Shorrock)

CONCLUSION

The contributors to this chapter have raised a spectrum of issues for HF/E practitioners working in the four contexts considered. What are we to make of these reflections? What are the themes? Some common factors, expressed in difference ways, are issues that are rarely taught on HF/E courses, and rarely discussed in the HF/E literature. They include the following:

- *Social issues* such as getting access to key stakeholders, building awareness, gaining recognition, finding your place and getting involved, building a network, maintaining relationships, having influence, and navigating the politics
- *Personal issues* such as maintaining integrity and independence, maintaining competence and interest, managing capacity and workload, and managing family and work needs
- *Organizational issues* such as responsiveness, innovation, location, and intellectual property
- *Economic issues* such as getting funded work, managing work in progress, and future work priorities

Some of the issues are generic to a range of professions, but for HF/E practitioners in industry they are integral to their day-to-day practice.

There are several trade-offs, such as breadth versus depth, project work versus end-to-end/lifecycle involvement, theoretical versus practical, and macro versus micro, and every context offers upsides and downsides. But what stands out is the collective diversity of HF/E practice in different types of organizations, in various departments within organizations, at every hierarchical level, in almost every industry. This diversity offers significant potential for influence and is a real strength for our profession, but perhaps also a challenge to our identity.

REFERENCE

Wilson, J.R. 1994. Devolving ergonomics: the key to ergonomics management programmes. *Ergonomics*. 37, 579–594.
Wilson, J.R. 2014. Fundamentals of systems ergonomics/human factors. *Applied Ergonomics*. 45(1), 5–13.

9 Integrating Research into Practice in Human Factors and Ergonomics

Amy Z.Q. Chung, Ann Williamson, and Steven Shorrock

CONTENTS

PRACTITIONER SUMMARY

This chapter explores the perceived barriers to the use of research for human factors and ergonomics (HF/E) professionals, examines the implications of the research–practice gap for HF/E as a profession and discipline, and suggests ways to strengthen the value of HF/E research for practice. Barriers for practitioners to actively apply research evidence in their everyday work relate to access and time for research, relevance of research, and organizational pressures. Barriers for researchers to conduct and apply research relate to access to operational settings that allow research to be implemented, lack of appreciation of the usefulness of research, and translation of research findings into practice. However, the development of the discipline needs research and practice to be integrated. Suggestions from practitioners to overcome these barriers include: forming partnerships; networking; reading, applying, and publishing more research; and promoting the need for HF/E research in organizations. The effectiveness of any efforts to improve the application of research into practice will depend on the motivation of both researchers and practitioners.

INTRODUCTION

There has been much discussion about the research–practice relationship in the HF/E discipline since the 1960s. For example, Waterson and Sell (2006) point out that tension between the theoretical subject matter of ergonomics (primarily academic and research based) and its practical application in industry (led by practitioners) has existed since the 1960s as evidenced by minutes of annual general meetings, correspondence, and debates of the UK Ergonomics Society (now the Institute of Ergonomics and Human Factors) in the 1960s, 1970s, and 1980s. Similar observations were made by Green and Jordon (1999) on the basis that "academics regard industrial approaches as sloppy and lacking in rigour and validity, whilst industrialists regard academic practice as over-complex and impractical" (p. 113). Much of the commentary refers to the lack of research application in practice—that practitioners do not implement research findings and that researchers do not address questions relevant to practitioners. In Meister's (1999) analysis of the history of the discipline and the US Human Factors and Ergonomics Society (HFES), he found through surveys of 46 most experienced members of the Society that HF/E professionals recognized the need for both research and practice, but did not understand how they relate or the relative importance of each. Meister also reported that HF/E professionals felt that HF/E research was overly academic and largely irrelevant to practitioner problems, that "HFE research has little to offer application" and "significant problems in application are not themes for HFE research" (Meister, 1999, p. 75).

A recent international survey of almost 600 HF/E professionals identified perceived barriers to research application in practice (Chung and Shorrock, 2011). The 10 barriers most frequently identified by HF/E professionals related to the following:

- The difficulty in obtaining journal articles due to dispersed locations and limited availability or accessibility
- Low applicability of research due to unclear research implications, lack of relevance to practice, limited generalizability to the organizational environment, and the view that the organizational environment was not adequate for the application of research findings
- Lack of time to read and apply research
- Lack of awareness of the research
- Methodological inadequacies in the research

When the survey respondents were divided into practitioners and researchers, perceived barriers to application of research differed but also showed some similarities. For both groups, most respondents noted the relevant literature not being compiled in one place as a barrier. Lack of clarity of the implications for practice was also in the top five barriers for both groups. Practitioners, meanwhile, specifically identified as barriers a lack of time, perceived lack of relevance of research to their work, and lack of availability or access to journals. Barriers identified by researchers were lack of access or opportunity to apply and implement research findings in organizations, and methodological inadequacies of the research.

Clearly, there is a pervasive view within the field of HF/E that a research–practice gap exists and practitioners and researchers have somewhat different views about the reasons for the gap. It is important to recognize, however, that the research in this area has looked primarily at the perceptions of HF/E professionals about the place and relationship of research and practice in this field. There has been little study of whether a gap actually exists between research and practice. While it would be useful to have evidence on the existence and nature of the gap, it could be argued that this issue is less important as perceptions often become, or even determine reality. This means that HF/E professionals simply tolerate the status quo and even use the gap between research and practice to justify their approaches. Practitioners blame their failure to keep up with and use research on the existence of the gap, and researchers fail to try to ensure their research is used because of the gap. The objective of this paper is to explore in more detail the perceived barriers to the use of research separately for HF/E practitioners and researchers, look at the implications for HF/E as a profession and discipline, and suggest some approaches to strengthening the value of HF/E research in practice.

PRACTITIONERS' VIEWS ON BARRIERS

Taken as a whole, the commentary on the research–practice gap falls into three main clusters of reasons for practitioners not actively using research or recent research evidence in their everyday work. These relate to access and time for research, relevance of research, and the influence of organizational factors such as lack of acknowledgment of the role of HF/E in solving problems as well as pressure to produce results.

LACK OF ACCESS TO AND TIME FOR RESEARCH

Compared to researchers, practitioners are less likely to have access to online databases of scientific literature (e.g., Web of Science, Scopus) that are commonly used in academia. The expense of purchasing subscriptions to journals may be seen as prohibitive in organizations, even where there are several HF/E specialists. For many practitioners, a society journal may be their only access without personally purchasing articles or journals. Even if access is available to a database of scientific and research articles, the time required to browse, search, and retrieve journal articles of unknown applicability may be perceived by some practitioners as unjustified by the benefits, especially if the time is not accounted for in project time. Indeed, practitioners and those without society memberships are less likely to spend time reading research. Chung and Shorrock (2011) found that those working in academic/research institutions read more articles (2–5 articles per month) over the most recent 12 months than those working in other organizations (1 article per month), although professional society members read significantly more and found journal articles significantly more useful than nonmembers.

While lack of access is an issue that could be addressed successfully by societies and publishers, the issue of time is directly controllable by practitioners. Obviously, practitioners will allocate time to the activities that they think are most important and perhaps those that require least effort (Meister, 1999). The argument of a lack of time however may reflect, at least in part, a lack of commitment to good practice and to the profession. Rice et al. (2006) argued for the importance of HF/E professionals to master and maintain competence in a complex body of knowledge and skills used in the service of others.

As a discipline, we need to facilitate access to new and emerging evidence relevant to practice. This would include very practical solutions such as professional societies making available state-of-the-art reviews on the broad range of HF/E issues in order to improve and centralize access to research findings for their members. Strategies that increase collaborations between researchers and practitioners will also be of benefit, such as requiring partners in research grants. Solutions to the problem of allocating time to seek out and read research are not easily found. Professional societies can again play a role through setting continuing education and professional development requirements that encourage practitioners to make the effort to keep up with new research.

LACK OF RELEVANCE OF RESEARCH

The issue of relevance is the barrier most often identified and discussed when the problem of the research–practice relationship is broached. For example, Chapanis (1967) stated, "one of the most difficult tasks for the ergonomist or human factors engineer is to find and identify that very small percentage of information that will really contribute to the solution of whatever problem he may have at hand" (p. 9). Part of the criticism relates to the research rationale and method. Based on his content analysis of over 621 research papers published in *Human Factors* and the annual meetings of the HFES, Meister (1999) found that the research stemmed mostly from universities, mostly used experimentation, involved students as participants, was more problem oriented than technology oriented, and was largely unrelated to theory or modeling. Only about 10% of all 368 studies analyzed by Meister utilized the operational environment as a measurement environment. On this basis, Meister argued that HF/E specialists make little use of research data. He put forward the view that this was because the research is not applicable, and pointed out the artificiality of the studies, and the decreased representativeness, and therefore relevance. In addition, he argued that the increasingly academic nature of the research has tended to alienate the discipline from the system development and application questions that were originally the source of the discipline.

To a large extent the origins of the problem of relevance have been laid at the door of researchers. Meister (1999), who has had the most to say on this topic, argued that psychology's control over many HF/E concepts has led to a disconnect between research (primarily human centered) and application (primarily equipment centered). He asserted that without this influence, the explanation of human responses to technology might have become more important in research than technology design. Meister maintained that the reality of practice pushes HF/E professionals into design (i.e., engineering), whereas many researchers do not possess this orientation. When research ignores system development, Meister argued, practitioners might not see value in it.

Salas (2008) raised the issue of the "translation problem," which he characterized as the lack of clear practical implications of research findings reported in journal articles. Salas (2008) recalled that, as an editor, as much as he pushed some authors to address the practical repercussions of their research, "they were resistant to the problem or apathetic or lacked know-how" (p. 353). Meister also took up this point and argued that the lack of practical implications is due to researchers not being able to come up with any, along with the publication-centric nature of research.

This question of the relative importance of fundamental or applied research plagues many disciplines. The solution is almost certainly not one or the other, but achieving a good balance between them. Furthermore, some problems are ripe for application but others simply languish for lack of fundamental knowledge about the problem. For example, before the significant growth in sophisticated techniques for analyzing narrative data, HF/E evaluation of internal human phenomena such as attitudes, beliefs, and values was extremely limited.

The barrier of lack of perceived relevance of research reports will always be viewed through the eye of the beholders. Again, involving practitioners in the research process may help to increase the breadth of the relevance of the research and reciprocally make researchers more aware of how the research applies in the real world.

Organizational and Contextual Pressures

Organizational pressures combined with the general resistance to HF/E in organizations and allied disciplines represent strong barriers for practitioners to apply research. These are likely reasons for the differences in rigorousness of approaches between practitioners and researchers. Organizational barriers to the application of HF/E research were reflected in Meister's (1999) survey. His respondents reported being "heartily sick of having to explain and justify HFE" (p. 220). Meister argued that this illustrates the historic tension between HF/E and management and suggested that this might contribute to practitioners being less research oriented. In addition, stakeholders (e.g., management, engineers) in organizations may not view HF/E as important; it may even be perceived as unnecessary (Meister, 1999). High pressure in organizations to produce rapid results at minimal cost may also lead to the continued use of methods and techniques that are, or were, in fashion rather than those that have solid theoretical underpinnings (Anderson et al., 2001).

RESEARCHERS' VIEW ON BARRIERS

Barriers reported by researchers relate to opportunity and access to real-world settings that allow research to be implemented, lack of appreciation of the benefits and utility of research, and translation of research into practice.

Lack of Opportunities to Implement Research Findings

Chung and Shorrock (2011) found that for researchers and those working in research institutions the most commonly cited barriers to applying research related to characteristics of the organization and support for putting research into practice. These barriers included decision makers not allowing the research application, key stakeholders not cooperating with the application of research findings, not having enough authority to change current organizational processes, and insufficient time on the job to apply new ideas. These all relate to the lack of opportunities or access for researchers to implement research findings in organizations. Kirwan (2000) asserted that HF/E is ultimately rooted in context. Without sufficient access to the organizational context, it is difficult for the work of researchers to make substantial impact in the workplace, both in conducting the research in operational settings and even in simply getting workplaces to pick up and apply new research evidence.

PRACTITIONERS LACK APPRECIATION FOR THE USEFULNESS OF RESEARCH

Meister (1999) also concluded that many practitioners do not recognize or appreciate that theory and research underlie their everyday practice and only see them as background influences. He argued that practitioners are not asking the question of whether there is a better way. Similarly, respondents in the survey by Chung and Shorrock (2011) suggested that there was a need for better understanding of the value of research for practice and better evaluation of the quality or usefulness of research.

Without an appreciation of the need for research, it clearly is unlikely to be seen as useful and certainly will not be used. In addition, criticisms that HF/E research is not useful or relevant enough implicitly assume that research should necessarily be directly applied or applicable to practice now. However, what is the purpose of research? If one argues that research should be conducted only if it generates "useful" and "relevant" findings at the present time, then much research would not be conducted since it may not be immediately useful. Although the immediate application may not yet be obvious, all research is potentially useful. What is not useful to some may be useful to others or useful in the future. There is, however, a lack of studies examining how research is actually used in practice.

TRANSLATION OF RESEARCH INTO PRACTICE

The question of whether results from laboratory or field research can be translated effectively into useful practice and whether it can map onto how things are in the real world has been raised by a number of authors. For example, Meister's (1999) survey of the most experienced members in HFES found that although the respondents recognized the need for research and practice, they did not understand how they relate or the relative importance of each. Meister made the point that HF/E research should add to the knowledge base describing human–technology relationships and so greater focus is needed in research on translation of the behavioral domain to the physical domain. He argued that currently there is a heavy focus on the human side of the human–technology relationship in HF/E whereas research conclusions should be interpreted in terms that relate to the effect on the technology, otherwise the research only has psychological meaning. An explicitly broader way of looking at the translation of research into practice would describe the relationship between humans and other elements of a system, including technology (Dul et al., 2012). The responsibility of translating research into practice should be with both researchers and practitioners.

It seems that in HF/E, as in a range of other disciplines, there is a separation between research and practice or at least in the perceptions and values of researchers and practitioners. What does this mean for the field? Is it important? would it matter if research and practice drifted into separate silos within HF/E? Is it possible that we can carry on HF/E research and application without each other? It seems that we often do. Many HF/E professionals carry out their day-to-day work with little acknowledgement of the wider implications of what they are doing. To what extent do practitioners use methods to solve human–technology–system problems with consideration and concern for their origins, reliability, validity, or theoretical implications? To what extent do researchers conduct studies with appropriate regard for real-world implications such as context and opportunities for application? The drift into silos must not be accepted as normal. It threatens the mutuality of research and practice, and thus the integrity of the discipline and profession of HF/E.

The question then is: what can be done to encourage a more holistic approach in HF/E in which research and practice are intertwined?

HOW TO OVERCOME THE BARRIERS? SUGGESTIONS FROM HF/E PROFESSIONALS

The barriers identified by practitioners and researchers and discussed earlier provide some leads about how we might overcome the problem. The barriers identified are of two main types: "surface" barriers, such as simply getting access to journals for practitioners and to organizations to do or implement research for researchers, and "deep" barriers, such as commitment and motivation to overcome access obstructions. Surface barriers may be overcome through some active changes to the way HF/E creates and shares knowledge. Deeper barriers will be overcome mainly through motivation and commitment to strengthen the whole discipline and increase value for creating effective and safe human–technology–system interactions, along with any associated incentive systems.

The following suggestions are those offered by HF/E professionals in the survey by Chung and Shorrock (2011). As such, these are strategies and tactics that survey participants from the HF/E community believe are necessary. Each of these is illustrated by some comments regarding two industrial projects:

1. The *EUROCONTROL European Air Traffic Management Safety Culture Program* (see Kirwan and Shorrock [2014] for an overview), as an example of an attempt to combine research and practice in a major program over several years.
2. The *Advanced Fatigue Management (AFM) accreditation program for the National Heavy Vehicle Regulator in Australia* (see NHVR, 2015a), as an innovative program offering an alternative compliance approach to fatigue risk management in the long-distance road transport industry in Australia. Commencing in 2014, this program integrates research evidence to improve practice of fatigue risk management for transport companies.

1. **Form Alliances and Partnerships:** A large subgroup of suggestions was aimed at developing stronger connections, alliances, and partnerships between researchers in academia and practitioners in industry to enhance collaboration, to more directly relate research to practice, and to allow mutual understanding between communities. This means getting industrial and practitioner requirements into the research agenda. Some suggestions concerned the crossover of roles, for example, researchers working in industry or playing a role in applied projects, and practitioners working in the academic arena, for instance via sabbaticals. Other suggestions focused on forming links between academic institutions and practice organizations to supervise and mentor postgraduate students working on research projects in industry. Co-supervision of postgraduate students helps organizations to obtain access to the research, and academic supervisors and students can achieve their research and educational goals. By collaborating with postgraduate students who are practitioners, academics can learn more about the field of practice (Jarvis, 1999).

Other suggestions included practitioners working in laboratories, partnerships with Centers of Research Excellence, being involved in research studies (for instance as an external advisor), communication on real-world practical problems, current gaps, solutions and future directions, and more domain-specific research. Apart from ensuring that research efforts are directed at practical needs in the industry, suggestions also illustrated greater practitioner–researcher collaboration during planning and designing of research, conduct of research, writing and publishing of research, and application and applied validation of research. In the healthcare literature, Sobel (1996) argued for the need to learn from successful businesses, that in order to produce products that practitioners want to use, researchers need to understand practitioners' needs and foster close working relationships with practitioners to give them a feeling of ownership of the product. Evidence from the healthcare, industrial, work, and organizational psychology literature suggests that practitioner involvement in the research process makes practitioners more likely to find the research useful and implement it, and other academics are more likely to cite the work (Bostrom and Sutter, 1993; Rynes, 2007).

a. An ongoing EUROCONTROL safety culture program involves a collaboration between EUROCONTROL, two UK universities (initially the University of Aberdeen and subsequently the London School of Economics and Political Science [LSE]), and a number of consultancy organizations (including NATS). The program began as a research project in 2003 (led in EUROCONTROL by Barry Kirwan until 2014, and Steven Shorrock from 2014 onwards), and continues until the time of writing, combining research and practice. The program is one of the largest safety culture programs in the world, and has been made possible only via a collaborative research–practice relationship.

b. The AFM accreditation program was established based on evidence from peer-reviewed research on what constitutes good fatigue management. The program involved active researchers in the area, in the development of a formal Risk Classification Framework that provides clear guidance to industry on designing alternative work–rest schedules that would not comply with standard working hours limits. The researcher involvement has continued with the establishment of a Fatigue Expert Reference Group (FERG), which is an expert advisory body that supports the National Heavy Vehicle Regulator (NHVR) in making decisions related to AFM accreditation. FERG members include Drew Dawson, Philippa Gander, Narelle Haworth, and Ann Williamson with Carolyn Walsh as Chair.

2. **Informal Personal Networking:** Another large subset of suggestions involved increasing communication and networking between practitioners and researchers, both in-person and online. Many respondents noted that this allows greater sharing of experience, knowledge, research findings, or implementation strategies and approaches. It was noted that this enables learning and discussion about the latest research and facilitates interaction and cross-fertilization of ideas between practitioners and researchers. Suggestions related to networking mentioned professional development meetings, workshops, and conferences. Many respondents suggested that there should be applied conferences specifically for practitioners, or more practice-oriented conferences with a focus on research/practice crossovers. The CIEHF has increasingly arranged and co-arranged conferences on HF/E in various fields, such as oil and gas, healthcare, manufacturing, and recently aviation, both as discrete conferences and streams within larger conferences. Conference programs may well need to move away from traditional presentation formats to more interactive exchanges. Again, the CIEHF Ergonomics and Human Factors conference has done this for the past few years, with many discussion workshops in the recent programs.

a.   The EUROCONTROL safety culture program has been discussed in networking groups via multi-industrial workshops, at an annual European workshop involving representatives from up to 30 countries, and in various conferences and as papers and workshops. The Ergonomics and Human Factors 2015 conference included an open workshop with the title "Is safety culture still a thing?" where some of the issues arising in research and practice were discussed. Interactive workshops give greater opportunity for informal discussion, compared to one-way presentations with limited time for more formal questions.

b.   Establishment of the AFM accreditation scheme involved an extensive program of workshops with industrial groups including those representing driver and transport operators and specific sectors of the industry. These workshops involved presentations by researcher advisors who were involved in development of the scheme and discussions with workshop participants.

Other suggestions associated with networking involved online groups or forums to discuss case studies and research strategies, methods, and processes. Several such forums and discussion boards now exist, for instance on LinkedIn.

3. **Read More Research:** The most frequent suggestion aimed at practitioners was incorporating more reading and awareness of current research into their practice. Recognizing this need, some suggestions were focused on increased reading to keep up with the latest research for self-development. Survey respondents suggested forming research discussion groups in their organizations, such as a journal club. Other suggestions were focused on reading research to inform internal stakeholders or clients, or to develop reviews of relevant research to synthesize findings and draw design-relevant conclusions.

Many respondents commented on the need for more time to read research at work, but found this difficult. For example, one respondent commented that the time spent reading research could not be billable. A solution suggested was to build literature review into project planning time and as part of billed work activities.

Related to this, other suggestions emphasized the need to critically evaluate the quality of research. Suggestions for practitioners included the need to distinguish good evidence from opinion, fads, or unfounded claims, and the need to understand the limitations of research results and avoid overgeneralizing.

a. The safety culture literature continues to expand, and time is built into the safety culture program for this to be reviewed—to an extent. However, the difficulty lies in identifying from the mass of research those findings and methods that make a real difference when it comes to working with organizations. To help with this, the EUROCONTROL safety culture program includes an annual session where key papers are selected and summarized, and implications for practitioners working in air navigation service providers are discussed.

b. An integral aspect of the AFM accreditation scheme is the development of a Risk Classification Framework and supporting evidence documents, which are freely available on the NHVR website (see NHVR, 2014, 2015b).

4. **Apply More Research Findings:** Another frequent (and obvious) suggestion aimed at both researchers and practitioners concerned the need to apply more research findings in practice—to support design, as a basis for recommendations in project reports, or to make research-informed decisions. To facilitate the application of research findings in practice, respondents suggested that practitioners could collaborate with researchers to identify problems and find and test solutions. Meister (1999) suggested that the specialist working in system development must become a translator of research into the specifics of the individual design problem, but stated that little research is already translated into design specifics. Norman (2010) argued "between research and practice a third discipline must be inserted, one that can translate between the abstractions of research and the practicalities of practice. . . . We need translational developers who can act as the intermediary, translating research findings into the language of practical development and business while also translating the needs of business into issues that researchers can address" (p. 12).

a. Part of the EUROCONTROL safety culture method involves a questionnaire (the other parts being focus groups, interviews, and informal observation). The questionnaire has been developed iteratively based on research and practice. Over the years, it has been subject to several published validation exercises. Recent research has refined the factor structure and question set, for example. This new structure has influenced not only analysis, but also communication with stakeholders. Other ongoing research concerns understanding differences in response patterns between different regions of Europe. Another application of the research has been the development of "safety culture discussion cards" (Shorrock, 2012), a practical tool to facilitate discussions, derived from the themes in the questionnaire. These have been published in multiple languages in a physical card form, and for viewing and print on www.skybrary.aero.

b. The AFM accreditation scheme includes evidence-based supporting materials that bring the most up-to-date research evidence into decision making on work–rest schedules that can be allowed under this scheme. The Risk Classification Framework provides guidance to transport operators on how to establish an alternative fatigue management program rather than being forced to work to prescriptive, one-size-fits-all standard hours. The FERG also ensures that new research findings can be included in decision making on what is judged to be safe and good fatigue risk management for this highly vulnerable industry.

5. **Promote the Need for HF/E Research in Organizations:** There is a need to raise awareness of research and gain support to apply and conduct research, both within organizations (i.e., management and colleagues) and outside organizations (i.e., customers, government, professional associations, and the public). The suggestions for marketing the benefits of HF/E research exemplify each of the three approaches observed by Catterall and Galer (1990) (see also Chapter 7 on HF/E practitioner roles):

1. *Practitioner as expert*—conducting cost–benefit or return-on-investment analyses, being more assertive, and communicating the long-term benefits of research application.
2. *Practitioner as provider*—promoting published material or providing news bulletins of current research, translating research findings into everyday language, educating stakeholders on the science, and sharing ergonomics solutions more openly.
3. *Practitioner as enabler*—gaining understanding and commitment from top management in the implementation of ergonomic designs, feeding HF/E into the tools and techniques that other technical specialists use, and encouraging the participation of these specialists from other divisions in HF/E projects.

Catterall and Galer (1990) argued that the "practitioner as enabler" approach allows the recipients of tools and techniques to practice ergonomics and demonstrate the benefits for themselves. This promotes the individual and organizational adoption of ergonomics principles, and the understanding of the benefits of ergonomics expertise as an in-house resource. Hence this approach, used in conjunction with the first two approaches, may be an effective strategy for the future expansion of the uptake of HF/E.

a. The EUROCONTROL safety culture program has promoted the three perspectives of practitioner as expert, provider, and enabler. The need for a solid research base has been argued and given support via a number of contracts with universities to act as scientific support providers over several years. This support has included questionnaire design support, independent questionnaire data collection and analysis, support at focus groups, support at European workshops (e.g., to provide updates on research or workshops on key issues in the research), and broader research using large data sets. The independence of the scientific support provider has been an important factor in the program, and the insights gained from research have helped to make broader links that may otherwise not have been possible, such as the interpretation of questionnaire data in national cultural contexts. Findings from the European safety culture program have fed into speeches by the EUROCONTROL Director General to CEOs (e.g., https://www.eurocontrol.int/speeches/operational-safety-needs) and the need for research in embedded in a EUROCONTROL Safety Culture Charter, which forms an agreement with air navigation service providers. Developments such as white papers and wiki pages (www.skybrary.aero) translate research findings into everyday language, available openly, while freely available safety culture discussion cards have encouraged the participation of many professions.

b. The AFM accreditation scheme and its strong research and evidence base providing clear guidance to industry is a first in the Australian transport industry and is novel internationally. It demonstrates that research and practice can come together to allow a more flexible approach to fatigue management that can take into account operation demands so making it much more practical. Success in this program will motivate duplication in other areas so encouraging integration of research into practice in other areas.

6. **Publish Your Work:** Many suggestions were directed at practitioners to conduct applied research to evaluate the usefulness of theories and methods, for example, via internal experiments to validate applicable research. Research methods suggested include case studies, evaluation studies within their own organization, or comparative studies with other organizations, which would be immediately relevant for other practitioners. Respondents also suggested that practitioners need to find

time to publish the work being done. Many respondents would like to see more practitioner publications that share information about how organizations have applied theory and research in the real world, for example, processes of ergonomic design, successful applications of research findings, replicated results, and best practices. Norman (1995) agreed that there is a need for practitioners to publish more in order to allow researchers to better understand the design and product process, but that the lack of time, inclination, or reward means that practitioners seldom conduct research or publish papers.

Meister (1999) asserted that HF/E is divided into two major groups: researchers and practitioners, with an intermediate group of practitioner-researchers whose research is of an applied nature. With rapid changes in the global marketplace and the need for continuing education, there is a demand for practitioners to become practitioner-researchers (Jarvis, 1999). Indeed, respondents' suggestions indicate a need for practitioners to move toward becoming practitioner-researchers to produce more relevant, real-world research. Better support structures and incentives from organizations, such as through increased funding and building more time as part of billed work activities, may be needed to encourage more research and publications from practitioners.

a. Several formal research articles have emerged from the EUROCONTROL safety culture program, published in journals including *Risk Analysis* (Reader et al., 2015), and *Journal of Occupational and Organisational Psychology* (Noort et al., 2016). In addition, many practitioner papers have also emerged, including various conference papers, EUROCONTROL White Papers, professional association magazine articles, and blog and wiki articles. The combination of formal research papers and wider outreach articles has combined to give the program credibility.

b. The rationale and background to the AFM accreditation scheme is available on the NHVR and has been referenced in peer-reviewed journal publications. A description of the approach has been submitted to a peer-reviewed journal. These publications ensure that this work is made available to a broader audience both within and outside the transport sector.

CONCLUSION

The effectiveness of any efforts to improve the application of research into practice will depend on the motivation of both researchers and practitioners. In the long term, better integrated research and practice in HF/E will benefit the whole discipline and profession. Continuing to develop both theory-driven and practice-driven underpinnings, with a dynamic interplay between science, craft, and engineering, will increase appreciation of the contribution of HF/E to system performance and human well-being.

REFERENCES

Anderson, N., Herriot, P., and Hodgkinson, G.P. 2001. The practitioner-researcher divide in industrial, work and organizational (IWO) psychology: Where are we now and where do we go from here? *Journal of Occupational and Organizational Psychology.* 74(4), 391–411.

Bostrom, J. and Sutter, W.N. 1993. Research utilisation: Making the link to practice. *Journal of Nursing Staff Development.* 9(1), 28–34.

Catterall, B.J. and Galer, M.D. 1990. Marketing ergonomics—What are we selling and to whom? *Ergonomics.* 33(3), 301–308.

Chapanis, A. 1967. The relevance of laboratory studies to practical situations. *Ergonomics.* 10(5), 557–577.

Chung, A.Z.Q. and Shorrock, S.T. 2011. The research-practice relationship in ergonomics and human factors—Surveying and bridging the gap. *Ergonomics.* 54(5), 413–429.

Dul, J., Bruder, R., Buckle, P., Carayon, P., Falzon, P., Marras, W.S., et al. 2012. A strategy for human factors/ergonomics: Developing the discipline and profession. *Ergonomics.* 55(4), 377–395.

Green, B. and Jordan, P.W. 1999. The future of ergonomics. In: Hanson, M.A., Lovesey, E.J., and Robertson S.A. (eds.). *Contemporary Ergonomics 1999.* London, UK: Taylor & Francis, pp. 110–114.

Jarvis, P. 1999. *The Practitioner-Researcher: Developing Theory from Practice.* San Francisco, CA: Jossey-Bass Publishers.

Kirwan, B. 2000. Soft systems, hard lessons. *Applied Ergonomics.* 31(6), 663–678.

Kirwan, B. and Shorrock, S.T. 2014. A view from elsewhere: Safety culture in European air traffic management. In: Waterson, P. (ed.). *Patient Safety Culture: Theory, Methods and Application.* Farnham, UK: Ashgate, pp. 349–369.

Meister, D. 1999. *The History of Human Factors and Ergonomics.* Mahwah, NJ: Lawrence Erlbaum.

National Heavy Vehicle Regulator (NHVR). 2014. Risk classification system for advanced fatigue management evidence statement. Version 1.0, June 2013. Fortitude Valley, Australia: NHVR. Accessed March 28, 2016. Available at: www.nhvr.gov.au/files/201402-150-risk-classification-system-for-afm-evidence-statement.pdf.

National Heavy Vehicle Regulator (NHVR). 2015a. Advanced fatigue management. Fortitude Valley, Australia: NHVR. Accessed March 28, 2016. Available at: https://www.nhvr.gov.au/safety-accreditation-compliance/fatigue-management/work-and-rest-requirements/advanced-fatigue.

National Heavy Vehicle Regulator (NHVR). 2015b. Risk classification system tool. Fortitude Valley, Australia: NHVR. Accessed March 28, 2016. Available at: www.nhvr.gov.au/files/201503-0152-risk-classification-system.

Noort, M.C., Reader, T.W., Shorrock, S.T., and Kirwan, B. 2016. The relationship between national culture and safety culture: Implications for international safety culture assessments. *Journal of Occupational and Organizational Psychology.* 89(3), 515–538. doi: 10.1111/joop.12139. [pdf]

Norman, D. 2010. The research-practice gap: The need for translational developers. *Interactions.* July and August, 9–12.

Norman, D.A. 1995. On differences between research and practice. *Ergonomics in Design,* 3(2), 35–36.

Reader, T.W., Noort, M.C., Shorrock, S.T., and Kirwan, B. 2015. Safety san frontières: An international safety culture model. *Risk Analysis.* 35(5), 770–789.

Rice, V.J., Duncan, J.R., and Deere, J. 2006. What does it mean to be a 'professional' ... and what does it mean to be an ergonomics professional? Bellingham, WA: Foundation for Professional Ergonomics. Accessed September 1, 2014. Available at: http:// ergofoundation.org/images/professional.pdf.

Rynes, S.L. 2007. Editor's afterword: Let's create a tipping point: What academics and practitioners can do, alone and together. *Academy of Management Journal.* 50(**5**), 1046–1054.

Salas, E. 2008. At the turn of the 21st century: Reflections on our science. *Human Factors.* 50(**3**), 351–353.

Shorrock, S.T. 2012. Safety culture in your hands: Discussion cards for understanding and improving safety culture. In: Anderson, M. (ed.). *Contemporary Ergonomics and Human Factors 2012.* London, UK: Taylor and Francis, pp. 321–328.

Sobel, L.C. 1996. Bridging the gap between scientists and practitioners: The challenge before us. *Behavior Therapy.* 27(**3**), 297–320.

Waterson, P. and Sell, R. 2006. Recurrent themes and developments in the history of the Ergonomics Society. *Ergonomics.* 49(**8**), 743–799.

10 The Challenges of Practice-Oriented Research

Claire Williams and Paul M. Salmon

CONTENTS

PRACTITIONER SUMMARY

Practice-oriented research aims to produce knowledge that practitioners and others can do something with, rather than simply to further describe a problem. However, using research to generate instrumental knowledge is challenging; there are constraints impacting practice-oriented research and they are imposed by both research and practice sectors. The challenge is to help solve practical problems while ensuring that the research undertaken is publishable. Issues that make practice-oriented research problematic concern funding, framing the problem for industry, choosing an appropriate methodology, publishing, and practitioner access to academic literature. Solutions can be found by open and honest dialogue up-front to ensure expectations are managed, offering postgraduate research posts to provide longer term studies at reduced costs, building in contingencies from the start, and remaining flexible and adaptable throughout. This chapter explores some of these challenges, outlining some of the major constraints concerning during practice-oriented research and offering practical advice for researchers and practitioners interested in practice-oriented research projects.

INTRODUCTION

One definition of research is "the systematic investigation into and study of materials and sources in order to establish facts and reach new conclusions" (Oxford Dictionary, 2015). By this definition, much of what academics and practitioners do can be classed as research. In human factors and ergonomics (HF/E) practice-oriented research, our "materials" and "sources" may range from journal articles to HR policies, senior managers to frontline operators, and oxygen levels to interview answers, but we endeavor to establish facts and reach new conclusions through our workplace investigations. We view the line between research and consultancy as a blurred one, with no consistent divide between methods or question types allowing for certain categorization. So, for example, the question "Is this work too physically demanding?" may be answered quickly by simple interviews; could use further psychophysical tools such as the Borg RPE scales; may combine these with heart rate measures for a more thorough answer; or could involve the use of Douglas Bags for even more accurate data on oxygen uptake. Each and any of these approaches may count as "research" by the preceding definition. We value research as "discovery" as well as research as "utility" (Buckle, 2011). However, we look to produce what Meister (1992) describes as "instrumental knowledge" rather than simply "explanatory knowledge" through this process; knowledge that practitioners and others can do something with, not just a further description of the problem.

But using research to generate instrumental knowledge is challenging; there are significant constraints impacting practice-oriented research and they are imposed by both research and practice sectors. This chapter brings together the perspectives of two practitioner-academics: one based in a university (Salmon) and the other predominantly based in a consultancy (Williams). It outlines the challenges of practice-oriented research under the headings of a typical systems analysis. We discuss the goals, demands, resources, methods, outputs, trade-offs, compromises, and constraints of practice-oriented research for each of our situations. In each part, we reflect on examples from our own experience and outline two case studies that draw the different strands together. Finally, we propose some conclusions and advice for HF/E practice-oriented researchers.

GOALS—WHY ARE WE DOING IT?

For most researchers, the main goal underpinning their research activities is a desire to positively influence practice in their area of interest. For those working in HF/E, this normally equates to a desire to improve safety, health, and well-being or performance. The ultimate goal is safe and efficient systems. On top of this, however, all researchers working within academic institutions are faced with various requirements to which they should adhere when conducting research. Although there are many of these, the majority relate to the need to publish the findings of their research. This need cannot be overstated.

Articles published in peer-reviewed journals represent one of the most important metrics of scientific achievement and are used to evaluate researchers' contributions to a particular discipline (Dul and Karwowski, 2004). To the researcher working within a university environment, these outputs are essential for career progression (Harris, 2012) and to the universities themselves they have significant financial implications. Publishing in high-quality journals (Buckle, 2011) enhances reputation, which in turn brings about increases in funding and students (Harris, 2012). The bottom line is that if the work isn't publishable, it has little benefit to the researchers and universities involved.

This, of course, does not mean that it does not have impact in practice—often research that might not be appropriate for publication is highly effective in solving the real-world problem that it was designed to solve (Buckle, 2011). So alongside the academic need to publish is the goal of answering the specific question asked by the client or customer. The challenge for practice-oriented research then is to help solve practical problems while ensuring that the research undertaken is itself publishable.

DEMAND—WHAT IS THE TRIGGER FOR IT?

As discussed earlier, in academia, publishable research begets more research, because of the funding attached to the publication profile of the University. The subject matter for the research is often defined by funding body priorities [see, for example, in the United Kingdom the Economic and Social Research Council's (ESRC) funding priorities or those of The Institution of Occupational Safety and Health (IOSH) and in Australia the Federal Government's strategic research priorities].

These areas will tend toward issues of global or societal importance for the larger funding bodies, but may have a more industry-specific or parochial focus for professional bodies such as IOSH. This means that for externally funded research, the HF/E question for investigation must fit within the framework of priorities for the funder.

Take, for example, the successful funding application to IOSH for the Move More Study (Williams et al., 2014), which investigated behavior change techniques to support postural breaks for musculoskeletal disorder (MSD) prevention. This application fit with the 5-year focus for that funder. MSDs were in the top five priorities in IOSH's 2008 publication "Workplace health issues" and behavior change was a key area identified in which practitioners would need knowledge and skills (Leka et al. 2008, p. 3). So for academia-based research, the triggers are often relatively high-level issues.

In a practitioner setting, however, research is triggered by more immediate business needs and questions. This can mean that research into these problems is far less fundable and publishable, though its application to practice is much more apparent. "Why do we keep having manual handling injuries?" sounds far less compelling to a journal than something like "A systems analysis of the contributory factors involved in manual handling incidents." However, to the organization with the problem, the question is the trigger for the research. For research funding bodies, the issue may be further removed still from the "sharp end," as strategic research priorities such as "Promoting population health and well-being" come into play. This leads to questions being framed in a broader occupational health setting to enhance the opportunities for funding outside the organization itself (e.g., from a national research council).

RESOURCES—WHAT DO WE NEED FOR IT?

The need for financial funding for the research work is implicit in the discussions above, but additional resources are needed in practice-oriented HF/E research. Access to the appropriate work environment and the right people (participants and stakeholders) with enough time to undertake the work is key, and will include all manner of "field experts" (EUROCONTROL, 2014) depending on the research area.

In addition, existing data on practice and the problem being tackled are almost always useful in practice-oriented research. For example, documentation describing aspects of practice such as standard operating procedures, training programs, and previous incidents is all useful. Accordingly, most practice-oriented research projects will have an initial documentation review phase. Not only will this provide useful information for the research questions being tackled, it is also useful for refining the methodological approach and identifying key aspects of the system to examine.

In conjunction with practice documentation, access to the extant academic literature is a requirement for most types of research, even where the focus is on collecting new data to answer an industry-driven specific question. Again, this enables the researchers to understand what issues have been identified in the past and to refine their approach (often there is an appropriate methodology or study design out there). Unfortunately, such access is more difficult for those researchers not affiliated with a university (Buckle, 2011; Chung and Shorrock, 2011).

Further tools and equipment will vary from project to project, but may well include structured HF/E methods (e.g., hierarchical task analysis, cognitive work analysis, the technique for the retrospective and predictive analysis of cognitive error, the human factors analysis, and classification system), standardized assessment instruments (such as the SF-36 Health Survey, often used for pain measures), proprietary surveying tools (such as the Health and Safety Laboratory's safety climate tool), video and audio recording devices, and other measurement technologies (such as heart rate monitors or movement sensors). Various software support tools are also used to reduce and analyze the data (e.g., ObserverPro, NVivo, SPSS).

METHODS—HOW DO WE DO IT?

The research methodology is important. For publication, it has to be rigorous and stand up to academic scrutiny. While this is a requirement of all research to ensure that the findings are valid, again it relates primarily to the need to publish the work. Without a clear, rigorous, and repeatable methodology, it will not be possible to publish the research. The difficulty here is that there can often be a huge difference between a methodology desired by industry and one that is acceptable to academia.

Industry partners often want a problem solved as quickly and as cheaply as possible, emphasizing efficiency. Academics often want to solve the problem with an intricate methodology or even a new approach and then gather more data to demonstrate they have solved the problem, emphasizing thoroughness. This increases the timelines and the resources required. To quote the title of the 2010 Institute of Ergonomics and Human Factors Annual lecture, in these instances, "The perfect is the enemy of the good" (Buckle, 2011).

In HF/E, there are various examples of where the methodology needed to solve a practical problem may differ from the methodology required to be able to publish the work. A good example of this is in error prediction studies, where an organization requires an assessment of the errors likely to be made with a new procedure or piece of technology (e.g., Salmon et al., 2011). Here the organization typically wants an experienced HF/E analyst to work with subject matter experts to predict the likely errors, using an accepted methodology such as the systematic human error reduction and prediction approach (SHERPA; Embrey, 1986) or the technique for the retrospective and predictive analysis of cognitive error (TRACEr; Shorrock and Kirwan, 2002). This exercise will tell them what safety critical errors are likely and provide suggestions for preventing them. Their needs are met by the analysis described.

To ensure that the work is publishable in a reputable journal, however, further work is needed. Predicting the errors that occur for a particular task or with a piece of technology is no longer publishable on its own. Often multiple analysts need to predict errors for the same task independently from the other analyst, and the predictions are required on different occasions from the same analysts. While multiple analysts are often used in industry, these additional activities can be used to enhance the chances of publication by enabling the calculation of inter- and intrarater reliability statistics (e.g., Stanton et al., 2009), which in turn can be used to support the reliability of the method (reliability is a critical issue for HF/E methods, see Stanton and Young, 1999). In addition, data on the actual errors made in practice are often required, or alternatively an expert panel is assembled to predict errors on the same tasks in order to provide a "gold standard." These activities enable the validity of the error predictions to be assessed. The final extra task involves applying appropriate statistical tests to the data in order to calculate these reliability and validity measures. All of this extra work equals more time (and money), and in most cases they represent lines of inquiry that are not a priority for organizations.

Finally, for publication, the methodology employed should be as novel and innovative as possible, that is, unpublished. Again this is tricky as often industry partners do not want something highly innovative; they want a problem solved as quickly and efficiently as possible. Tried and tested is much more desirable, though using the standard methods in novel ways is possible (see the UPLOADS case study, Table 10.1).

OUTPUTS—WHAT DOES IT GENERATE?

We have already discussed the need for journal publication output for research undertaken in a university setting. The choice of journal may be dictated as much by the journal's impact factor [a measure of the frequency with which the "average article" in a journal has been cited in a particular year or period (Buckle, 2011)] as by likely readership among practitioners. For practice-focused research, however, journals are not necessarily the most useful location for findings that will impact practice (Buckle, 2011; Chung and Shorrock, 2014). Nor are journal papers generally of interest to industry customers.

Instead, outputs can range from publically available research reports (sometimes lengthy documents such as Williams et al., 2014) or summary versions (IOSH, 2015). Such reports will often have practitioner-focused sections with headings such as "What does this research mean?" or executive summaries in which the application to practice is made clear.

Where industry customers have commissioned the research, the outputs can be even more parsimonious. One such customer demands a "summary on a page" (SOAP) covering the research carried out for them—which is often the distilled "so what?" from a literature review of a hundred articles or an organization-wide survey with 500 respondents covering 30 or more question areas. It is designed to inform the most senior management, show that the money was well spent, and provide a case for further work. This ability to summarize and apply the research findings is a key attribute of good HF/E advisors, taking the knowledge from theory to practice (Williams and Haslam, 2011). We have at times been required to interpret the research findings that other organizations have carried out in our customers, giving them the "so what?" which was not evident from the research undertaken by others.

Ensuring an academic audience for the practitioner-oriented findings can also be tricky if the research is not sufficiently robust to warrant journal publication. However, poster presentation of findings at conferences is a good option as a research output. This allows academic researchers to see what issues and questions are being tackled in practice, to inform their future research endeavors and to reduce the academic–practice gap. In addition, books are subject to less-rigorous peer-review procedures, providing another useful avenue for disseminating practice-oriented research. Finally, shorter articles summarizing research outputs are becoming popular via websites such as www.theconversation.com, white papers on company websites, and also blogs.

CONSTRAINTS—TRADE-OFFS AND COMPROMISES—WHAT ARE THE LIMITATIONS AND HOW DO WE DEAL WITH THEM?

The major challenge faced when conducting practice-oriented research is to convince those in practice that research is the right approach to solving the problem of interest. This can often take some convincing for industry; industry partners usually want a form of investigation that is quick and provides a straightforward answer. They are also less concerned about methodological rigor. In some cases, the very concept of university-driven research can ring alarm bells. The timelines are much longer, the methodologies are intricate and can be confusing, funders may lose intellectual property to the university, and researchers want to tell everybody about the work and evaluate its impact. In addition, when seeking additional research funding from industry-research schemes the proposal may not be successful; while some schemes have around a 40% success rate, some are as low as 15%–20%.

On the plus side, if successful the project becomes far more prestigious for the industry partners involved and the funding pot is increased significantly. The key at this stage is to communicate honestly and clearly with the industry partners: Tell them exactly what the research will be and what the outputs will be. Don't skate around the timelines associated with obtaining ethics approval and recruiting participants. Clarity and honesty at the outset enables industry funders to make informed decisions about whether they want to go down the research path. If funding a longer term approach is an issue, it can be useful to encourage the organization to support a postgraduate research student. This increases the available resources and can support the production of useful, practice-oriented research findings.

Another compromise approach is for the researchers to be creative in how they think about and express the problem. Because industry partners may want their own study to tackle a problem that others have already "solved" in the literature, researchers will have to work hard to come up with a new approach to solving the problem or to emphasize reasons why the work is still of interest. Applying a different methodology is one way that the study can become more publishable.

One important aspect of practice-oriented research is the need to deliver as promised. Quite often, researchers struggle to deliver the project as promised. There are various reasons for this, such as restrictions on access to required data, researchers with key skills leaving the project, equipment not being available or suitable, lack of access to industry personnel, or industry champions moving on. Planned research activities will become unfeasible for various and most often unforeseen reasons. For example, in Salmon's first position as a research assistant he was all set to fly all over Europe in cockpits to observe pilot errors—then 9/11 happened and restrictions on entering cockpits were immediately put in place. Despite issues such as this research teams are still expected to conduct the innovative research project as promised, and it is imperative that industry partners' needs are still met. Typically the proposed research therefore requires significant revision to stay within scope—but still achieving what was proposed originally and still meeting both the industry partner's needs (a solution to a problem) and researcher's needs (including publishable work).

In the case of unforeseen constraints such as this, it is important to have contingency plans around all data collection and analysis activities, for example:

- Work on the premise that they will not run as planned
- When planning the project try to have contingency plans in place
- Consider what other data collection method could be adopted
- Consider what other source of data can be utilized
- Consider who else could you potentially involve in the project

A final challenge relates to ensuring that the industry partners are heavily involved throughout the research project. This is one of the major challenges associated with practice-oriented research. It is imperative that funders are involved every step of the way; however, often this is surprisingly difficult. It can be helpful here to:

- Have a project steering committee comprising researchers and industry stakeholders
- Have regular project steering committee meetings (quarterly if possible)
- Provide regular updates via email or progress reports

In addition, it is important that researchers have a highly visible presence at industry events such as practitioner conferences, workshops, and committee meetings.

CASE STUDIES

Table 10.1 provides an overview of how these challenges have emerged and been responded to in recent projects involving the authors. Salmon's case study (UPLOADs) was undertaken in the led outdoor activity sector in Australia. Led outdoor activities represent an educational form of active recreation, and have been formally defined as facilitated or instructed activities within outdoor education and recreation settings that have a learning goal associated with them (Salmon et al., 2010). Williams' example outlines a new approach for a food production customer whose accident rates were no longer reducing in line with health and safety interventions.

TABLE 10.1
Summary of Two Case Studies

Case Study	UPLOADS	Food Production Behavioral Safety Project
Goals	Develop an incident reporting and learning system in the led outdoor activity sector in Australia that will enable sector to better understand and prevent injury incidents.	Reduce the plateaued number of lost time accidents and the costs associated with them.
Demand	Following a series of major injury incidents, a working group within the led outdoor activity sector discussed the problem of injury-causing incidents and how to respond to it, identifying the need for research on the causal factors underpinning such incidents. An initial industry-funded project involving a literature review and case study analyses highlighted the need for an incident reporting system based on systems thinking (see Salmon et al., 2009).	The health and safety department had identified there was not a simple fix to the continuing accident profile and had secured funding from their senior team to investigate why and intervene.
Resources	Industry funding. Australian Research Council funding. In-kind contributions from industry partners. Access to existing incident data. Access to organizations existing incident report systems. Access to incident reporting systems from other domains. Access to literature on accident causation and analysis. Software developer. Researchers with experience in systems thinking and accident reporting and analysis. Practitioners with experience in led outdoor activities and reporting incidents.	Access to people, the shop floor, meeting rooms, and funding. In-house ownership of the project. Senior manager sponsorship. Facilitation and listening skills. Systems thinking skills. Safety climate assessment tool license. Time for workshops. Stakeholder steering group meetings.

(*Continued*)

TABLE 10.1 (CONTINUED)
Summary of Two Cases Studies

Case Study	UPLOADS	Food Production Behavioral Safety Project
Methods	Accimap (Svedung and Rasmussen, 2002), HFACS (Wiegmann and Shappell, 2003), STAMP (Leveson, 2004). Delphi questionnaire (to determine stakeholder needs for incident reporting system). Ravden and Johnson's HCI checklist (to assess usability of incident reporting system). Reliability and validity studies. 6-month trial of incident reporting system. 12-month national trial.	HSL safety climate questionnaire and analysis. "What Good Looks Like" (WGLL) workshops with representative stakeholders to dig deeper into the survey findings. Facilitated sense checking meetings and round table feedback sessions.
Outputs	UPLOADS incident reporting system (now in use by over 50 organizations). Website (see www.uploadsproject.org). UPLOADS training manuals and videos. UPLOADS training workshops. Report describing analysis of all led outdoor incidents occurring over first 6 months of national trial. Multiple journal and conference articles describing development and national trial results.	10-page summary report for project team. One-slide summary to prompt discussion at WGLL workshops. One-page summary report for roundtable and sense checking meetings. Handwritten agreed interventions emerging from the roundtables. Joint consultant/industry funder case study poster at EHF 2015 conference, describing the link between HF/E and behavioral safety.
Constraints, Trade-Offs, and Compromises	Initial planned project activities had to be scaled back due to only 50% of requested research council funding being awarded. PhD funding also had to be sourced from elsewhere. Limited availability of incident data led to requirement for analysis of led outdoor activity incident data from elsewhere (e.g., New Zealand, see Salmon et al., 2014).	Project plan has to be developed *en route*; in this instance the funding stopped, restarted then stopped. Originally we wanted to go to all sites to get a full picture—we had to start with one and get a full enough picture for that site.

CONCLUSIONS

Practice-oriented research raises significant challenges for HF/E researchers and practitioners. These relate to the engagement of industry partners, alignment of research goals with funding priorities, acquisition of research funding, the methodology, publication of the work, access to resources, and production of useful outputs.

Thankfully, there are many practice-oriented research success stories from which to draw practical advice, including two of the authors' own projects as detailed in

Table 10.1. This advice includes the involvement of industry partners throughout the entire process (from proposal development and methodological design to data analysis and project delivery), being creative during study design, the development of contingency plans to deal with events impacting data collection and analysis, open and honest communication with industry partners, the use of postgraduate programs as part of practice-oriented research, and the production of a range of practice-friendly outputs (such as poster presentations, industry articles, and short communications).

Solutions to modern-day problems can only be arrived upon through practice-oriented research. We hope that researchers and practitioners will find this chapter useful when designing, conducting, and delivering such research. While there are many difficulties, the sense of reward when research impacts practice is hard to put into words. We encourage researchers and practitioners to continue their efforts. Likewise, we encourage industry to continue to engage with our discipline.

REFERENCES

Buckle, P. 2011. 'The perfect is the enemy of the good' – ergonomics research and practice. Institute of Ergonomics and Human Factors Annual Lecture 2010. *Ergonomics*. 54(1), 1–11.

Chung, A.Z.Q. and Shorrock, S.T. 2011. The research-practice relationship in ergonomics and human factors – surveying and bridging the gap. *Ergonomics*. 54 (5), 413-429.

Dul, J. and Karwowski, W. 2004. An assessment system for rating scientific journals in the field of ergonomics and human factors. *Applied Ergonomics*. 35(3), 301–310.

EUROCONTROL. 2014. *Systems thinking for safety: Ten principles (a white paper)*. Brussels: EUROCONTROL.

Harris, D. 2012. *Writing Human Factors Research Papers*. Aldershot, UK: Ashgate.

IOSH, 2015. *Move More: Encouraging Postural Breaks – Behaviour Change in the Office. Research summary*. Wigston, UK: IOSH. Accessed March 12, 2016. Available from: http://www.iosh.co.uk/movemore.

Leka, S., Khan, S., and Griffiths, A. 2008. Exploring health and safety practitioners' training needs in workplace health issues. Research report 08.2. Wigston, UK: IOSH.

Leveson, N. 2004. A new accident model for engineering safer systems. Safety Science. 42(4), 237-270.

Meister, D. 1992. Some comments on the future of ergonomics. *International Journal of Industrial Ergonomics*. 10(3), 257–260.

Salmon, P.M., Goode, N., Lenné, M.G., Cassell, E., and Finch, C. 2014. Injury causation in the great outdoors: A systems analysis of led outdoor activity injury incidents. *Accident Analysis and Prevention*. 63, 111–120.

Salmon, P.M., Williamson, A., Lenne, M.G., Mitsopoulos, E., and Rudin-Brown, C.M. 2009. *The role of human factors in led outdoor activity incidents: Literature review and exploratory analysis*. Monash University Accident Research Centre Report, November 2009.

Salmon, P.M., Williamson, A., Lenne, M.G., Mitsopoulos, E., and Rudin-Brown, C.M. 2010. Systems-based accident analysis in the led outdoor activity domain: Application and evaluation of a risk management framework. *Ergonomics*. 53(8), 927–939.

Salmon, P.M., Young, K.L., and Regan, M. 2011. Distraction 'on the buses': A novel framework of ergonomics methods for identifying sources and effects of bus driver distraction. *Applied Ergonomics*. 42(4), 602–610.

Shorrock, S.T. and Kirwan, B. 2002. Development and application of a human error identification tool for air traffic control. *Applied Ergonomics*. 33(4), 319–336

Stanton, N.A., Salmon, P.M., Harris, D., Demagalski, J., Marshall, A., Young, M.S., et al. 2009. Predicting pilot error: Testing a new method and a multi-methods and analysts approach. *Applied Ergonomics*. 40(**3**), 464–471.

Stanton, N.A. and Young, M.S. 1999. *Guide to methodology in ergonomics: Designing for human use*. London: Taylor and Francis.

Svedung, I. and Rasmussen, J. 2002. Graphic representation of accident scenarios: Mapping system structure and the causation of accidents. *Safety Science*. 40, pp.397–417.

Williams, C. and Haslam, R. 2011. Exploring the knowledge, skills, abilities and other factors (KSAOs) of ergonomics advisors. *Theoretical Issues in Ergonomics Science*. 12(**2**), 129–148.

Wiegmann, D. and Shappell, S. 2003. A human error approach to aviation accident analysis: *The human factors analysis and classification system*. Aldershot, UK: Ashgate.

Williams C.A., Denning, E., Baird, A., and Sheffield, D. 2014. *The Move More Study: Investigating the impact of behaviour change techniques on break taking behaviour at work*. Wigston, UK: IOSH. Accessed March 12, 2016. Available from: http://www.iosh.co.uk/movemore.

11 Human Factors and Ergonomics Methods in Practice
The Right Tool for the Right Job

Matthew Trigg and Richard Scaife

CONTENTS

PRACTITIONER SUMMARY

Tool selection in human factors and ergonomics (HF/E) is very important, but the right knowledge of the person using the outputs from the tool is critical for success. Understanding the client's need requires investigation before selecting the tool(s) to be used—clients often think they know what will work, but part of the practitioner's job is to critically assess this and assist in formulating the right approach. HF/E tools are generally good for describing issues, but weaker where they attempt to link to a hard threshold for action. There will be times when one tool is not sufficient to find the answers to the questions posed. Learning to use tools in combination and to apply experience and knowledge is at least as important as learning to use the tools individually.

INTRODUCTION

Tool (noun): Anything used as a means of accomplishing a task or purpose.

By ergonomics "tools" we mean those fairly rigid methodologies that ask for the same steps and data points for every application. In practice, these probably blur with what one might call "approaches"—semi-structured interviews and observations of work, tailored usability questionnaires—where the data collection may be consistent, but that rely on the skill and interpretation of the practitioner to add value (see, for example, Marras and Karwowski, 2006; Stanton et al., 2004).

Every HF/E practitioner, whether directly employed or contracted, faces some of the same challenges over the selection and use of tools. We will demonstrate through this chapter how, regardless of role, tools have their place in practice. However, we will also make an argument for the avoidance of over-reliance on tools; the practitioner's own skill in making sense of the output from tools, in collaboration with stakeholders such as frontline personnel, is key to successful use.

WHY SHOULD WE USE TOOLS?

By applying a tool correctly, the user can claim some measure of diligence in their planning and design, as tools add varying degrees of rigor and structure to the analysis. Tools with objective thresholds, although set from someone else's rules, allow organizations (and project teams) to coalesce around an agreed approach. A tool that shows your work makes it easy to communicate your findings and therefore allows effective criticism of your methods, observations, and interpretations by peers and customers alike. Furthermore, a good tool allows more people to contribute to the HF/E program, including nonspecialists (e.g., incident investigators, designers, end-users). This allows for greater engagement in human factors activities but also facilitates the formation of multidisciplinary integrated development teams.

Ergonomics tools used properly at the appropriate phase of a project can reduce lifecycle costs of the design. The cost of change after a product has entered service is many times higher than the cost of change at the concept phase. Using the right tools at the right time by the right people can therefore have major financial benefits (Bias and Mayhew, 1994).

THE PITFALLS OF HF/E TOOLS

There are potential pitfalls associated with the use of any tool. Some of the common issues that practitioners will often come across are as follows.

First, you need to understand the procedure. We have encountered many cases where the instruction on how to use a tool misses out basics such as how to conduct a basic task analysis to support the use of the tool. Whether a tool is designed for an expert or a novice, the relevant background skills required to organize the data before the application of a tool must be in place. For example, in postural analysis, the size of the work chunk defines the outcome of the analysis. Failing to agree simple and appropriate starting and stopping rules, whether to include occasional movements or exceptions, and whether to look at everyone or a defined number and type of user, seems to get lost in the desire to get on with using the tool.

Second, there will be variance between observers and the users of tools. The reliability of tools among users is available for some tools, but this is not always the case (Olson, 2013). Some tools have been shown to be less reliable when used by trained novices who are not connected with the developers of the technique (Olson, 2011; Olson and Shorrock, 2010). In the selection and use of tools, these factors need to be taken into account in both application of the tools and in the interpretation of their results.

Third, people can change their behavior when they are under observation. They can also change their behavior based on who else is being observed at the same time. For example, observing shop floor workers with their supervisor present will likely yield a different set of results to the same measurement without the supervisor present. This sounds obvious, but such effects need to be accounted for when using a tool. Even when the plan for the measurement accounts for these effects, the practitioner needs to be alert to the fact that the client may not be. The aforementioned example happened in a workshop run by one of the authors where the workforce had been separated into management and worker groups that would attend different workshops to avoid this effect. However, managers wanted to hear what their people were saying about them and sneaked into the workshops. The author spotted the problem relatively early and asked for the group composition to be changed before continuing with the workshop.

WHO CHOOSES THE TOOLS?

Whether an external consultant, internal consultant, or academic, every HF/E practitioner is under pressure to deliver something that adds value to the project, research, or work in which they are involved. How this pressure is applied differs between roles. As practitioners, it is up to us to try to guide those who are seeking our services in the right approach; right for the job in that it will also results to the customer who can be used to add value. One way in which this guidance can be given is in demonstrating the validity of the approach that is being used. This can depend on the maturity of the organization seeking the HF/E service. Some organizations with a greater understanding of the importance of HF/E will require high-validity tools and methods to help solve their problem. Such organizations have recognized the benefits of HF/E and want to do the most robust job possible. They are attempting to "cross the chasm" (Moore, 1991) from theory to the application of theory. They will put effort and money into ensuring that they get the best solution possible, and are often "early adopters" of new approaches and tools. Organizations with a high level of maturity in relation to HF/E will similarly place a great deal of effort into ensuring that what they get from their consultants is best practice, and proven to be valid.

Those with a lower level of maturity will be more likely to seek an easier way of achieving a result, with less rigor being applied in terms of the selection of practitioners and/or the tools to perform the task, and will often wait to see how other organizations get on with the solution before they adopt it. An organization, for example, that has been told by their Regulator that they must do something will often find the least expensive way of meeting the requirement. This may also result in them asking that a certain approach be adopted, which may not be the approach that an HF/E practitioner would advise. Practitioners have an important role in guiding customers at this stage. For example, in a recent project, a customer approached one of the authors with a request for a lighting survey for a control room. The stated objective was to identify ways of bringing the control room lighting in line with an internal company standard. As an experienced consultant, an initial discussion was held to understand the tasks to be completed in the control room, staffing levels, systems to be installed, and furniture to be used. Having gathered all of the available data on the requirements and the customer's objectives, it became clear that the initial request of measuring against the internal company standard was flawed, as the standard had been written with reference to a different type of control room where operators were performing a significantly less-demanding task.

The approach that was actually required was to integrate the lighting survey with a workload analysis and link analysis and design the control room and its environment around the tasks and the operators performing them. There is a danger that as a practitioner still building experience, one could be encouraged to adopt an approach that is not the most appropriate for the task, but the most convenient for the customer, and the investigation of the true requirements may not be as thorough.

There is a possible conflict here for the practitioner: Either (a) to provide a service that the practitioner recognizes is not appropriate, but will help the project in question to at least address part of the problem and improve usability to some degree; or (b) to only engage with organizations that request the kind of tools that the practitioner believes are appropriate for the issues at hand. Most practitioners would want the latter, but pragmatism must be applied. Influencing a low-maturity organization can be worth the effort, but this may require patience to help improve their understanding over time.

In consultancy, we seldom end up using the gold-plated "best practice" approach (if such a thing exists). The more usual practice is a compromise solution that will meet the client's requirements. Practitioners have a role in guiding clients in the most appropriate use of tools, including helping them to understand why their preferred tool may not be the most appropriate approach in every case.

But those entering the discipline with a view to combining consultancy with "best practice" brought forward from the latest research should not be discouraged. There are organizations with higher levels of maturity that will spend the time on research. These are typically the "early adopters" of new methods and tools, but this level of insight requires a good level of internal capability to achieve. For example, one of the major oil companies became concerned with the fact that despite doing thorough incident investigations they continued to have repeat incidents. They had identified a gap in their approach in that they treated "human error" and "noncompliance" as root causes, and therefore kept making the same recommendations whenever an incident involved "human failure." They invested a great deal of time and effort in developing a set of analysis tools for their incident investigators that has since been adopted by a wide range of organizations in other industry sectors as well (Lardner and Scaife, 2006).

HOW DO WE USE TOOLS AND EXPERIENCE IN COMBINATION?

In HF/E, practitioners come from a range of different backgrounds (e.g., psychology, engineering, physiology, as well as more operational backgrounds), and learn about a wide range of tools during training. Some of the tools learned during training (particularly early training) are discipline specific, and later in the practitioner's training (and sometimes not until they enter practice) they begin to learn about tools that span several disciplines (Buckle, 2011; Chung and Shorrock, 2011).

We also tend to learn about tools in isolation—the trainee will spend time, for example, hearing about the origins of task analysis, then practice using it, and then move on to a different tool. We learn the textbook way to use each of the tools in the toolbox.

When we get into the practical setting we find that—for many tasks—one tool is not enough; we need to use them in combination. In some cases, we even need to find new ways of using tools that are not described in textbooks. For example, one of the authors needed to adapt an approach to safety-critical task screening to ensure that the customer was able to begin using the results as quickly as possible to solve some high-priority problems. The approach documented by the Health and Safety Executive (Brazier et al., 1999) was used as the starting point with the addition of two further levels of screening in order to refine the assignment of priorities to individual tasks. This was done using expert judgment coupled with consultation with end users to ensure that the results of the process would be valid and would meet the end users' requirements. We either work this out from the application of our knowledge, or from other practitioners who have already done something similar. These adaptations are rarely mentioned in research, or spoken about in conferences, though there are exceptions (e.g., Jenkins, 2015). In practice, it is not only important to know how to make adaptations, but it is also important to be able to demonstrate to others (peers and customers alike) why what you have done is valid, and how you arrived at your conclusions.

As well as developing knowledge and expertise in order to be able to use tools in a more flexible manner, both individually and in conjunction with other tools, understanding the work setting is crucial. A newly qualified practitioner can develop a deep understanding of a tool and how to use it from the textbook. Take, for example, the concept of situational awareness (Endsley, 1995). A number of tools are available to measure this in experimental, simulated, and real-world settings. But to measure situational awareness in fighter jet pilots and draw sensible conclusions on how this should be addressed through design, understanding the work setting is essential. For task analysis, this will usually involve a combination of procedure reviews and job observation.

TOOLS FOR SPECIALISTS AND TOOLS FOR NONSPECIALISTS

Many customers have a "preferred tool" that they would like to use or have someone else use to support their own HF/E approach. Some tools require expertise—for example, many undergraduate and even postgraduate degree courses do not go into topics such as human reliability assessment or user needs analysis in enough depth to use related tools; it takes an experienced practitioner to perform this well. Other tools are much easier to understand and use by those without formal training in HF/E (e.g., a number of postural assessment tools that draw from biomechanics).

HF/E practitioners will often question anything that tries to take years of experience and turn it into a tool that can be used by a nonspecialist (e.g., incident investigator, product designer). It may be possible to use a very structured tool as a novice, but very difficult to make sense of the data. Some tools clearly require skill and experience to use them properly, often because they are less structured and require a great deal of in-depth understanding to even deploy them.

One good example is the Manual Handling Assessment Charts (MAC) tool (Health and Safety Executive [HSE], 2014), a simple method to help nonspecialists decide where to put their attention in preventing injuries through manual handling at work. It is easy to use with minimal training, and provides a set of thresholds that are clear. It is a simple tool—but is based on a very complex combination of data sources (e.g., the National Institute for Occupational Safety and Health [NIOSH] "lifting equation") and is therefore flexible in the hands of a specialist practitioner. Unfortunately it tends to be used for applications where the only question the user wants answered is "Is this safe?"

The developers of the MAC tool almost certainly never intended it to present *de facto* predictions of musculoskeletal risk. But for all sorts of users—employers, employee groups, enforcement agents—it is tempting to treat the resulting colors definitively, rather than look at the other sources of data available as well, which is best achieved using expertise.

However, there is a place for tools that help the novice to understand the most important issues and become an "intelligent customer" who can then seek more in-depth analysis. Even with an intelligent customer, the skill of the practitioner lies in the sense-checking of their espoused requirements and selecting the best tool or tools for the job, alongside convincing the customer with robust arguments of the best approach. There are many cases where a tool can be used by a novice. Tools such as the Human Factors Analysis Tools (HFAT®) are designed to be used by people without specialist training in human factors or ergonomics. The guidance on what the results of using the tool mean is provided by the tool itself. Other tools, for example, safety culture assessment tools, might be used by a novice, but the novice will need help to work out how to use the results of the assessment. This is the purview of the HF/E practitioner.

Just because tools are designed to be used by nonspecialists, it does not always follow that they have the background knowledge and skills to use the output from the application of the tool correctly. Interpretation is often required, and the skills required to perform this interpretation need careful consideration before the tool is used. For example, due to increased regulatory interest in the extent to which organizations in the high-hazard industries are measuring the influence of human performance on major accident hazards, many companies are training personnel to perform human reliability analysis using a range of different tools. But when it comes to making recommendations on how to mitigate some of the errors predicted, specialist knowledge and judgment is required. If people are only trained in the use of the tool, they need to recognize when they are operating outside their capabilities, and seek appropriate support for using the results in an effective way.

There is also a difference between a tool that will allow analysis and then informed, nuanced decision making, and a tool that locks on to a specific action as a result of its use. For tools with hard rules at thresholds ("if [level] then [action]"), this has the potential to complicate decision making even though it may seem to simplify it, especially when working with other sources of data. An example of this is "just culture algorithms" that can seem to dictate certain responses to certain situations depending how some standard questions are answered. As practitioners we are often in a position to be able to accept less-definitive answers and use multiple data sources more effectively to arrive at our conclusions.

Having said all of this, we are yet to use a tool that is a better predictor of whether a job was OK than talking to and observing the people doing that job, and checking some of the other markers for "good" ergonomics such as injury rates, staff retention, or the job satisfaction and motivation of people at work. Increasing the number of measures can take the practitioner further from reality, and this suggests a greater need for direct observations and discussion in combination with the use of tools. Dekker (2015) makes this point with regard to workload-rating tools versus discussion in context. Nothing beats the experience of real users with views they can share. Facilitating the collection and discussion of these views in the design or development process goes beyond the ability to apply relevant and appropriate tools (Dempsey, 2007; Annett, 2002).

TOOL SELECTION IN THE REAL WORLD

As a practitioner, it can become tempting to use the same tools as a level of expertise with them develops, especially when a client requests a specific tool to be applied. As practitioners we need to reevaluate the approach as project progresses. Defining the most appropriate approach comes down to planning and reevaluating the project at regular stages.

Defining a Human Factors Integration Plan for larger projects is helpful. In the early stages, the plan is used to define the human factors activity, and linking this to the overall project plan. This must be kept as a living document and updated as the project progresses to ensure that the approach and the tools remain valid.

For example, for a recent control room project it was identified at the outset that a link analysis would be conducted of the existing control room to inform the design of the new one (by identifying the existing equipment usage, communications paths between operators, etc.). As the project progressed it emerged that a number of errors using the existing systems had led to incidents. The plan for the work was therefore updated to include further task analysis and human reliability analysis, used in combination with the link analysis, to provide requirements for the layout and design of equipment components in the new control room.

HF/E concerns interactions between factors and disciplines (e.g., psychology, physiology, engineering). We are yet to come across an ergonomics tool that fully accounts for these complex interactions. Tools exist that cover the characteristics of people, or tasks or the organizations, but few, if any, cover all three in a comprehensive manner. The reason is probably because such a tool would be unusable. The practitioner needs to integrate different tools and the results of their application, particularly for more complex projects.

In practice, the tools we use most often—in different aspects of HF/E consulting—include the following:

- Human factors tools for incident investigation—understanding why people behaved the way they did leading up to an incident is essential to formulating recommendations.
- Task analysis—whether it is used as the basis for human reliability analysis or for the design of procedures or training.
- Safety culture measurement tools—for examining some of the influences of the way the organizational factors can influence the people in the organization in relation to safety.
- Stakeholder analysis—this might include "who uses, who's affected by, who influences" questions, but helps prevent hidden stakeholders from appearing at the end of a project.
- Postural/biomechanical analysis—the MAC process provides an ability to code quite complex handling tasks well, and against what increasingly seems a pretty robust interpretation of available data. The observational ratings of RULA and REBA are easy to agree between observers and help to explain joint displacements. Moore and Garg's Job Strain Index (JSI) is useful to analyze fine distal postures (Moore and Garg, 1995).

It is almost always the case that as practitioners, we are competing with productivity goals, costs, time pressure, etc., that require us to be inventive and move away from the usual means of data gathering to a more pragmatic approach (e.g., observational analysis over workshops or simulations). See, for example, Stanton and Young (1998).

In some organizations that we have worked with (e.g., the military and civil service) there are situations where the application of tools can be close to the ideal, there is the willingness to remove barriers and make sure the project is done as per our advice (e.g., releasing large numbers of military personnel to achieve a representative sample for user testing of mission-critical human–machine interface). However, we have also experienced the opposite of this in some of the more commercial organizations that use ergonomics consultants. For example, during the design of a control room, it may only be possible to speak to three control room operators at a time to do some human reliability work due to the cost of releasing them.

Often we as practitioners need to weigh up the result of not doing something against the consequences of doing something but not to the same level of rigor that we would like. For example, if we can only run an HRA with half a dozen control room personnel, identify a number of major design requirements to reduce the potential for errors, we are much better off than doing nothing at all because we can't meet the standard recommendations on sample size.

There are times when constraints reduce the robustness of the analysis, up to a point where it becomes impossible to meet client requirements. Usually a way can be found to set expectations with the client to ensure that something beneficial can be achieved, even if it is not the most robust solution.

WHERE ARE HF/E TOOLS USED IN PRACTICE?

Some industries are more mature in terms of the application of HF/E than others. As stated earlier in this chapter they are often early adopters of HF/E tools and approaches, and they recognize the benefits of using them not just because they need to demonstrate compliance with good practice, but because they want to adopt best practice.

Such organizations will invest more heavily in HF/E, for example, in simulations where participants can be observed performing the task, but not in a live environment. This is especially useful for proposed changes to equipment and procedures. Because the simulation is virtual, those involved can try new things without fear of negative impacts. This can provide a useful opportunity to test new systems in an environment somewhere between theory and real life.

There is also benefit when applying some tools to do so when people are actually completing the task that is being measured. For example, when performing task analysis the opportunity to gather the data while receiving a walkthrough of the task can be significantly more beneficial than sitting around a table with the procedure receiving a talk-through of the task. Pragmatism is required however, as in some cases the level of interference introduced by attending a walkthrough can introduce an additional level of risk that would be unacceptable.

CONCLUSION

Tools help the specialist and novice to perform a particular type of job. Where the tool is right for the job, careful consideration of whether the user of the tool has the right skills and knowledge to apply the results to address the issue in hand is required. This is true of experienced practitioners and novices—some tools are designed for use in a niche area, and a general ergonomics or human factors practitioner may struggle if they have not specialized in this area. So the right tool for the right job is an important consideration, but so is the right practitioner for the tool.

The results of the use of the tool need to be examined to ensure that they make sense. All practitioners can probably cite a situation where they have applied a tool, but when they look at the results with reference to a bigger picture they question whether what the tool produced is useful. Blindly using the results of applying a tool is not wise, hence the need to ensure the right practitioner for the right tool.

There are times when only experience and skill will do the job, and a tool used on its own will not be sufficient. This may mean more effort, analysis, and application of skills. Although it may seem easier to use a tool to help achieve this, first principles may need to be applied. This may result in using tools in combination and the results being integrated, or it may result in the application of expertise alone. Either way, these are definite situations where an experienced specialist practitioner is going to be required, and not just someone who knows how to use the tools.

REFERENCES

Annett, J. 2002. A note on the validity and reliability of ergonomics methods. *Theoretical Issues in Ergonomics Science.* 3(**2**), 355–365.

Bias, R.G. and Mayhew, D.J. 1994. *Cost-Justifying Usability.* Orlando, FL: Academic Press.

Brazier, A., Richardson, P., and Embrey, D. 1999. *Human factors assessment of safety critical tasks. HSE Offshore Technology Report OTO 1999 092.*

Buckle, P. 2011. 'The perfect is the enemy of the good'—ergonomics research and practice. Institute of Ergonomics and Human Factors Annual Lecture 2010. *Ergonomics.* 54(**1**), 1–11.

Chung, A.Z.Q. and Shorrock, S.T. 2011. The research-practice relationship in ergonomics and human factors—surveying and bridging the gap. *Ergonomics.* 54(**5**), 413–429.

Dekker, S. 2015. What is your maximum workload? *HindSight* 21, 8–9. Brussels: EUROCONTROL.

Dempsey, P.G. 2007. Effectiveness of ergonomics interventions to prevent musculoskeletal disorders: Beware of what you ask. *International Journal of Industrial Ergonomics.* 37(**2**), 169–173.

Endsley, M.R. 1995. Toward a theory of situation awareness in dynamic systems. *Human Factors.* 37(**1**), 32–64.

Health and Safety Executive (HSE), 2014. Manual Handling Assessment Charts. INDG 383 (Rev 2). ISBN:9780717666423 Available at: http://www.hse.gov.uk/pubns/indg383. htm.

Jenkins, D.P. 2015. Specifications for innovation. In: Sharples, S., Shorrock, S., and Waterson, P. (eds.). *Contemporary Ergonomics and Human Factors 2015.* London: Taylor & Francis.

Lardner, R. and Scaife, R.G. 2006. Helping engineers to analyse and influence the human factors in accidents at work. *Process Safety and Environmental Protection.* 84(**3**), 157–233.

Marras, W.S. and Karwowski, W. 2006. *The Occupational Ergonomics Handbook: Fundamentals and Assessment Tools for Occupational Ergonomics,* Second Edition. Boca Raton, FL: CRC/Taylor & Francis.

Moore, G. 1991. *Crossing the Chasm.* New York, NY: Harper Business.

Moore, J.S. and Garg, A. 1995. The Strain Index: A proposed method to analyze jobs for risk of distal upper extremity disorders. *American Industrial Hygiene Association Journal.* 56(**5**), 443–458.

Olson, N.S. 2011. Coding ATC incident data using HFACS: Inter-coder consensus. *Safety Science.* 49(**10**), 1365–1370.

Olson, N.S. 2013. Reliability studies of incident coding systems in high hazard industries: A narrative review of study methodology. *Applied Ergonomics.* 44(**2**), 175–184.

Olson, N.S. and Shorrock, S.T. 2010. Evaluation of the HFACS-ADF safety classification system: Inter-coder consensus and intra-coder consistency. *Accident Analysis & Prevention.* 42(**2**), 437–444.

National Institute for Occupational Safety and Health. 1994. *Applications Manual for the Revised NIOSH Lifting Equation.* NIOSH (DHHS) Publication no. 94–110. Washington, DC: U.S. Government Printing Office.

Stanton, N.A., Hedge, A., Brookhuis, K., Salas, E., and Hendrick, H. 2004. *Handbook of Human Factors and Ergonomics Methods.* London, UK: CRC Press.

Stanton, N. and Young, M. 1998. Is utility in the mind of the beholder? A study of ergonomics methods. *Applied Ergonomics.* 29(**1**), 41–54.

12 Becoming a Human Factors/Ergonomics Practitioner

Andrew Baird, Claire Williams, and Alan Ferris

CONTENTS

PRACTITIONER SUMMARY

An effective human factors/ergonomics (HF/E) practitioner has a blend of knowledge, skills, abilities, and other factors (KSAOs) that allow for the development of solutions to what are often complex client needs within a variety of real-world

165

constraints. The development of these KSAOs occurs as part of initial professional development (IPD) and continuous professional development (CPD) throughout a practitioner's career. IPD is a fundamental step toward formal professional recognition and demands input from academic institutions, trainee and mentor practitioners, and employers. Academic institutions try to ensure that practitioners develop skills and underlying knowledge such that graduates are useful to employers from the outset. But academic institutions work within constraints placed on them by academic regulation and available time, and cannot provide everything required. HF/E students must be active learners and look to develop real-world skills. Mentor practitioners need to help develop skills and knowledge, particularly in light of the realities of particular market requirements. Employers must accept that students will not emerge from university courses as fully capable for independent HF/E work, and will need time and resources to continue their development.

INTRODUCTION

Becoming a HF/E practitioner is a process in which there are multiple stakeholders with various interests, including employers, the public, product and service users, and other professionals. Society requires us to be capable of helping to address human problems and opportunities both now and in the future. Employers requires practitioners to be capable of some useful practices as they enter the workforce. HF/E educational institutions have to ensure that academic standards are maintained and academic skills are acquired while delivering courses from which students often enter industrial practice directly. In light of the wide range of goals for HF/E training and practice, it is unsurprising that there are differences in opinion about the extent to which these goals are achieved.

Training to be an HF/E practitioner is a process that should continue through the whole of a professional career. For simplicity, the training is sometimes split into two parts. Initial professional development (IPD) involves education and "on-the-job" development to the point that officially recognized registration, certification, or chartership can be achieved. Continuing professional development (CPD) then supports the maintenance of competence and thereby ongoing recognized status.

This chapter will bring evidence from the literature and the experience of two of the authors (Baird and Williams) in running a University MSc in Ergonomics (a qualifying course) and the third author (Ferris) recently responsible for chairing the Professional Affairs Board of the Chartered Institute of Ergonomics and Human Factors (CIEHF), the body whose role is to set and oversee the professional standards of the CIEHF membership. Juxtaposed with this will be the thoughts of a number of other practitioners as hirers into the HF/E profession. We will use these reflections to help draw some conclusions for academics, students, employers, and professional bodies. The aim is to outline the demands, constraints, and opportunities that impact HF/E IPD.

> Being a practitioner of human factors ... demands a much broader understanding of the discipline than ever before. ... Today's practitioner will most likely be called upon to support the integration of people and technology into complex and safety critical systems. They must be able to comprehend the full range of psychosocial and

technological influences that will affect human performance and be able to guide and advise engineering, safety and operational professionals on the optimization of human-centered design. ... Students should be well prepared to enter the world of professional human factors practice.

(Ian Hamilton, ERM Ltd., UK)

INITIAL PROFESSIONAL DEVELOPMENT

IPD often starts with a "qualifying course." In the United Kingdom, this is a course accredited by the CIEHF to show that it covers the full breadth of educational requirements set out by the CIEHF and similar bodies around the world. Graduate membership is available to those who successfully complete these courses. To attain professional standing, a practitioner subsequently needs to demonstrate competence in the practical application of HF/E knowledge to achieve chartered membership. In the United Kingdom, the CIEHF asks for a period of IPD to be recorded and demonstrated through the production of a logbook of projects, completed across a 3-year, mentored development period after qualifying. Similar systems exist in other national professional bodies.

A qualifying course, then, is only the first part of a journey toward professional practice. Developing other characteristics for effective performance is an ongoing process, which we will consider next.

KNOWLEDGE, SKILLS, ABILITIES, AND OTHER FACTORS

One way to characterize effective performance is in terms of KSAOs an individual possesses (Kierstead, 1998; Landy and Conte, 2004). KSAOs have been defined for a range of professional disciplines and HF/E professionals are sometimes tasked with defining KSAOs for particular contexts when matching people to jobs. However, HF/E as a discipline has rarely considered the KSAOs of HF/E professionals themselves. Williams and Haslam (2011) therefore explored these factors in terms of ergonomics advisors, and many of the comments from practitioners cited in this chapter can be viewed in this context.

THE PROFESSIONAL DEVELOPMENT PROCESS

I see HF/E practitioners journeying through four stages in their professional development. 1. Content learning—learning the sciences that contribute to the development of competency in the understanding of the core disciplines that make up an HF/E practitioner. 2. Skill development—developing key skills to collect information, analyze, and synthesize information to make a decision. This often comes in the form of a postgraduate level degree. 3. Craft refinement—through workplace practice the professional develops their craft. They learn the art of ergonomics/human factors. 4. Facilitator of Change/Educator—with senior practice and years of experience the HF/E becomes a facilitator of change. They pull on many skills and variety of bases of knowledge and experience to refine and facilitate effective sustainable change.

(Linda Miller, EWI Works, Canada)

KNOWLEDGE AND EDUCATIONAL STANDARDS

Reflecting on Linda Miller's remarks above, initial training might be viewed as supporting the first two steps of professional development—content learning and skill development. The first, foundational step therefore relates to knowledge and it is the job of academic institutions to help students attain that knowledge. In doing so, there are certain standards that constrain the options available to universities.

Given that many HF/E professionals worldwide and most in the United Kingdom come through a Master's Degree, they must follow certain standards for Master's-level education defined by The Quality Assurance Agency for Higher Education (QAA) in their framework for higher education qualifications (QAA, 2014). In this they explain the following:

Master's degrees are awarded to students who have demonstrated

- A systematic understanding of knowledge, and a critical awareness of current problems and/or new insights, much of which is at, or informed by, the forefront of their academic discipline, field of study or area of professional practice
- A comprehensive understanding of techniques applicable to their own research or advanced scholarship
- Originality in the application of knowledge, together with a practical understanding of how established techniques of research and enquiry are used to create and interpret knowledge in the discipline
- Conceptual understanding that enables the student
  - To evaluate critically current research and advanced scholarship in the discipline
  - To evaluate methodologies and develop critiques of them and, where appropriate, to propose new hypotheses

(QAA, 2014)

This indicates that Master's students should demonstrate they have current knowledge of the field of HF/E and have the skills to evaluate and undertake research. There is no mention here of practitioner skills—the only techniques referred to are those which are "applicable to ... research or advanced scholarship" (see the following).

KNOWLEDGE AND HF/E PROFESSIONAL STANDARDS

The aim of the MSc courses is to prepare the students for careers in any facet of ergonomics—research or practice (or both)—while satisfying national and international requirements of professional bodies. This means that training involves necessary compromises. Overall, HF/E course providers face the difficult aim of ensuring that their offerings get the right balance of knowledge and skills to allow individuals to become a researcher or practitioner. This compromise means that it is not possible fully to meet the needs of both sets of potential professionals within the constraints of a 1-year (UK) qualification.

In addition to meeting the overall educational requirements for any Master's-level science subject, course providers must also meet the requirements set out by professional bodies such as the CIEHF and Centre for Registration of European

Ergonomists (CREE). The "Harmonising European Training Programmes for the Ergonomics Profession" (HETPEP) initiative produced the following key areas in the early 1990s:

1. *Ergonomics principles*: Introduction to the "ergonomics approach" and its relation to sciences.
2. *Human characteristics*: Basic knowledge from disciplines such as human biology and psychology that has particular relevance for ergonomics.
3. *Work analysis and measurement*: Techniques and methods for analysis, measurement, and computation.
4. *People and technology*: Applied knowledge from engineering and human sciences that has particular relevance for applying ergonomics.
5. *Applications*: The integrative, interactive, social, and iterative nature of applying ergonomics in the context of a structured and concrete research or design project.
6. *Professional issues*: Legislation, economics, "politics" of ergonomics investigations, ethics, organization.

The current CIEHF (2015) interpretation of these requirements provides a series of overarching "knowledge areas" and a series of "themes," both of which must be adequately covered within a qualifying course. The Centre for Registration of European Ergonomists (CREE) has taken a slightly different approach to the application of the original HETPEP definitions, but it is broadly comparable in terms of overall requirement to the CIEHF approach (see Table 12.1).

In effect, this lays out that there are key subject areas that all students must cover, but they must also understand why the knowledge is useful to ergonomics and how they can integrate the knowledge from individual subject areas to address complex real-world problems.

Coverage of these professional body requirements is mandatory for course providers—failure to meet requirements would lead to an unaccredited course, with obvious effects on recruitment of students. Competency in research methods and a clear understanding of the evidence base around ergonomics is therefore essential for all ergonomics students regardless of eventual career path.

SKILLS

I find that students come in with strong quantitative skills, but often lack the ability to refine solutions so that they can be implemented.

(Linda Miller, EWI Works, Canada)

As described, academic and professional requirements focus heavily on knowledge, with skills receiving comparatively less attention. Research skills are, however, a clearly defined requirement, so it is no surprise that research methods tend to form a mandatory part of ergonomics syllabi. For example, the University of Derby's syllabus of six taught modules includes "Investigations and Analysis," a module covering qualitative and quantitative research methods (University of Derby, 2016). Similarly,

TABLE 12.1

Chartered Institute of Ergonomics and Human Factors (CIEHF) and Centre for Registration of European Ergonomists (CREE) Knowledge Areas

CIEHF Requirements	CREE Requirements
The knowledge areas are defined as follows:	CREE defines 10 knowledge areas:
1. Human mental characteristics	1. Principles of ergonomics
2. Human physical characteristics	2. Populations and general human characteristics
3. Systems ergonomics	
4. Ergonomics methods	3. Design of technical systems
5. Research methods	4. Research, evaluation, and investigative techniques
The key themes are defined as	5. Professional issues
• Ergonomics approach characteristics: user-centered design, participatory, holistic and integrated essential knowledge areas	6. Ergonomics: activity and/or work analysis
	7. Ergonomic interventions
• Ergonomics knowledge attributes: theoretical and practical knowledge, scientific basis, integrated, evidence based	8. Ergonomics: physiological and physical aspects
• Ergonomics knowledge aims: health, safety, wellbeing, efficiency, effectiveness	9. Ergonomics: psychological and cognitive aspects
• Ergonomics knowledge application: product design, workplace interventions, systems design, and evaluation	10. Ergonomics: social and organizational aspects
• Professional skills: legislation, standards, project management, ethics, communication, interdisciplinary work	

the University of Nottingham's Distance Learning MSc has six modules, one of which (Ergonomics Methods) deals with research methods and statistics (University of Nottingham, 2016). Loughborough University's course has (at the time of writing) the more traditional eight taught modules, one of which is "Data Collection and Analysis." It is apparent that between an eighth and a sixth of the taught part of these CIEHF qualifying courses comprises research method and statistical analysis training. This level of coverage is required even though many practitioners are unlikely to need some of these academic research methods after graduation.

This issue is not limited to the United Kingdom. The CIEHF requirements broadly follow the HETPEP guidance. It is also a requirement of CREE that a significant proportion of an applicant's education must have covered research and analysis if they are to be awarded "Eur.Erg." status, despite the fact that CREE is specifically for practitioners and not researchers. In other words, HF/E bodies devoted to practitioners recognize the value of research.

This focus on research skills may detract from practical skills required of HF/E practitioners but the "Ergonomics Methods" element of CIEHF requirements should ensure that practical skills are acquired while studying on a qualifying course. Within the time available, it is likely that students gain knowledge of various methods, their history, and development, but lack proficiency in using them, especially to meet client needs. Ultimately, HF/E practitioners must use their range of skills and knowledge to provide solutions to what are often complex client needs. This can be difficult to mimic in an academic environment, even when practical, scenario-based assessments are employed. This relates to the issue of "craft refinement" as described above and is something that is ideally developed as part of mentored work activities. This can be a source of frustration to employers, who would like HF/E graduates to be billable from the outset.

> I like generalists who have at least analysed and designed one interface; conducted one environmental survey and conducted a basic physical ergonomics assessment. As an applied discipline based on science, that's what I would like to see reflected in the courses.

> **(Nigel Heaton, Human Applications, UK)**

The issue of skill development must also be seen in the context of the changing nature of HF/E education where full-time, "hands-on" students are now in the minority with most students studying part-time and at a distance or with block study setups.

ABILITIES AND OTHER FACTORS

Regardless of academic ability, successful HF/E graduates require some broader abilities and personal characteristics if they are to be effective HF/E practitioners, as Nick Taylor makes clear:

> ... the main requirements for graduate employment were whether their personality fit with the existing team and they had the organizational skills to be self-sufficient and effective whilst on the road ... if they couldn't get from A to B, know how to solve problems on the hoof, and represent themselves and the company professionally – then their technical knowledge was wasted.

> **(Nick Taylor, Hu-Tech, UK)**

Arguably one of the key attributes of effective HF/E practitioners is empathy (Shorrock and Murphy, 2007) as it is vital that practitioners understand human activity (e.g., work) from the point of view of those doing the activity, along with the systemic context and impact of work.

> I've come to believe that the most important attribute of an HF/E practitioner in general and one that works for a small consultancy in particular is the ability to be empathetic.

> **(Ron Gantt, SCM, USA)**

> I think it is about being able to apply an ergonomics framework to problem solving ... getting good at getting inside the heads and skins of other people.

> **(Nigel Heaton, Human Applications, UK)**

Allied to this are some general "people skills" not unique to HF/E, but vital if workers are to be put at ease and practitioners accepted into a workplace as positive agents of change. This is the difference between someone seen as being there to help workers and someone who is seen as a "time-and-motion"-type extension of management. Again, these are attributes that can be discussed in a classroom, but are not easily taught away from the realities of a workplace and indeed are often more related to personality or character than knowledge. This demands an effective selection and training process to ensure a good fit between practitioners, the organization, and the sort of work it undertakes.

BEING A USEFUL GRADUATE

The key is that graduates aren't just "knowledgeable," but also "useful." It is hard to define the totality of what "useful" means in terms of graduates, but it is probably best seen as an *ability to meet client needs*, which will require a blend of KSAOs. For consultancies, having graduates with enough knowledge and skill to be able to earn fees, almost from the outset, is certainly one metric. In other contexts, making a "value-added" contribution, however value is defined, is probably the equivalent. Understanding that at least three years of mentored practice is required before professional status might be gained (Chartered Membership), what is reasonable and possible to expect of those coming straight from qualifying courses? This must take into consideration the broad range of experience of people going into qualifying courses and the industries and roles that they may take up on leaving them.

While there are core approaches across the domains (human centered, systems focused) and even general methods that most practitioners will use (e.g., task analysis, stakeholder analysis, participatory approaches), there are hundreds of HF/E methods used in different domains for many different purposes. HF/E university degree programs, as opposed to PhDs, have to focus on breadth at the expense of depth, partly because of the constraints of the time available, and partly because too narrow a focus would preclude a graduate from taking up a position in any domain. This means that graduates may be aware of certain techniques without having used them in a realistic way, or in any way. Their general understanding should, however, ensure that specific tools or techniques can be picked up quickly with suitable supervision/mentoring.

THE EDUCATION–PRACTICE GAP

The easiest way to sniff out a graduate or academic author of a report, is when they demonstrate a lack of understanding of how a commercial organization will view, respond to, or implement recommendations for change.

(**Nick Taylor, Hu-Tech, UK**)

There may be a disconnect between what employers want and what qualifying courses provide in terms of both knowledge areas and skills. The disconnect is not simply a case of requirement versus provision, as it will also depend a great deal on the nature of the student, their current work (if any—some courses have students who have not yet entered the workplace), and whether they wish to continue to work in their current sector/area.

For example, the majority of students undertaking the University of Derby course are employed and tend to stay with their current employer, in a similar role, working on similar projects. This gives these students the ability to initiate and participate in discussions on the practical use of theory and methods, which ensures that everything they are learning can be applied to their workplace and role requirements. If, however, a student is not employed and/or wishes to move to a completely different work activity or work sector, then those questions require more imagination.

Something that can easily be forgotten is that, typically, only two-thirds of a Masters is "taught" (which again reinforces the constraints that universities face) with the remaining third being the students "thesis" or research project. Some may argue that this proportion of "research" is at odds with what employers need. However, in Masters programs "applied" research is common, particularly with employed students studying part-time. This allows the students both to study a technique or intervention in great depth and to consider its evidence base within the real-world context of application.

WHAT MORE COULD UNIVERSITIES DO?

While different universities will offer a different flavor of HF/E with variations in optional modules and content packaging, the constraints described above mean that wholesale changes to programs are not possible. That is not to say that nothing can change; educators can think about how subjects are taught and the way in which students are likely to learn.

Applied, Scenario-Based Assessment

It is well recognized that students tend to focus effort on what will generate most reward, that is the module assessment (Gibbs and Simpson 2004). Knowing that most effort will be made around assessment activities, educators need to think very carefully about how they assess subjects, particularly with summative (end of module) assessments, but also any formative (during module) activities. Assessments should have clear and direct relevance to work activities. For example, anatomy and physiology needs to be taught and assessed, but the assessment can be made in terms of applied issues such as musculoskeletal disorders (MSDs), physical capacity, or fatigue. An approach that is popular in some applied disciplines is scenario-based assessment. This approach produces many benefits (UNSW, n.d.) including the following (Beaubien and Baker, 2004):

- Engaging students in research and reflective discussion
- Encouraging clinical and professional reasoning in a safe environment
- Encouraging higher order thinking and creative problem solving without risk to third parties or projects
- Allowing students to develop realistic solutions to complex problems
- Developing students' ability to identify and distinguish between critical and extraneous factors
- Enabling students to apply previously acquired skills

- Allowing students to learn from one another
- Providing an effective simulated learning environment even though they are considered low-level fidelity

A development at the University of Derby to facilitate scenario-based assessment was the production of a virtual workplace (Richardson and Baird, 2008). Despite having a low level of fidelity, as Beaubien and Baker found, this is not a major barrier to the way in which it can be used. The virtual workplace has been used with both formative and summative assessment and provides the important "context" for assessment activities. The facility has also proved popular with students.

Developing Professional Skills

Beyond situating assessments in terms of real-world issues and within realistic scenarios, a proportion of assessments must assess the professional skills in a particular area. Sometimes these skills are general to all professionals, such as the ability to write well, and use software that they will need to use as practitioners:

> I would like to see a professional writing course and software familiarization. A number of students have very poor technology skills limiting their professional skills when they get out.
>
> **(Linda Miller, EWI Works, Canada)**

Other skills are more commercially focused, such as: the need to meet deadlines and for peer reviews before the deadlines; how to actually scope and price a job; how to manage time and stay in budget.

This leads to a problem that universities may face when an academic is reticent to engage in discussion around practitioner activities as it is "not their area" or they have never undertaken the work in question. Making use of guest lecturers and contributors to any program, with good representation of practitioners from different domains who bring current workplace case studies, helps support in-house academics in this way.

Careful Project Selection

Finally, academics should give great thought to the projects that they approve for students' dissertations. Projects should benefit them students and fit with their desired career path as well as developing a range of skills. It may be reassuring for both academic and student to stick close to the academic's experience, but unless it genuinely develops the student in their chosen area this should be avoided. The supervisor doesn't need to be an expert; the student will become that. The academic needs to be a wise sounding board. This will inevitably develop the academic as well as the student over time.

WHAT MORE COULD EMPLOYERS DO?

> We don't employ graduates any more. Overall the investment of time and resources has not shown a good return over the mid to long term and so has been stopped in favor

of looking for folk with their work skills already in place. The industries that we have been working in are intolerant of recent graduates being proposed for contract work, so graduates are difficult to keep chargeable and financially viable.

(**Nick Taylor, Hu-Tech, UK**)

Codesigning Content, Methods, and Assessments

If particular employers have particular needs with respect to graduate attributes, they could engage with institutions to discuss these needs. Employers could help by providing case study material, or by codesigning assessment activities.

Sponsorship and Work Placements

If an employer is very keen to ensure that they get the "best" students available, then they could consider sponsoring students and/or providing work placement opportunities. This would ensure that the student was able to answer the important "How does this fit with …?" questions mentioned above, but would also give the employer the chance to see the employee in action in a way no interview can allow.

Project Opportunities

Related to the preceding discussion would be to offer project opportunities. This has great potential for both student and employer. The student can undertake and/or evaluate a real-world, commercial ergonomics intervention. This will help the student to develop expertise around the area while also ensuring that the employer gets their work evaluated to a level that would never be commercially viable.

We provide on-the-job mentoring and coaching to bring the new start up to speed with our ways of working and our products. This typically involves co-delivery of a project with an experienced consultant at least twice before being allowed to lead under supervision, finally going solo on a project of a similar nature after we are confident in them and they are confident in themselves.

(**Richard Scaife, Keil Centre, UK**)

If "missing" elements are common to all graduates, then the production of in-house resources (training notes, etc.) seems sensible and the graduates themselves could be involved in much of the resource production for one another. It would also be appropriate to provide some form of on the job training. This would all be part of a graduate's ongoing IPD.

Ultimately employers need to ensure that developing practitioners have the skills that clients need, developed via both initial training and ongoing development. If employers want more developed practitioners from day one of employment, then it is necessary to employ professionally recognized (e.g., Chartered or Certified) HF/E practitioners.

WHAT MORE COULD EARLY CAREER PRACTITIONERS DO?

Make Use of Opportunities Afforded by Affiliation with Academic Institutions

While one is a student, access to materials and academic support is vast compared to the limited resources available to most practitioners. Utilizing these university resources is essential to the development of competent professionals. Fellow students within this subject area are likely to provide a wealth of experience and insights that can be drawn upon. Discussion between students and academics and between students and students is key to bringing the subject to life and contextualizing knowledge.

Get a Mentor

Universities can only provide the core elements that allow students to go on into any aspect of ergonomics. Wherever a graduate chooses to work, there will be additional elements to learn throughout one's career. IPD should be approached with the same vigor and commitment as the formal university education. To that end, other individuals, as well as the student-practitioner, have an important role to play; line managers and mentors in the early years of practice are integral to the development of competent practitioners.

Keep Current

Once professionally recognized, practitioners may feel that they have a "badge of approval," which can be sold as "expertise." Successful projects and repeat business may then lead to a degree of complacency among practitioners. There is always a need to critically evaluate and update the way we work; just because there is work doesn't mean the work is appropriate. The role of some ergonomics practitioners in the "explosion" of MSD cases is a case in point (Baird, 2008; Lucire, 2003).

Ergonomics and allied disciplines will change with time, and this will impact the services offered in particular areas of ergonomics. A good example of this is the change in our understanding of chronic musculoskeletal pain. While the evidence base around chronic musculoskeletal conditions was emerging, ergonomics was slow to respond or recognize the new evidence (Baird, 2013). Such evidence changes inevitably lead to the need for a change in approach if a problem is to be successfully addressed. When thinking about CPD, therefore, developing practitioners can reflect on how other disciplines can contribute to our understanding and/or activities, and vice versa. As well as contact within places of work, social media (e.g., Twitter) can be a great help in understanding what is going on in allied disciplines and in industry as a whole.

CONCLUSIONS

Effective HF/E practitioners require a broad range of KSAOs that will vary depending on the sector in which they work. Accredited university courses must ensure that all students gain sufficient knowledge to be able to work in any sector and to go on to become professionally recognized (e.g., chartered). Although universities could do more to develop practical skills, it must be accepted that universities work under constraints such that they cannot produce the complete HF/E practitioner and they must ensure academic quality and focus on research evidence. Most students are studying at Master's level with significant work experience behind them; utilizing this peer experience can provide significant practical insights that will help when discussing skills applications.

Employers must recognize that graduates will require guidance and mentoring to develop the skills and abilities required to work in a particular sector. Employers could work more with universities to try to narrow the gap that employers currently perceive exists by working on case study materials and offering work placement opportunities. Professional bodies must ensure that accredited courses continue to produce graduates capable of work in a full range of potential sectors and must ensure that mentoring programs have the desired effect of helping graduates achieve registered status.

REFERENCES

Baird, A. 2008. Teaching about musculoskeletal disorder – are we barking up the wrong tree? In: Bust, P.D. (ed.). *Contemporary Ergonomics 2008*. London: Taylor & Francis, pp. 441–446.

Baird, A. 2013. Musculoskeletal disorder – are we better off seeing them as pain conditions rather than injuries? *Association of Canadian Ergonomists 44th Annual Conference on From Sea to Sky: Expanding the Reach of Ergonomics*, 8–10 October 2013, Whistler, BC.

Beaubien, J. and Baker, D. 2004. The use of simulation for training teamwork skills in health care: How low can you go? *Quality and Safety in Health Care*. 13(s1), i51–i56.

CIEHF. 2015. Degree course accreditation. Accessed April 29, 2015. Available at: http://www. ergonomics.org.uk/degree-courses/degree-course-accreditation/.

Gibbs, G. and Simpson, C. 2004. Conditions under which assessment supports student learning. *Learning and Teaching in Higher Education*. 1, 3–31.

Kierstead, J. 1998. *Competencies and KSAOs*. Accessed March 13, 2016. Available at: https://wayback.archive-it.org/3608/20080201031859/http://www.psagency-agencefp. gc.ca:80/research/personnel/comp_ksao_e.asp.

Landy, F. and Conte, J. 2004. *Work in the 21st Century: An Introduction to Industrial and Organizational Psychology*. Boston, MA: McGraw-Hill.

Lucire, Y. 2003. *Constructing RSI: Belief and Desire*. Sydney, Australia: UNSW Press.

QAA. (2014). UK Quality Code for Higher Education. Part A: Setting and Maintaining Academic Standards. Gloucester, UK: QAA. Accessed April 29, 2015. Available at: http://www.qaa.ac.uk/en/Publications/Documents/qualifications-frameworks.pdf.

Loughborough University. 2016. Ergonomics/Human Factors Programmes. Accessed March 13, 2016. Available at: http://www.lboro.ac.uk/departments/lds/pg/ergonomics/.

Richardson, M. and Baird, A. 2008. Teaching ergonomics at a distance: A virtual workplace. In: Bust, P.D. (eds.). *Contemporary Ergonomics 2008*. London: Taylor & Francis, pp. 34–39.

Shorrock, S.T. and Murphy, D.J. 2007. The role of empathy in ergonomics consulting. In: Bust, P.D. (eds.). *Contemporary Ergonomics 2007*. London: Taylor & Francis, pp. 107–112.

University of Derby. 2016. Ergonomics MSc. Accessed March 13, 2016. Available at: http://www.derby.ac.uk/online/course/ergonomics-msc.

University of Nottingham. 2016. Applied ergonomics and human factors (distance learning) MSc. Accessed March 13, 2016. Available at: http://www.nottingham.ac.uk/pgstudy/courses/mechanical-materials-and-manufacturing-engineering/applied-ergonomics-by-distance-learning-msc.aspx.

UNSW. (n.d.). Assessment by case studies and scenarios. Accessed May 4, 2015. Available at: https://teaching.unsw.edu.au/printpdf/610.

Williams, C. and Haslam, R. 2011. Exploring the knowledge, skills, abilities and other factors (KSAOs) of ergonomics advisors. *Theoretical Issues in Ergonomics Science.* 12(**2**), 129–148.

Part III

Domain-Specific Issues

Part III has 15 chapters, which summarize some of the key, industry-specific issues and challenges for practitioners. Working as a human factors and ergonomics (HF/E) practitioner in different industries provides an incredibly varied set of challenges and rewards and requires skills, knowledge, and ways of working that differ from domain to domain. Part III is not intended to cover every industry, and within each there is much more that could be said by many others, also very differently, but it gives a flavor of HF/E in practice in a wide range of industrial contexts.

We begin with Chapter 13 (Ken Catchpole and Shelly Jeffcott) on HF/E in healthcare. Ken and Shelly describe the healthcare system as complex, opaque, and prone to accidents, and as such there are challenges for the HF/E practitioner. There is, however, considerable interest in HF/E, and significant potential.

The next chapters consider two modes of transport. Chapter 14 (Ben O'Flanagan and Graham Seeley) explores HF/E practice in the rail industry. Ben and Graham address the unique characteristics of this sector, which make it a fascinating and challenging place to be as a practitioner. Chapter 15 (Jean Pariès and Brent Hayward) concerns aviation, a sector with a long tradition of human factors in many aspects of the sector. Jean and Brent reflect on their diverse experience with regard to issues of HF/E practice in aviation personnel selection, in aviation training, in equipment design and certification, focusing especially on aviation safety management, and aviation safety occurrence investigation. Chapter 16 (Ben Cook and Ryan Cooper) goes on to consider aviation in a military environment. Ben and Ryan discuss the tactical applications of HF in military aviation and the array of challenges and opportunities to enhance operational effectiveness and safety.

The so-called "process industries" also often have significant HF/E involvement. Some of these are also "high hazard" industries. Chapter 17 (Rob Miles and Ian Randle) discusses HF/E practice in one such industry: oil and gas. Rob and Ian consider the role of HF/E in major capital projects followed by a detailed and very practical consideration of human factors safety issues in operations. Chapter 18 (Clare Pollard) turns our attention toward HF/E in the nuclear industry, with its focus on helping to deliver safety. Clare describes the multiple features that characterize this industry and the opportunities to design systems and processes to help ensure the safety of workers and the public. Chapter 19 addresses HF/E in manufacturing (Caroline Sayce and Fiona Bird), in the context of defense (submarines) and rail (rolling stock). Caroline and Fiona consider HF/E as part of a multidisciplinary design process within a requirements-driven engineering environment.

Chapter 20 (John Wilkinson) stays within the field of process industries, but takes on a regulatory perspective on human and organizational factors, with respect to the formation and compliance with legislation to protect health and safety. John gives an experiential account of working in this field as a former regulator, considering the needs, challenges, and opportunities in this field.

We then turn to HF/E in a range of consumer product and digital service contexts. Chapter 21 (Daniel Jenkins) considers human factors and ergonomics practice for consumer product design. Daniel describes how, unlike the high-hazard process industries, few standards or regulations mandate the involvement of HF/E in this sector, so it must be justified by helping to design products that are more usable and desirable, and have the greatest chance of commercial success. Accessibility is one aspect that is the focus of legislation and standards, but these can have unintended consequences. Chapter 22 (Edward Chandler and Phil Day) discusses how the needs of disabled people can be included in mainstream digital and technological solutions, avoiding a "race to the bottom" or building niche solutions for separate user groups. Many suppliers of products designed for human interaction offer "ergonomic products" or refer to the "ergonomic features" of their products. Chapter 23 (Guy Osmond) considers the issues involved in selling "ergonomic products," an area where there is only superficial understanding of ergonomics and what it can achieve. Guy emphasizes different approaches for different applications.

The subsequent chapters address two very different aspects of digital services, both fast-growing and limited in regulation and standards. The user experience (UX) field has grown considerably over the past decade. Chapter 24 (Lisa Duddington) considers how this increasing demand, along with financial and time constraints, may contribute to weaker connections with the discipline of HF/E. Lisa describes issues of limited UX practitioner competency and research. At the other end of the digital spectrum is the fast-growing field of web engineering and operations, the subject of Chapter 25 (John Allspaw). John considers the possibilities for HF/E in a complex and opaque domain that crosses geographic and geopolitical boundaries yet has no singular overarching framework or body for regulations and standards.

We end this part by moving from the newest domains to two of the oldest. Chapter 26 (Daniel Hummerdal and Stuart Shirreff) considers HF/E in the construction and demolition industry. Daniel and Stuart outline dimensions that may be relatively unique to the construction and demolition industries with each aspect providing a challenge and an opportunity for HF/E. Finally, Chapter 27 (Dave O'Neill and Dave Moore) provides a comprehensive overview of HF/E practice in a diverse and complex sector of worldwide and fundamental relevance: agriculture.

13 Human Factors and Ergonomics Practice in Healthcare
Challenges and Opportunities

Ken Catchpole and Shelly Jeffcott

CONTENTS

PRACTITIONER SUMMARY

The healthcare system is complex, opaque, and prone to accidents. This provides challenges and opportunities for the human factors and ergonomics (HF/E) practitioner. The current challenges faced by the healthcare HF/E practitioner are primarily of legitimacy. The needs, benefits, and value of the HF approach are frequently unrecognized, and the business case can be challenging to justify without explicit regulation. Furthermore, those familiar with some HF/E practices may have a limited understanding of the breadth and depth of the discipline. Communication and collaboration with clinicians and other stakeholders are, therefore, prerequisites for success. To understand this complex environment and work, direct observation, listening, and knowledge elicitation from clinicians and managers are vital. Since the role of HF/E is not necessarily understood (Russ et al., 2013), it is also important to explain in simple terms what your role is and how you can help the delivery of frontline care. In a hospital environment, this tends to involve assisting with field investigations, incident analysis, and delivering interventions or training. In each of these areas, the HF/E practitioner can help clinicians to think in new ways about the relationship between humans and other aspects of the system.

THE HEALTHCARE SYSTEM

For the HF/E practitioner, healthcare offers a huge range of application opportunities and is among the most challenging environments for HF/E practice. With initial interest focused on safety, there has been considerable growth in the awareness and activity of HF/E in healthcare in recent years. This was helped in the United States primarily through growing Food and Drug Administration (FDA) regulation and in the United Kingdom through a National Health Service concordat. There are many ways in which HF/E can be deployed successfully to improve patient and practitioner experience, reduce costs, and improve outcomes. However, no application of HF/E or change is simple or easy. In this chapter, we will focus on fieldwork practice, incident analysis, and the process of making change, including specific interventions related to teamwork, technology, and task redesign, largely concentrated in hospital care. Though there is considerable relevance for other healthcare settings, such as the home, they offer other challenges that we do not address.

About 5%–15% of hospitalized patients worldwide are accidentally injured, and about 1 in 300 hospitalized patients has an "error" that contributes to their death (Brennan et al., 1991; Leape and Berwick, 2005; Vincent et al., 2001). Cost estimates for this range from $98 billion to $980 billion per year in the United States alone (Andel et al., 2012). Of the care that should be given, based on a specific diagnosis, only around 55% is successfully delivered (McGlynn et al., 2003).

Although there is a tendency to view events as being more predictable than they really are (i.e., the "hindsight bias"), it is clear that there is a degree of preventable error in the healthcare system, which had previously not been the subject of any major effort or investment from a human-centered perspective. Subsequently, interest in HF/E has been growing at a rate that has outpaced the available expertise.

Unlike virtually all other industries in which HF/E has been developed and practiced, healthcare is not substantially technologically mediated. Most goals are delivered directly by humans, to humans, with technology assisting rather than enabling. Like other safety-critical industries, it has multiple, diverse, conflicting goals that can make an already complex system resemble a chaotic system.

Rather than being engineered in the last 200 years, healthcare has grown organically over the last 2000 (or more) years. Uncertainty about the goals, processes, risks, and workload means that it is not always possible to conduct clinical work methodically or as planned. There is variability in the needs of patients, requirements of treatments, and other work demands over time, and evidence is constantly changing. There is also a very large variety of different professionals.

Since most hospitalized patients are sick, the default condition for any patient is to be "at risk." Furthermore, there are many treatments and approaches, and the ability to predict the effect of a treatment upon an individual patient is unclear. Some are sensitive to certain drugs; some patients respond well to treatment or are otherwise resilient. Others can respond in highly divergent ways, especially where they have other (known and unknown) conditions.

Healthcare is a highly professionalized sector, with an emphasis on individual accountability and self-determination and a general lack of understanding about the systemic contributions to performance and outcome. There is still a mistaken impression that the outcomes are primarily a function of the individual patient and the individual provider (physician, nurse, or other care deliverer).

This is in spite of the fact that surgical care might start with the primary care doctor, through diagnostic specialists, processes and technologies, appointment booking systems, and administrative clerks, to operating teams, recovery nurses, intensive care and ward nurses, pharmacists, and physiotherapists. A key relationship in the delivery of hospital care is the physician–nurse relationship, which is complex and frequently challenging. The role of senior managers in mediating how frontline work happens is also integral, since they can shape culture.

Equipment is often poorly designed, maintained, and integrated, and environments are noisy, with frequent interruptions. Information technology is often designed and introduced with little involvement from frontline staff, or understanding of the so-called "ironies of automation" (Bainbridge, 1983). Contrary to what might be imagined, healthcare practitioners do not spend the majority of their time practicing their skills in diagnosis or treatment. Rather, most time is occupied by communicating and working with other disciplines, and navigating patients safely through complex bureaucratic and care systems to receive the correct diagnosis and treatment.

As an HF/E practitioner, you will likely be engaged in one of three different activities: (i) field investigations, (ii) incident analysis, or (iii) delivering interventions or training. You may also be involved in product design, development, and usability evaluation—typically outside of a hospital environment—but this will not be covered here. Work in these latter areas is more developed in other industries. Healthcare has been slow to adopt user-centered design thinking.

FIELDWORK IN HEALTHCARE

The complexities of understanding work demands mean that HF/E practice requires fieldwork in healthcare environments. Direct observation and discussions with practitioners are essential, even if you are not being directly engaged to do so. When entering a clinical domain, you need to establish: (i) who you are and what you are doing, (ii) what HF/E is and how your expertise can help frontline clinicians to deliver better care, and (iii) the value of clinical partnership and nature of that cooperation (Blandford et al., 2015).

It can be intimidating to be in clinical spaces without having clinical expertise. While it is important to show willingness to understand the technical language and processes, the broad and holistic system approach that typifies HF/E expertise provides an ability to see many things that clinicians are blind to. (However, avoid describing your work as "holistic," which has a specific and not well-regarded meaning in medicine due to its association with "alternative.") A productive approach is the naivety that allows us to ask what might seem to be stupid questions ("Why is it like that?") that quickly reveal truths, mistaken perceptions, cultural norms, and tensions. Furthermore, provided we establish trust with the people we work with, they can provide insight that would not be possible if we were seen to come from "within" the system.

Healthcare is a busy, sometimes chaotic, environment. You are ethically bound not to interfere or to assist, and it is easy to find yourself in the way of clinicians obtaining equipment, moving patients, or simply trying to conduct their everyday work. Interrupting clinicians is difficult, even though it will often be necessary, so watching, waiting, and learning allow you to identify when and where you need to create the least amount of disruption. Using opportunities when multidisciplinary teams assemble, such as handovers, will allow you to make a broad introduction, which can be cascaded to others. Information sheets on notice boards or email communication (with a photograph) can then reinforce who you are and the purpose of your interactions with them. You will never reach everyone but clinicians are accustomed to observers—especially relatives, administrators, and students—and if in doubt of your identity they will invariably ask directly who you are. Having a concise response takes practice but is vital to establish rapport and trust, allaying concerns about your presence.

A simple illustration that facilitating human performance (not constraining it) leads to improved system performance is a good place to start. You could illustrate the ways that the organization, environment, and culture support or obstruct the delivery of safe, effective, and person-centered care; the purpose of understanding demand and workload; the design and use of equipment; the quality of team interactions; or how work flows and processes are designed. Given that most improvement solutions in healthcare have focused on training, direct behavioral change, raising awareness, checklists, or "trying harder," illustrating the broader range of influences on human performance can go a long way. Making it clear that you are not interested in an individual's performance should be emphasized; it puts clinicians at a far greater ease once they know that they are not being watched, but that you are trained to explore their experience of the system, and you are there to help them.

As rapport and trust builds there will be a corresponding openness to let you see and understand the system as it really is (i.e., the "system-as-found"). You will find that people will seek you out to describe the difficult circumstances that they have to work in—pointing out bad designs or difficult working conditions. Many problems are ubiquitous, which can be a good entry to a conversation. Communication, interruptions and distractions, throughput or financial pressures, and equipment availability can be good places to start a discussion. This can quickly lead to discussions of workarounds, and adaptation to local conditions. You will quickly find that there is a difference between policy and practice, or work-as-imagined and work-as-done (see Hollnagel, 2014), and that administrators may not be aware of the latter. Direct observation usually illustrates a further difference between what is said and what is done.

HF/E practice also requires talking to those working across all levels in the system (Rasmussen, 1997) and observing what happens, in as many different scenarios as possible. It is important not to simplify before the complexity of a system is well-understood. The variability from day to night shift and from week to weekend routines in relation to demand, staffing numbers, and skill mix, fatigue, and workload must be considered if solutions are to be adaptable to the 24/7 working environment.

If the HF/E practitioner cannot be present out of hours then there needs to be a way of gathering and reviewing data from these periods. Eventually, you will be able to build partnerships. The clinician provides the focus on the appropriate aspects of the domain, clarifies domain semantics and culture, and helps gain access to frontline workers and the work processes themselves. You, as the HF/E practitioner, define the boundaries of projects, interpret the results, and link your work to the existing knowledge bases. You will always find a far greater range of potential approaches than you can possibly employ, make sense of, and address, so another important skill is to prioritize your work on the most productive lines of inquiry to deliver something actionable of importance and relevance to your stakeholders.

With regard to working across the system levels, from organization to team to individuals, it is important to get the commitment and permission from those senior in the organization, and also to encourage the middle management level to be part of driving change. Ultimately, nothing will work without the support of those at the frontline, so sense-checking any proposed changes with them, before investing time and energy, is critical. Undoubtedly, there will be differences of opinion and resistance to new ideas and approaches.

Some clinicians may appear (to you) to be deliberately obstructive toward your attempts to bring in improvements. However, understanding their motivations and experience will help you to maintain the dialogue to obtain their respect, and provide you with new knowledge and approaches. Being aware of the innovation adoption curve and how people often sit in distinct categories (innovator through to laggard) can help you to assess attitude to change (Rogers, 2003). Your observations of the "as is" system will also help to understand the reasons for the various attitudes. However, to reduce the subjectivity of your observation, a process called "respondent validation" should be used to improve the accuracy, validity, and transferability of any findings by taking them back to original participants for checking (Mays and Pope, 2000).

Eventually, this open and honest dialogue will establish frontline cooperation, assistance with data collection, and clinician advocacy. In return, your commitment is to be an advocate for improved working conditions for clinicians and care for patients. It is important to bear in mind that you will have more time and a different perspective from clinicians to suggest new approaches, but must be prepared to listen to criticism. You, as an HF/E practitioner, are an expert in HF/E method, but not the work, the system, or the environment. For further reading on the challenges and rewards of fieldwork in healthcare, Furniss et al.'s work (2014) is worth exploring, and reflects our experiences.

IMPROVING RESPONSE TO INCIDENTS

There is now a wealth of nomenclature ("adverse event"/"significant adverse event"/"never event") to define a range of specific undesirable outcomes. Some also argue that any "complication" is, in essence, a form of adverse outcome. Root cause analysis (RCA) is the de facto process that is almost universally used in healthcare to reconstruct serious events and incidents. It is usually conducted as a single meeting—typically an open committee, chaired by a patient safety administrator in one 1–3-hour session. It usually occurs some time after the event (it can take several weeks to get everyone involved together). If feedback is to be given to frontline staff then this time delay creates a missed learning opportunity.

The purpose of the RCA meeting is to decide upon the causes of, and solutions to, the incident. It is often conducted specifically for regulatory purposes to demonstrate "response to an incident." Most are carried out by healthcare practitioners with limited training in human factors, systems safety, or other investigative skills, with most conducted internally, and often opaquely. However, a powerful element of the RCA meetings is that a multidisciplinary team brings together their thoughts on the same event, from different professional and system-level perspectives. Arguably, healthcare provides limited opportunities for such multidisciplinary interaction although in situ education is slowly moving toward this. There is a realization that we must teach and learn as teams, not as single professionals.

There are many problems with the current RCA process. The notion of one "root cause" is the first since each part of each multicausal event has its precursors (Hollnagel, 2014). However, the use of RCA is at least attempting to track upstream from the event to the contributory factors (e.g., organization, team, working conditions, patient, task, and individual). The RCA meeting itself can be a threatening environment for people involved and is often dominated by senior opinion leaders, who may mean well, but can stifle subsequent discussion. Information on these events is only available through limited medical records, and the memories of those who were there. There can also be a "self-preservatory" interest in keeping quiet—especially to avoid litigation or ridicule from colleagues—while professions can close ranks, or indeed, inadvertently reinforce the culture of blame. In the complex politics of hospitals and healthcare, this remains a considerable impediment to organizational learning when things go wrong. Whatever the method used, it is important to bear in mind that the discovery of "causes" is more of an issue of post hoc social consensus, than investigatory fact finding.

As HF/E practitioners, our roles include supporting better recording of incidents, better analysis of incidents, and better solutions. Ultimately, rather than working out a sequence of "causes," reconstructing the unfolding mindset (Dekker, 2002), including discovering why actions made sense to people at the time, should be a broader aim. Though you will find this a challenge, good questions to ask include: (i) Why did the person making the erroneous decision think this was the right thing to do? (ii) How many other patients were being looked after at the time? (iii) Was all the equipment serviceable? (iv) Were the staffing levels appropriate? and (v) Were there any precursor events?

More generally, it is worth thinking about things from a Safety-II perspective (Hollnagel, 2014), considering systems thinking principles concerning people, system conditions, system behavior, and system outcomes (e.g., Shorrock et al., 2014). You may be on your own defending a different approach, but this could make a fundamental shift in the outcome of an investigation. Incident reporting systems are also prone to a range of biases—who reports (nurses report more than doctors), when they report (injuries are reported more than near misses), and what they report about (falls are reported more than complications), so finding a broad range of information sources is important for organizational learning and systems understanding.

Eventually, HF/E may help move the focus from "What went wrong?" to "How things work" and so from "human error," to recovery from errors and, more generally, trade-offs and performance adjustments (Hollnagel, 2014). The notion of error recovery is important in healthcare since many errors are caught before they reach the patient, yet we have little information on the system and process-related factors that help correct these errors. The newer way of thinking, concerning performance adjustments and performance variability, goes further beyond the binary notions or erroneous and successful behavior (Jeffcott et al., 2009), and is gradually gaining acceptance.

INTERVENTIONS

Given the complexity of healthcare, problem definition and causation analysis can be challenging. Problem definitions are frequently imprecise and the influences of context, performance shaping factors, and complexity are often opaque. The biggest mistake made in quality improvement is not understanding the problem before identifying the solutions. Most healthcare practitioners are both anxious to reach a solution and unaware of the true complexity of human behavior and clinical systems. Indeed, some may consider a continued focus on the problem as procrastination; that you should "just do it." Many conversations about improvement lead to "we need more staff," "people should just do their jobs," or "we need a checklist." Though people will talk of "low hanging fruit," in fact there are few easy solutions. If there are obvious known problems, it is important to understand why they haven't been successfully resolved before. What seems simple from the outside is far more complex in practice. The source of a problem may be spatially and temporally distinct from the manifestation of that problem. Thus, a commonly identified cause does not equal a practical or valid solution. This is where a systems approach is extremely valuable.

Although the technical challenges are large, the opportunities for improvement using HF/E techniques are also frequent and varied. Taking the systems approach—considering the wide range of influences on human performance—immediately gives many more dimensions on which change is possible than "being aware," "trying harder," "more training," or "more people." Human factors can help prevent the risk of failing to define the problem before implementing the solution and counter the history of nonevidence-based safety interventions. It is also crucial to consider the cost and efficiency in any evaluations. The more costly and less efficient a solution, the less likely it will be sustained in the face of future financial and production pressures. Again, HF/E can not only help with these calculations, but also help with ways to reduce costs by embedding system level rather than behavioral-level changes. Once again, you may well be the lone voice, so you may need to expect only limited progress initially. Learning how to convince people that there may be other ways to think about improvement is a skill that needs constant practice. Fortunately, there is a growing cadre of engaged and enthused clinicians with experience in HF/E approaches who will be supportive. Many more have negative experiences of seemingly cheap and quick (and typically behavioral) fixes that only provide a temporary alleviation of symptoms. Acknowledging and providing tools to support and communicate more complex approaches, HF/E practice advocates and assists longer term cures. To realize the benefits requires steadfastness, a willingness to challenge and be challenged, and the ability to influence, win over, and eventually partner with people who may be initially deeply (and quite rightly) skeptical.

QUALITY IMPROVEMENT

Within healthcare, there is a considerable and growing interest in management or improvement science, commonly referred to as "quality improvement" or QI. The de facto gold standard for change using QI methodology is based on Deming's Plan-Do-Study-Act (PDSA) cycle of iterative improvement (e.g., see Schriefer and Leonard, 2012). However, this can lead to an overreliance on data, rather than the gathering of intelligence (such as direct observation, task analysis, or interviews). While PDSA cycles in theory should be ongoing, the limited timescales of many improvement projects (frequently driven by demands for specific improvements) mean that the nature of PDSA cycles can be curtailed. The "Plan" part in the original cycle must be informed by data since it is impossible to explore ways to improve without a clear idea of the current state. This is particularly important given "confirmation bias" pressures to demonstrate project success and refute counter-indicative findings, reducing the ability to identify unsuccessful interventions. This may perpetuate ineffective solutions, compounding cost, quality, and safety issues rather than addressing them.

The QI movement is beginning to embrace the need for, and value added by, HF/E. Integrating the worlds of QI and HF/E may be a way to stimulate system change, leveraging on the fact that QI is already, in many places, an established part of practice. HF/E can help address the need for evaluation, to provide better iterations or add systems understanding in successive development cycles, and also preempt or predict (and measure) emergent properties. These two communities are beginning to work together, representing a strong alliance for patient safety (Hignett et al., 2015).

TEAMWORK INTERVENTIONS

Probably the most frequent proximal "cause" cited for so-called medical errors is communication. Communication is required to provide individualized care across multiple professionals, multiple shifts, and multiple patients, with constantly changing work and organizational demands. Teamwork is rarely taught, and contextual conditions (such as fatigue, tasks, equipment, environment, and organizational stressors) are rarely discussed. Thus, it was perhaps inevitable that exposure to HF/E concepts in many healthcare systems occurred initially through crew resource management (CRM) training. There is reasonable evidence that this approach works if sufficient resources are dedicated to it, if staff are willing to accept the practices, and if there is a strategy for recurrent training. The TeamSTEPPS approach (Clancy, 2007; Guimond et al., 2009) is generally accepted to be the de facto standard, although nontechnical skill frameworks also exist for surgeons (non-technical skills for surgeons [NOTSS]) (Yule et al., 2008), anaesthetists (anaesthetists' non-technical skills [ANTS]) (Fletcher et al., 2003), and scrub techs (scrub practitioners' list of intra-operative non-technical skills [SPLINTS]) (Mitchell and Flin, 2008).

Simulation facilities are now widely available and can be extremely powerful, if time and resources allow. Since behavioral solutions dominate in healthcare safety, CRM-type training has often been equated as HF/E, to the neglect of design or other system-level interventions. Always take the opportunity to emphasize the value of CRM as part of a wider approach to systems safety.

TECHNOLOGICAL SOLUTIONS

Although healthcare will remain predominantly driven and delivered by people, technology is playing more of a role. "Smart" infusion pumps, barcode systems, electronic health records, simulation facilities, and surgical robotics have all driven changes in performance, behavior, and system configurations. There is relatively little formal recognition of the value of HF/E in healthcare. However, the United States FDA, which is responsible for the regulation of most medical devices, previously demanded limited human-centered design considerations. A more rigorous approach is now available (FDA, 2016). It may be possible to evaluate devices before purchase, or explore their integration with other aspects of clinical care. Again, maintaining the view that healthcare is opaque and complex may help to identify problems before significant investments are made. In particular, in situ (or at least high-fidelity simulation) trials are required as part of the design or procurement activities, and the future may see an increased emphasis on postmarket evaluation and monitoring.

CHECKLISTS

Checklists have been extremely popular and are often seen as the default way to change behavior or embed safety practices. This has led to a proliferation of checklists for a whole range of purposes—some appropriate and some not. Some clinicians like them but many do not (even when they can be helpful). The evidence for their effectiveness is bound up with naïve views about how they work. In particular, the method of implementation, and level of involvement of stakeholders in the design, can have huge influence on whether a checklist is used or not. Many require the lead clinician's signature as a way to signal responsibility for audit purposes, even though this can be impractical (e.g., when scrubbed for surgery) and is not at all an effective guarantee of use. Much has been written for and against the use of checklists in healthcare, with a reasonably succinct summary by Catchpole and Russ (2015).

Relying on a checklist, when there is still huge systemic predisposition to error through design or process, is a strategy that will only be met with limited success. The introduction of care bundles and the WHO checklist has not been met with universal approval because other changes to the work system are required. For example, one criticism is that even though the checklist helps identify deficient equipment, it occurs late in the process, and the problem has a set of complex causes, so it is not easy to rectify. A further criticism is that by being designed to be universally applicable, it needs further local adaption to be demonstrably effective.

HF/E practitioners can help through the following:

- Helping to elucidate the problem that needs solving
- Identifying a range of ways to improve performance that may negate the need for a checklist

- Helping to identify what a checklist would offer as a solution and where and when it could be applied in the process
- Designing it to have a small number of vital components on it
- Considering ways to implement and monitor use

MULTIPLE DIMENSIONS OF CHANGE

It has become increasingly recognized that safety and quality interventions need to fit with local demands and engage local behavioral change. Quality improvement approaches such as Lean (e.g., Ben-Tovim, 2007 ; McCulloch et al. 2010) have been popular and can be extremely useful if conducted appropriately. Lean is a collection of management practices derived from the processes used in the Toyota Production System to manufacture cars cheaply and reliably. Though the translation of concepts from motor vehicle manufacturing to healthcare is perhaps controversial, the use of Lean in healthcare has achieved considerable penetration and early success.

Working from the bottom up engages staff (since it directly improves their day-to-day work), and helps address the genuine complexity of the healthcare system in a way that top-down management interventions do not. The best ways to carry out a task cannot be identified in a meeting or remotely from the shop floor; it needs to be tested with the people doing the job and within the context of each individual area. Interventions that ignore the importance of engaging and utilizing the expertise of frontline staff are unlikely to generate sustainable or optimal improvements. Moreover, addressing a particular problem using a number of different methods may be more successful than just one (Morgan et al., 2015).

CONCLUSION

At the time of writing, there is a considerable interest in the application of HF/E to healthcare, and excitement of the benefits that it might bring. HF/E practice in healthcare is complex, challenging, but extremely rewarding for an HF/E practitioner who is prepared to observe and learn about frontline clinical care. Clinicians and managers have little training or time to think about interactions between people and systems, so understanding the language and environment can place you in a unique position to work with many stakeholders at every level of the system. Emphasizing your willingness to help improve the working conditions at the frontline, and the benefits to safety and performance that this will bring, will establish trust and foster new learning and collaboration across an often "siloed" workplace. A wise practitioner maintains and spreads enthusiasm for HF/E, while remaining realistic about the challenges to overcome, and the timescales required, to realize the benefits.

Interventions are challenging to implement and benefit from clinical involvement, thorough problem definitions, and where possible, multiple dimensions of change. Success is most likely to be achieved by collaborating with a broad range of clinicians, administrators, and safety and quality professionals, advocating for an HF/E view of the world that will be new to many of your colleagues. As an HF/E practitioner, it is possible to blaze a trail for future generations, contribute to the development of HF/E expertise, and deliver tools and improvements specifically tailored to this complex sector. This can fundamentally impact frontline care delivery and help save lives.

REFERENCES

Andel, C., Davidow, S.L., Hollander, M., and Moreno, D.A. 2012. The economics of health care quality and medical errors. *Journal of Health Care Finance.* 39(**1**), 39–50.

Bainbridge, L. 1983. The ironies of automation. *Automatica.* 19(**6**), 775–779.

Ben-Tovim, D.I. 2007. Incident reporting—Seeing the picture through "lean thinking". *British Medical Journal.* 334(**7586**), 169.

Blandford, A., Berndt E., Catchpole, K., Furniss, D., Mayer, A., Mentis, H., et al. 2015. Strategies for conducting situated studies of technology use in hospitals. *Cognition Technology & Work.* 17, 489–502.

Brennan, T. A., Leape, L.L., Laird, N.M., Hebert, L., Localio, A.R., Lawthers, A.G., et al. 1991. Incidence of adverse events and negligence in hospitalized patients: Results of the Harvard Medical Practice Study I. *The New England Journal of Medicine.* 324(**6**), 370–376.

Catchpole, K. and Russ, S. 2015. The problem with checklists. *BMJ Quality and Safety.* 24(**9**), 545–549.

Clancy, C.M. 2007. TeamSTEPPS: Optimizing teamwork in the perioperative setting. *AORN Journal.* 86(**1**), 18–22.

Dekker, S. 2002. *The Field Guide to Human Error Investigations.* Aldershot, UK: Ashgate.

Fletcher, G.C.L., Flin, R.H., Glavin, R.J., Maran, N.J., and Patey, R. 2003. Anaesthetists' non-technical skills (ANTS): Evaluation of a behavioural marker system. *British Journal of Anaesthesia.* 90(**5**), 580–588.

Food and Drug Administration (FDA). 2016. *Applying Human Factors and Usability Engineering to Optimize Medical Device Design: Guidance for Industry and Food and Drug Administration Staff.* 3 February 2016. Rockville, MD: FDA. Accessed March 3, 2016. Available at: http://www.fda.gov/downloads/MedicalDevices/.../UCM259760.pdf.

Furniss, D., Randell, R., O'Kane, A., Taneva, S., Mentis, H., and Blandford, A. 2014. *Fieldwork for Healthcare: Guidance for Investigating Human Factors in Computer Systems.* San Rafael, California: Morgan & Claypool.

Guimond, M.E., Sole, M.L., and Salas, E. 2009. TeamSTEPPS. *American Journal of Nursing.* 109(**11**), 66–68.

Hignett, S., Jones, E.L., Miller, D., Wolf, L., Modi, C., Shahzad, M.W., et al. 2015. Human factors and ergonomics and quality improvement science: Integrating approaches for safety in healthcare. *BMJ Quality and Safety.* doi:10.1136/bmjqs-2014-003623. Available at: http://qualitysafety.bmj.com/content/early/2015/02/25/bmjqs-2014-003623.full.

Hollnagel, E. 2014. *Safety-I and Safety-II: The Past and Future of Safety Management.* Aldershot, UK: Ashgate.

Jeffcott, S.A., Ibrahim, J.E., and Cameron, P.A. 2009. Resilience in healthcare and clinical handover. *Quality & Safety in Health Care.* 18(**4**), 256–260.

Leape, L.L. and Berwick, D.M. 2005. Five years after To Err Is Human: What have we learned? *Journal of the American Medical Association.* 293(**19**), 2384–2390.

Mays, N. and Pope, C. 2000. Assessing quality in qualitative research. *British Medical Journal.* 320(**7226**), 50–52.

McCulloch, P., Kreckler, S., New, S., Sheena, Y., Handa, A., and Catchpole, K. 2010. Quality improvement report effect of a "Lean" intervention to improve safety processes and outcomes on a surgical emergency unit. *British Medical Journal.* 341, c5469.

McGlynn, E.A., Asch, S.M., Adams J., Keesey, J., Hicks, J., DeCristofaro, A., et al. 2003. The quality of health care delivered to adults in the United States. *New England Journal of Medicine.* 348(**26**), 2635–2645.

Mitchell, L. and Flin, R. 2008. Non-technical skills of the operating theatre scrub nurse: Literature review. *Journal of Advanced Nursing.* 63(**1**), 15–24.

Morgan, L., Pickering, S., Hadi, M., Robertson, E., New, S., Griffin, D., et al. 2015. A combined teamwork training and work standardisation intervention in operating theatres: Controlled interrupted time series study. *BMJ Quality and Safety.* 24(**2**), 111–119. doi:10.1136/bmjqs-2014-003204.

Rasmussen, J. 1997. Risk management in a dynamic society: A modelling problem. *Safety Science.* 27(**2–3**), 183–213.

Rogers, E.M. 2003. *Diffusion of Innovations.* Fifth Edition. New York, NY: Free Press.

Russ, A.L., Fairbanks, R.J., Karsh, B.T., Militello, L.G., Saleem, J.J., and Wears, R.L. 2013. The science of human factors: Separating fact from fiction. *BMJ Quality and Safety.* 22(**10**), 802–808. doi:10.1136/bmjqs-2012-001450.

Schriefer, J. and Leonard, M.S. 2012. Patient safety and quality improvement: An overview of QI. *Pediatrics in Review: American Academy of Pediatrics.* 33(**8**), 353–359; quiz 359–360. doi:10.1542/pir.33-8-353.

Shorrock, S., Leonhardt, J., Licu, T., and Peters, C. 2014. *Systems Thinking for Safety: Ten Principles. (A white paper).* Brussels: EUROCONTROL.

Vincent, C., Neale, G., and Woloshynowych, M. 2001. Adverse events in British hospitals: Preliminary retrospective record review. *British Medical Journal.* 322(**7285**), 517–519.

Yule, S., Flin, R., Maran, N., Rowley, D., Youngson, G., and Paterson-Brown, S. 2008. Surgeons' non-technical skills in the operating room: Reliability testing of the NOTSS behavior rating system. *World Journal of Surgery.* 32(**4**), 548–556.

14 Human Factors/ Ergonomics Practice in the Rail Industry
The Right Way, the Wrong Way, and the Railway

Ben O'Flanagan and Graham Seeley

CONTENTS

PRACTITIONER SUMMARY

The rail industry offers huge potential for the practical application of human factors/ ergonomics (HF/E) to support human-centered design, to improve human performance, and to help shift mindsets away from "blame and train" toward an understanding of how work is done and how systems can be improved. HF/E is still somewhat misunderstood within the rail industry, but it is a discipline that is steadily maturing and one that is in increasing demand. This is thanks in no small part to the legislated requirements to consider HF/E that have increased following a number of significant rail incidents. The long history and traditions of the rail industry are, however, at times at odds with the current demands placed on modern rail networks; there is a very real pressure to modernize infrastructure, improve efficiency, and deliver increasing levels of service that have led to significant change and supported rapid growth. To achieve greater efficiencies, many rail operators now look closely for better ways of working, good practice industry standards, cutting edge technology, and even "off-the-shelf" solutions. These characteristics make the rail sector a fascinating and challenging place for an HF/E practitioner.

WHAT CHARACTERIZES THE RAIL INDUSTRY?

The rail industry is a largely male-dominated industry, with proud engineering roots, robust trade unions, and rule-based methods of working that tend to have evolved slowly over time. The industry faces important medium-term challenges to attract and retain younger workers, to address the gender imbalance, and to put in place strategies to retain knowledge and domain expertise.

As with other industries, the rail industry is moving away from an environment that afforded workers years to learn skills on the job in a very unionized environment, to one that maximizes return on investment, reduces waste, and focuses keenly on customer service delivery. For example, Australian rail patronage has experienced significant increases in passenger operations since 2008 that has resulted in a need to increase train services to maximize capacities on the network (Naweed et al., 2015).

Although the principles of rail traffic management are relatively simple, the industry has evolved toward more complex and layered bureaucratic systems. Where technological improvements have been adopted, the fundamental rules and the hierarchies of control have been much slower to evolve. The complexity associated with the layers of bureaucracy that come with an industry with such a deep seated history and culture is often considered to inhibit the take-up of new ideas and reduce the likelihood of successfully delivering change. As people are often quick to point out, "there's the right way, the wrong way, and the railway."

The rail industry is quite reactive, but the pace of change has increased rapidly. Modern technology is increasingly juxtaposed with centuries-old infrastructure, to deliver more frequent rail services, and meet growing customer demand—but at the same time meeting the need for a strong focus on safety.

The rapid pace of technological change is driven by increasingly competitive business environments, changed regulatory practices, and increased public pressures. These changing features of the rail domain can combine as "production pressures," which Naweed et al. (2015) highlight as potentially impacting safety through normalization of deviance.

Advances in safety, changing attitudes to HF/E, and the need to consider adopting good practice from other industries have been driven in response to investigations into a number of significant safety incidents. There are numerous examples of recommendations from rail accident investigations that have led to widespread positive changes. A few examples are as follows:

- The collision at Clapham Junction, UK, in 1988 (Hidden, 1989)
- The crash at Ladbroke Grove, UK, in 1999 (Cullen, 2001)
- The derailment at Waterfall, NSW, Australia, in 2003 (McInerney, 2005)
- The train collision at Chatsworth, LA, in 2008 (National Transportation Safety Board, 2010)

These and other significant rail incidents demonstrate how the safety and performance of high-reliability engineered systems—such as safety integrity level (SIL) 4 signal interlocking—can be undermined if there is inadequate consideration of the human interactions with that system in all modes of operation. Further safety improvements have been sought in industry through the introduction of systems such as automatic train protection, as well as through an acknowledgment of the need to develop HF/E awareness and capability.

Beliefs about the benefits of applied HF/E research are now widely held across the industry and HF/E maturity has increased significantly over the last decade. It is also now often a requirement of regulations to formally integrate HF/E into safety management systems and large projects. At the same time, however, there is also a belief that problems associated with "the human factor" and "human error" can be engineered out through increasing automation and that a tightly coupled and precision engineering system can deliver complete safety.

THE CHALLENGES OF APPLYING HF/E IN THE RAIL INDUSTRY

When you start out in HF/E in the rail industry, you face the same issues as in other industries—many of your coworkers will struggle to articulate what HF/E is or where its value lies. If it is not confused with "human resources" (not always an innocent slip of the tongue), many are happy to hold it in vague regard as a catch-all topic for "those-problematic-parts-we-can't-engineer-out." Without a sound understanding of the value of HF/E, it is also, unfortunately, too easily dismissed as a soft discipline. To one observer it can appear as "just common sense," whereas to another it is "too academic." Because of a tacit belief that nearly all "human error" can be designed out, trained out, or blamed out, there is not always acceptance that HF/E involvement is needed in day-to-day train operations.

However, things are changing and HF/E is maturing as a discipline within the rail industry, but its role within multidisciplinary teams needs to be more clearly and consistently understood. It does not help that there are a number of ill-defined relationships among the disciplines. There are some functional interfaces and overlaps between an HF/E specialist and a risk specialist, for example, or a business analyst, or a change manager. These other disciplines are, perhaps, more easily understood in lay terms and it seems they may have been more successful in establishing themselves as key ingredients for successful change initiatives. Many large projects invest in any number of change managers and business analysts, to map processes and churn out documents citing "functional and nonfunctional requirements." Yet those same projects are sometimes reticent to spend money on HF/E. If HF/E isn't perceived as adding value, it might not be appropriately integrated into the project manager's project.

With these challenges in mind and based on our work as HF/E practitioners in the rail industry, we provide in this chapter some observations on some key areas of focus for HF/E and some thoughts on future trends.

KEY AREAS OF FOCUS FOR HF/E IN RAIL

There are mandated requirements for HF/E required by law in many countries, and the opportunities and scope to add value are numerous. These include readily apparent goal conflicts, systems that cry out to be improved, and an appetite for modern technology that presents an interesting systems integration challenge. But taking

FIGURE 14.1   Key components of human factors/ergonomics within rail.

opportunities is not always easy; HF/E maturity varies among different organizations and you have to be prepared to be sufficiently persistent, persuasive, and often just in the right place at the right time in order to get projects off the ground.

Based on the United Kingdom HSE human factors' key topics (HSE, 2016) and the United Kingdom RSSB guide to understanding human factors (RSSB, 2008), we have summarized what we believe are some of the main areas that HF/E can contribute to the rail industry as shown in Figure 14.1. This model was developed for a rail operator to explain the breadth and focus of HF/E and to help people understand some of the key components of the discipline. Some further examples are provided to demonstrate how HF/E can add value in each of the different areas shown. We have highlighted three key areas in the following:

1. Continuing to develop a mature HF/E integration process to ensure the widespread adoption of human/user-centered design approaches
2. Better integration of nontechnical skills (NTSs) and applied human factors training to achieve greater resilience
3. Improved systems and procedures to optimize human performance in support of day-to-day train operations

DESIGNING FOR HUMAN USE

Despite increasing maturity there is still work to be done to improve understanding of human factors integration in the rail industry. Appropriate integration within project activities can be hard because of a belief that HF/E is something that can be provided in a meeting or through an expert review. The HF/E practitioner might be invited to fix a user interface, write a better procedure, design a control room, or help navigate acceptance problems simply by looking at a draft design and making suggestions based on a seemingly deep and mysterious understanding of human psychology. Projects are often keen for a solution and, if we are lucky, they know that a HF/E practitioner can help them to deliver. The key skills for any HF/E practitioner, therefore, include the ability to communicate in terms that resonate with your audience, to influence and persuade, to be patient, and pragmatic with your advice.

Helping project managers and senior leaders understand what it means to follow a user-centered design process is a recurring theme on projects big and small. Having an ISO standard or two in your back pocket helps, but even then it can feel like fighting against entropy or swimming against the tide (and project budgets) for quite a while. Once a project manager starts to see the output from the user-centered design process and realizes the benefits, they can become very useful HF/E champions who support the early introduction of HF/E on other projects.

Projects in the rail industry are often keen to get the benefits of an off-the-shelf solution, but the operating context in which the product needs to be applied tends to limit the extent to which this is possible, or else there is a lack of appreciation for a need to change operations that may result from a whole system approach. The original good intention—to adopt standard products and therefore to deliver value for taxpayer dollars, whether communication systems, signaling control systems, or incident management systems, for example—can result in either unworkable systems that create huge frustration, or contractual variations and product customization to fit with the operating context. Buying a solution can mean buying a different problem.

In the rail industry, HF/E can play a key role in procurement to ensure that the context of use is understood and that user requirements are identified and fed into the process. In essence, this requires commitment to embark on a human-centered process before going to market. One of the challenges for HF/E is to influence senior managers and decision-makers sufficiently early in the procurement process, which is also crucial in order to manage appropriate assurance expectations with product suppliers. HF/E must also be integrated into the system safety and safety change processes to make sure that adequate assurance is provided by system manufacturers during the procurement process to avoid costly late design changes or contractual variations.

DEVELOPING RESILIENT SYSTEMS

NTSs are defined as the interpersonal and cognitive skills that complement technical skills and are required for safe and effective job performance. In Australia, the Rail Safety Regulators Panel developed guidelines for rail resource management (RRM, Lowe et al., 2007) as part of a national RRM project (Rail Safety Regulators' Panel, 2007). This was an adaptation of crew resource management (CRM) from aviation for the rail industry to increase awareness and ensure greater emphasis is placed on NTSs, behavioral markers, and applied human factors training.

These tools and examples from other industries can be invaluable in helping to convince management to adopt programs to collect data on NTS performance and to ensure that applied human factors training is developed and follows RRM/CRM principles. Naweed et al. (2015) suggest that "Developing and applying rail crew resource management practices ... may be one way towards achieving a mature systems and organisational culture" (p. 110).

An integrated program for NTS development should include collation of predictive data from normal operations, to supplement proactive data from competence assurance activities and reactive data from investigations. Although threat and error management (TEM) techniques [such as line operations safety audit (LOSA)* and its various adaptations] are not necessarily easy to implement at first, their power lies in collecting aggregate observational data from normal operations (albeit with a focus on threats and errors) to understand work-as-done. This is in complete contrast to the punitive regimes of the past, waiting until something bad happens and seeking to blame transgressors who did not follow procedures.

The collation of predictive data from normal operations is consistent with the idea that we need to focus on learning from successful performance as well as unsuccessful performance (Hollnagel, 2008, 2009). The rationale for this perspective is that failures and successes result from the same underlying processes (Hollnagel, 2009), namely performance adjustment and variability.

The rail industry lags a long way behind aviation in this regard but we are slowly following suit. In Australia, Queensland Rail implemented a TEM program based on the LOSA model known as CORS (Confidential Observations of Rail Safety) and has seen positive effects on error management (Carter, 2012, 2013; McDonald et al., 2006). Once managers can see tangible data that demonstrate a perceived improvement (such as a reduction in certain incidents or disturbances) they are much more likely to understand the value of an HF/E–led predictive approach. The predictive approach goes hand in hand with improved competence assurance using simulators to test NTSs in degraded modes or emergencies, and collation of trends from investigations.

OPTIMIZING HUMAN PERFORMANCE IN TRAIN OPERATIONS

Train delays are a major priority in the rail industry; nobody likes delays, and railway senior managers have to answer to government ministers on matters concerning train running. Increasingly, railways are judged by a jury of commuters and taxpayers, armed with up-to-the-minute headlines and apps to inform them about the latest incident or delay. This is the reality of running a modern railway, which is reflected in the growing interest of rail operators to actively manage their media (and social media) presence.

All of this generates productivity pressures and managerial focus to meet punctuality and reliability targets. Accountability keeps an important focus on continuous improvement within the public sector and in some cases can reduce bureaucracy;

---

* Line Operations Safety Audit—Trained observers collect data about external threats and crew behavior during normal flight operations. The observations are conducted under nonjeopardy conditions by pilots trained in the use of specially designed observation materials. Data gained through LOSA can be used to guide improvements to training and operations. LOSA uses the TEM model as a framework for data collection and analysis.

however, it is also a pressure that can have unintended consequences. For example, when frontline staff are pressured and harried into making decisions, or fearful of making a decision, safety margins can be reduced and the likelihood of an incident such as a signal passed at danger (SPAD) increases.

With hindsight, it might appear clear-cut to management that a train driver or maintainer made a wrong decision, or that they violated procedures and that such an incident warrants more training, or even punishment. When there is pressure to *manage* the fallout from an incident, that is, when managers are under pressure to react in an appropriate way, a simplistic explanation of events, a label of "human error," and a default low-cost training "solution" offer a way of making the problem go away. The need is not always obvious to consider more complicated explanations, looking deeper for reasons with time-consuming investigations that generate difficult-to-resolve tactical and strategic recommendations. It is within these day-to-day decisions and organizational reactions that HF/E considerations add value.

Noncompliance with procedures is frequently cited as a contributory factor in many rail incident investigations. Rail industry traditions depend upon the development of detailed and numerous rules and procedures as an administrative risk control measure. This can backfire when inexperienced frontline staff are unfamiliar with all of the applicable procedures, when procedures are not usable, or, when faced with the day-to-day challenge to maintain on-time running, successful outcomes can't be achieved by following procedures to the letter and they are violated "to get the job done." Often the only way to achieve successful outcomes is to act in a way that is not wholly consistent with the operational procedures. There is often a gap between work-as-imagined versus work-as-actually-done (Dekker, 2006) and this can be a practical necessity or a mark of expertise. However, there are different reasons for the gap that exists and sometimes the motivations for the way that the work is actually done are not laudable.

HF/E can help rail organizations to understand the operational context and the underlying reasons why people act a certain way, and help to explain some of the reasons why people do not comply with a procedure. Detailed investigations and understanding of the context and conditions in which people work can reveal some of the issues related to scarcity, pressure, or multiple goals that may affect the likelihood of procedural compliance.

Gaining user input and involvement in procedure design can improve the development of practical and usable operational procedures. Combined with initiatives to improve system performance that addresses cultural issues, this can help to reduce the gap between work-as-imagined and work-as-actually-done.

Focusing on *how* the job gets done, and therefore placing less importance on the success of an outcome or the attribution of failure, is useful when tackling another key challenge within rail—the challenge of implementing and sustaining a safe and just culture. There is a critical need to raise and sustain understanding within workers and senior management about the importance of a culture that adopts more of a system-based approach when tackling incidents on the railway, encouraging safety-related reporting, and moving away from blaming delays or incidents on individuals.

A well-implemented just culture program is vital to reduce the likelihood of apparently poor postincident decision-making—and can be a monumental change of mindset for many rail operators. It should include usable tools to support managers with postoccurrence decision-making so that individuals can be treated fairly and consistently to improve the performance of the system. The big challenge here is the change of focus from the management of outcomes (e.g., there was a bad incident so someone "stuffed-up") to the appropriate and timely management of behaviors and decisions. That is, regardless of an outcome, the quality of behaviors and decisions should be considered within the constraints of the operating environment at the time; in some cases individuals should be recognized or praised for the way they managed incidents. To be effective, this approach also requires that managers and supervisors are equipped to identify and proactively address day-to-day behaviors and decisions that fall short of expectations, and instead ensure that behavioral norms support effective system performance. These sorts of changes require significant organizational commitment and can also be very difficult to sustain when faced with increasing competitive pressures. However, well-implemented just or fair culture programs combined with other initiatives to focus on positive behaviors are important and may act on some of the factors that influence the safety culture of the organization.

THE TRACK AHEAD

The rail industry is adapting to become leaner and more efficient, with greater reliance on technology and with a greater focus on customer service delivery. Advances in safety and legislative requirements to integrate HF/E into safety management systems have increased awareness and there are many useful tools that have been implemented to support improved HF/E integration. There is an ongoing need to drive a focus away from blaming individuals for delays or errors, toward adopting more of a system-based approach with the use of appropriate just culture tools, integration of NTSs, building an understanding of the work-as-done, and the application of human-centered design approaches. Using data and demonstrating improvements from the adoption of RRM programs such as CORS or other NTS training initiatives can help to increase awareness and demonstrate the value of HF/E to senior managers.

HF/E will increasingly play a key role in the development of more resilient systems that better harness the potential of the people who work in the rail industry and are designed with more careful consideration of human–system interactions through a greater understanding of everyday work.

REFERENCES

Carter, S. 2012. How tailored NTS training has shown positive effects on error management in the rail industry. PACDEFF 2012, Sydney, Australia. Available at: http://pacdeff.com/pacdeff-2012/.

Carter, S. 2013. An outline of an NTS program in rail. PACDEFF 2013, Gold Coast, Australia. Available at: http://pacdeff.com/pacdeff-2013/.

Cullen, W.D. 2001. *The Ladbroke Grove Rail Inquiry.* Sudbury, UK: HSE Books.

Dekker, S. 2006. Resilience engineering: Chronicling the emergence of confused consensus. In: Hollnagel, E., Woods, D.D., and Leveson, N. (eds.). *Resilience Engineering: Concepts and Precepts.* Aldershot, UK: Ashgate.

Health and Safety Executive (HSE). 2016. Introducing the key topics. Accessed March 13, 2016. Available at: http://www.hse.gov.uk/humanfactors/top-ten.htm.

Hidden, A. 1989. *Investigation into the Clapham Junction Railway Accident.* Department of Transport. London: Her Majesty's Stationery Office.

Hollnagel, E. 2008. Safety management: Looking back or looking forward. In: Hollnagel, E., Nemeth, C.P., and Dekker, S. (eds.). *Resilience Engineering Perspectives, Volume 1: Remaining Sensitive to the Possibility of Failure.* Aldershot, UK: Ashgate.

Hollnagel, E. 2009. The four cornerstones of resilience engineering. In: Nemeth, C.P., Hollnagel, E., and Dekker, S. (eds.). *Resilience Engineering Perspectives, Volume 2: Preparation and Restoration.* Farnham, England: Ashgate.

Lowe, A.R., Hayward, B.J., and Dalton, A.L. 2007. *Guidelines for Rail Resource Management.* Adelaide, Australia: The Rail Safety Regulators Panel.

McDonald, A., Garrigan, B., and Kanse, L. 2006. Confidential observations of rail safety (CORS): An adaptation of line operations safety audit. In: *Proceedings of the Swinburne University Multimodal Symposium on Safety Management and Human Factors.* Melbourne, Australia, 9–10 February.

McInerney, P.A. 2005. *Special Commission of Inquiry into the Waterfall Rail Accident.* Final Report, Volumes 1 and 2. The Honourable Peter Aloysius McInerney.

National Transportation Safety Board (2010). *Collision of Metrolink Train 111 with Union Pacific Train LOF65-12 Chatsworth, California September 12, 2008.* Washington, DC: National Transportation Safety Board. Available at: http://www.ntsb.gov/investigations/AccidentReports/Reports/RAR1001.pdf.

Naweed, A., Rainbird, S., and Dance, C. 2015. Are you fit to continue? Approaching rail systems thinking at the cusp of safety and the apex of performance. *Safety Science.* 76, 101–110.

Rail Safety Regulators' Panel. 2007. *Guidelines for Rail Resource Management. Fortitude Valley.* Queensland, Australia: Rail Safety Regulators' Panel.

Rail Safety and Standards Board (RSSB) 2008. *Understanding Human Factors, a Guide for the Railway Industry.* London: RSSB. Available at: http://www.rssb.co.uk/Library/improving-industry-performance/2008-guide-understanding-human-factors-a-guide-for-the-railway-industry.pdf.

15 Human Factors and Ergonomics Practice in Aviation

Assisting Human Performance in Aviation Operations

Jean Pariès and Brent Hayward

CONTENTS

PRACTITIONER SUMMARY

By nature a highly dynamic and intrinsically risky system, aviation has always manifested the need to consider human factors and ergonomics (HF/E) very seriously. First, it did this empirically, through trial-and-error, following intuitions, which naturally designated pilots or aircraft (and their designers and/or maintenance technicians) as primarily responsible for mishaps. It subsequently opened up to scientific rigor, and became the focus of much research. Over the past 30 years, the global footprint of aviation has facilitated rapid adoption and dissemination of quality HF/E rules and principles. Aviation remains relatively open to new ideas and is often venerated by other domains with regard to safety management matters. It is today a very demanding arena for HF/E practitioners, who are faced with high-level requests from demanding, experienced, and knowledgeable clients. This chapter attempts to reflect the authors' diverse experience with regard to issues of HF/E practice in aviation personnel selection, aviation training, equipment design and certification, aviation safety management, and aviation safety occurrence investigation.

INTRODUCTION

On October 30, 1935, at Wright Field in Dayton, Ohio, final evaluation flights were conducted under a US Army Air Corps tender for a long-range bomber. The performance of Boeing's four-engine Model 299 was so superior to its twin-engine competitors that many considered these final evaluations mere formalities. Two army pilots were at the controls of the 299, assisted by Boeing's Chief Test Pilot. Unfortunately, they forgot to disengage the elevator "gust locks," designed to restrict movement of control surfaces while the aircraft was parked. Hence just after take-off, the aircraft entered a steep climb, stalled, and crashed. Model 299 and Boeing were thus disqualified from the evaluation. Unsurprisingly, the official investigation identified "pilot error" as the cause of the accident. The idea that Model 299 was "too much airplane for one man to fly" took hold and almost terminated the prototype's future.

The Air Corps was, however, still enthusiastic about its potential, and after some politicking, gave Boeing another chance, ordering 12 aircraft to be delivered for "further testing." All crews involved knew their operations were under close scrutiny. They brainstormed how to make sure nothing would be forgotten in operating such a complex aircraft. The outcome was the pilot checklist, both alleviating memory load and establishing a cross-check process among the crewmembers (Schamel, 2012). It appears that the idea caught on. The squadron also implemented rigorous training and flight standards, and managed to fly the 12 aircraft thousands of hours without serious incident. The US Army eventually ordered 12,731 units of Model 299, renumbered the B-17, the "Flying Fortress," of the Second World War legend.

Considering this story today, it is fascinating to realize the similarity of practical issues behind different conceptual clothes. In contemporary language, while the official accident investigation blamed pilot error, the aviation community implicitly understood there was a *systemic* issue (aircraft complexity overwhelming human capacities). Additionally, the operations community reset the balance within the human–machine interface, introducing the checklist and better teamwork; the squadron added organizational compensations for the inexperience and learning needs associated with the unique introductory period; and the army found an acceptable trade-off between performance and safety, and could buy the superior performance they desired to achieve their strategic goals with a reasonable guarantee about safety concerns. All this happened in the 1930s.

As illustrated by this account, HF/E issues have always been an integral component of the aviation industry, from the selection of personnel, through training development and implementation, to equipment and procedural design and safety occurrence investigation. So what has changed since the early pioneering empirical trial-and-error efforts? Perhaps it is the way we make sense of the issues, rather than the issues themselves. In light of the long history of HF/E in aviation, this chapter will consider the *evolution* of the influence of HF/E on the various aspects of the aviation industry mentioned earlier, and some implications for practice from experience. The size and scope of the aviation industry precludes an examination of the full gamut of HF/E contributions, but these also include aircraft manufacturers, cabin design, maintenance, and in-service support or continued airworthiness.

PERSONNEL SELECTION

Looking back to the earliest days of military aviation, large numbers of aircrew were lost due to what can best be described as human performance issues. William H. Wilmer (1918) famously wrote that of every 100 British military pilot fatalities during the first year of the First World War, 90 resulted from "individual deficiencies," 8 from aircraft defects, and just 2 from enemy action. While these remarkable figures may not stand up to rigorous analysis, they nonetheless underline the importance of selecting appropriate individuals for potentially hazardous tasks.

Since those early days, considerable resources have been devoted to attempting to identify the essential physical and psychological characteristics of successful aviators. One of the earliest efforts comes from Rippon and Mannel (1918), who described the successful military pilot as "a high spirited, happy-go-lucky sportsman," who "seldom takes his work seriously, but looks upon 'strafing' as a great game ..." and returns after a day's flying to a life of theatre, music, dancing, and playing cards. In contrast, just 3 years later a study by Dockeray and Isaacs (1921) described the ideal military pilot as "quiet and methodical." Since that time many researchers have attempted to further delineate the characteristics of successful aviators, with similarly mixed results.

Pilot selection methods have, however, advanced considerably in the past 90 years and best practice systems today employ structured job analysis techniques (see Goeters et al., 2004) to identify the key knowledge, skills, abilities, and other traits (KSAOs), including team management skills and social competence (Hoermann and Goerke, 2014), that ought to be assessed in a balanced selection process. In 2014, the *International Journal of Aviation Psychology* devoted two volumes to a "special issue on pilot selection" and Damos (2014) provided an editor's overview of the current state of selection processes globally. The special issue also includes spirited discussion of some of the current "hot topics" in pilot selection (e.g., Eißfeldt, 2014; King, 2014; Weissmuller and Damos, 2014; Wiggins and Griffin, 2014).

Though HF/E focuses on design, one thing that has certainly not changed with time is the critical importance of having a clear understanding of the relevant job characteristics (of the pilot/air traffic controller/flight attendant/mechanic...) and selecting individuals who possess KSAOs appropriate for that job. While other elements of the aviation system are very important to safety and success, well-designed equipment, procedures, and training—the common purview of the HF/E practitioner—will amount to little if we don't start by selecting the right people, and of course training them appropriately.

HF/E PRACTICE IN AVIATION TRAINING

Perhaps, the greatest influence of HF/E practice in aviation in the modern era has been in the design, specification, and implementation of applied human factors training. Applied HF training arrived for aviation at a time when it was sorely needed.

It was an idea whose time had come. Commercial aviation expanded exponentially after the Second World War, but as the aviation industry grew so did public concern over what seemed to be the increasingly common occurrence of fatal aircraft accidents.

In 1975, the International Air Transport Association dedicated its landmark in the twentieth Technical Conference in Istanbul to examining the role of "pilot error" in accidents. The outcome was a more widespread recognition that rather than being the result of "negligence" or "carelessness" on behalf of flight crew, aircraft accidents were more frequently the result of a range of system deficiencies generating "errors" in crew performance. The conclusion was that unless HF/E was taken far more seriously, a major disaster would occur. Fewer than 18 months later, the catastrophic collision between two passenger-laden B747s on the runway at Tenerife followed. This remains the most harmful aviation accident to date with 583 fatalities and 59 serious injuries.

Tenerife and several other prominent disasters (including a DC-8 fuel exhaustion accident at Portland, Oregon) were a catalyst for development and implementation of applied HF training within aviation. Parallel research conducted at NASA demonstrated the negative effects of increased workload on the ability of flight crew to effectively manage internal and external communications, work together as a team, and prioritize tasks effectively (Smith, 1979).

The aforementioned accidents and the Ruffell Smith experiments were a precursor to the development of applied HF training for airline crews. In Europe, KLM led the way through development of the KLM human factors awareness course, which followed from the company's involvement in the Tenerife accident. In the United States, United Airlines developed their command/leadership/resource training program, aimed at assisting flight crew to better manage available resources in adverse situations. Once again the stimulus for this innovation was calamity: The DC-8 accident at Portland.

In 1979, NASA sponsored an aviation industry workshop on *Resource Management on the Flight Deck*, coupling the concerns emanating from industry research and aircraft accidents (Cooper, White and Lauber, 1980). As the technical and mechanical reliability of aircraft improved, it was observed that many aviation accidents were occurring for a broad range of "human factors" related to teamwork, including leadership, dealing with technical problems, communication, task allocation, judgement, and decision-making. As noted by Dr. John Lauber in opening the workshop: "One of the principal causes of incidents and accidents in civil jet transport operations is the lack of effective management of available resources by flight-deck crew" (Lauber, 1980, p. 3). This first NASA workshop provided the foundation and impetus for the development of "resource management" training solutions (studying how the people, procedures, and equipment are used and how this can be optimized and understood). A second NASA-sponsored workshop in 1986 (Orlady and Foushee, 1987) focused on discussion of the various "Cockpit Resource Management" training programs and techniques that had been developed and implemented to address these issues in the interim.

This form of applied HF training evolved considerably over the years and is today most commonly known as crew resource management (CRM) training. Helmreich et al. (1999) provide an overview of the evolution of CRM training during the 1980s and 1990s. Although it may not have seemed like it at the time, that period may be regarded as a "golden era" for the development of applied HF training in aviation. While CRM training is now mandated in one form or another for flight crew in aviation jurisdictions across the globe (Pariès, 1996), the battle for resources in the "CRM wars" that Bob Helmreich repeatedly referenced during those days (e.g., Helmreich, 1993) now seems more formidable than ever. This is reflected in the struggle for current HF/E practitioners to obtain and maintain adequate resources for the development and delivery of effective CRM training.

Although aviation is often regarded as a leading domain in the development and implementation of applied HF training, in recent years the innovation and enthusiasm of earlier times has been harder to detect. CRM training efforts have been extended to many more frontline aviation workers (pilots, cabin crew, maintenance engineers, and air traffic controllers), and the concept has migrated successfully to various other domains (Hayward and Lowe, 2010), such as shipping, rail, and healthcare. However, recognition of the key safety role of such training has sometimes been replaced by "tick the box" approaches aiming at superficial compliance to regulations.

Lamentably, examples of high-quality CRM training programs are becoming harder to find. Resources are scarce. Business cases are rare. With few exceptions (e.g., Edkins, 2002; Salas et al., 2006) the effectiveness of CRM training has not been subjected to rigorous evaluation or business-case analyses. This is an area of HF/E practice in aviation that still requires significant improvement in comparison to other safety-critical domains, and one where HF/E practitioners could and should have considerable impact.

HF/E PRACTICE IN DESIGN AND CERTIFICATION

Aviation is the realm of technicians, and for decades the priority has been to meet the challenges and threats of the physical world, with the implicit belief that human operators would eventually adapt to any resulting design. Recognition of the importance of HF/E to the design of human–machine interfaces in aviation has grown the hard way, with most progress resulting from catastrophic accidents, across the last century. Since the 1940s "hard" human capabilities and limitations (such as anthropometric size and strength, vision, hearing, and memory) in physical and perceptive interactions have been taken into account by designers when developing controls, designing displays, or arranging the workplace, particularly in the United States where HF/E practitioners and manufacturers seemed able to cooperate efficiently.

Cognitive interactions between human operators and systems began to be significantly considered in design processes in aviation around the end of the 1980s. In domains such as nuclear, chemical, and the offshore industry, the introduction of automation into control rooms in the late 1970s triggered the emergence of "cognitive ergonomics." Its main contributors were primarily from Europe (e.g., Bainbridge, 1987; Hollnagel, 1993; Leplat, 1985; Rasmussen, 1982; Reason, 1990), but it also evolved in the United States (Anderson, 1985; Billings, 1997; Norman, 1981; Sarter and Woods, 1992). Its preoccupation was to keep the human "in the loop" (through "human-centered design"), to improve human reliability (through user-friendly interfaces), and to reconcile safety with human variability (error resistant and tolerant design, error monitoring facilitation). However, aircraft manufacturers, especially those designing more highly automated aircraft, may have been reluctant to cooperate with those with rather critical attitudes toward automation.

It is recognized today that proper consideration of HF/E in the design of a cockpit or an air traffic control (ATC) tower or operations room could actually govern the efficiency and the reliability of the interaction between those systems and their human operators and maintainers; hence influencing productivity and safety. In theory, the perspective has almost been inverted, and many efforts are made to adapt the systems to the humans as well as the other way around. However, the road is not without bumps. Many obstacles need to be overcome. These include the following:

- Scientific knowledge about human performance and limitations is not stabilized for high-level cognitive functions.
- There is no such thing as an "average operator"; test pilots or controllers do not really represent typical workers.
- The systems designed will last for decades, and will be operated by two or more generations of people, with different contexts and habits.
- The systems and procedures produced will be used by people of various backgrounds, experience, and cultures from all over the world.

Industry certification requirements do not aim at a best possible design. They only intend to specify the minimum objectives to be matched by an applicable design; guarantee that the minimum crew is able to do the job (without exceptional skills, excessive workload, or fatigue); minimize the risks of error in the use of controls; minimize ambiguities in information displayed by instruments; provide crew with relevant warning information about unsafe functioning states of any equipment or system; and allow appropriate crew action.

The methodology used to check the compliance of a proposed design with a relevant airworthiness requirement depends on the nature of the requirement. When requirements are directly expressed in terms of design characteristics (means), the compliance is rather easy to check, and direct examination of descriptive material (drawings, scale models, and mock-ups) can be used. However, most of the HF/E-related issues are covered by requirements expressed in terms of the objectives. In this case, the *methodology* used to evaluate whether a proposed design can reach the objective *is a critical part* of the certification process. A first possible source of difficulty is the interpretation of the regulatory objective itself. A second possible source of difficulty is in the evaluation of an acceptable means of compliance (AMC). To guide these interpretations, requirements are complemented with advisory material, including *interpretation guidelines* and/or indications on *AMC*.

For many years, aircraft cockpits, ATC towers and control rooms have been designed with tangible success, although very little "HF/E" was explicitly integrated into the design process. While manufacturers have not always directly employed many HF/E specialists, they have nonetheless designed most systems with a great deal of innovative engineering, common sense and simplistic human–machine interaction models. Although some significant mistakes have been made, it would be somewhat arrogant for the HF/E community to claim that, had they been associated in a more systematic manner in the design process, they would have done significantly better in the creation process. Being able to criticize does not necessarily make someone a better designer.

Nowadays, the challenges have shifted again. Computer capabilities have reached a point that makes it possible to transfer higher level "cognitive functions" to automated systems, and to remove more human operators from the frontline. At the same time, the complexity of the aviation system has reached a point at which "fundamental surprise" (or startle) may be the main accident generator. The debate is no longer to design automated systems that can "keep humans in the loop," but to decide about the "reasonable" level of autonomy of the automated systems. The human capability to keep control (for frontline operators as well as for decision-makers) will be central for future effectiveness. This is clearly something of concern to HF/E practitioners who are engaged in the design of automated systems support.

HF/E PRACTICE IN AVIATION SAFETY MANAGEMENT

Safety management is another aspect of aviation that has advanced considerably in recent years. The concept of integrated safety management systems (SMS) has evolved and been gradually implemented to cover all areas of aviation operations over the past two decades. Since the early 2000s, the International Civil Aviation Organization (ICAO) has led aviation SMS developments. The release of ICAO Annex 19 on Safety Management (ICAO, 2013) complemented by a third and improved issue of the ICAO Safety Management Manual (Doc 9859; ICAO, 2012) may prove a watershed for the implementation of integrated SMS across all elements of the aviation system (including airline operations, airports, ATC, maintenance organizations, manufacturers, etc.). An SMS describes what an organization and its managers must implement and do to effectively manage their risks. It is the extension to safety of the traditional quality approaches, that is, the implementation of a closed control loop including an inventory of risks, the setting of goals, the implementation of reactive and proactive control levers (methods), the selection of sensors (the definition of leading and lagging indicators to assess the outcomes), and the implementation of a continuous improvement process. As such, it is a formal approach, defining the piping of the safety management structure rather than what should flow through the pipes.

The question for us is then: What can and should HF/E practitioners do regarding SMS implementation? The short answer is that we see the role of HF/E practitioners as providing reality checks, or lucidity data, about the percolation of the SMS to the base. On the one hand, they should bring their knowledge and analysis methodologies of shop floor activities and social realities into the development process of the "piping" framework. On the other hand, they should contribute to the sense-making process of the feedback from the field. However, in practice, they may feel uncomfortable in this role. Indeed, to follow on the previous metaphor, while an SMS is supposed to merely define the piping, the piping partially constrains what can flow through the pipes. The concept of SMS was born in process industries, where safety is based on very tight control of physical and chemical processes, and where the level of randomness and uncertainty about what will happen is very low. Consequently, it usually conveys a highly normative, prescriptive, and hierarchical model of safety management, in which managers and designers provide both the equipment and the usage rules, including safety rules that frontline operators must strictly respect, and are sometimes not even supposed to understand.

In our experience, one consequence of this historical heritage is a resistance of frontline aviation operators against what they perceive as an attempt to further limit their autonomy at work, and a growing gap between "paper operations" and real activities. Managers then tend to satisfy themselves by developing attractive Key Performance Indicator graphs, while the reality in the field may be, to say the least, different. This may put HF/E practitioners in a difficult situation at the interface between the managerial doctrinal certainty that there is no life anymore outside an SMS, and their knowledge of, and sensitivity to, real life and field operations. But they should not despair, as their contribution will be critical to SMS realism, and success. They should work hard at proposing methodologies to include end users into the design of operational processes and procedures, as well as at collecting field data to criticize, validate, or invalidate all the explicit or implicit behavioral assumptions hidden behind the "paper safety model." An SMS usually comes with a strong data-oriented culture, and its ambition is to establish evidence-based safety management. If they are to contribute, HF/E practitioners must learn to speak that language as well. One powerful approach to success is to think in terms of what creates reliability rather than failures, which is close to adopting a "Safety-II" mind-set (understanding successes rather than just failures; Hollnagel, 2014). This implies, among other things, collecting data on the denominators (the number of actions, or risk exposure situations) as extensively as on the numerators (number of failures) of the reliability equation. Failures are usually rather easy to count, and data on human "errors" are ubiquitous in organizational databases. Data on exposure rates are impressively lacking, not necessarily because they are more difficult to collect, but mainly because the need for them is not understood. Yet, this need is critical to establish a sound and efficient safety strategy, with relevant priorities, because they allow the computing of success rates (reliability). For example, in a recent consulting experience concerning around the clock operations, night activities were not felt to be a safety problem by the organization, considering the low frequency of incidents. However, once confronted by the actual number of day/night activities, the relative failure rate for a specific procedure turned out to be almost one hundred times higher during the night.

Associated with the development of SMS has been the rise of the concept of *safety culture*, referring to the "software" (values, beliefs, habits, and practices) in the collective mind of an organization, both underlying and overdetermining safety behaviors. In other words, safety culture is something that SMS implementation ultimately aims to improve, should it be effective. Consequently, SMS implementation processes more and more often include (or are accompanied by) a monitoring of safety culture evolution. Not surprisingly, this monitoring is based on data: Quantitative assessments of the safety culture, and/or quantitative assessments of the "maturity" of the safety culture. In-house HF/E practitioners will most probably be enrolled in the implementation of the corresponding methodologies. Safety culture and climate assessments are based on the use of questionnaires, interviews, and focus groups. Safety culture "maturity" assessments are mainly derived from the classification of organizational cultures (*pathological, bureaucratic,* and *generative*) originally proposed by Westrum (2004). Safety culture assessment grids and associated maturity scales typically seek to assess the following:

- The commitment of senior management to recognize and promote safety as a priority when it competes or conflicts with productivity
- The ability of the organization and its managers to produce and reinforce safety rules, and to learn from failures
- The commitment of frontline operators to follow safety rules and report their failures to do so, as well as other perceived anomalies (safety significant events, unsafe situations, ineffective design, etc.)

There are several safety culture assessment tools available "on the market" in aviation. Some are intended to be implemented by external practitioners in cooperation with the surveyed organization. Some are designed as self-assessment tools. In aviation, the air traffic management (ATM) domain was an early adopter of SMS and among the first to address safety culture. EUROCONTROL developed a safety culture assessment process tailored to ATM, which is made available to its European members, as well as a process to assess maturity of safety culture (see Kirwan and Shorrock, 2014, for an overview). In-house HF/E practitioners need not be overly impressed by safety culture/maturity assessment methodologies, even if they are not familiar with them. There is no rocket science behind them, and the raw data they produce will not avoid a long and controversial interpretation process in which HF/E practitioners can take an important role, once again based on their understanding of the human factors and "real life" within their organization. They should keep in mind that there is currently no consensual "scientific" vision of what makes a complex sociotechnical system, or an organization, safe. The notion of (good) safety culture is blurred by this lack of objective reference, and it might be fair to say that notional "safety culture" is what these surveys actually measure (just as IQ tests may measure notional intelligence), rather than a clearly defined concept allowing rational measurement.

The vision of a safe organization suggested by most SMS and some safety culture criteria is one of a fully controlled world, in which managers and designers anticipate every potential situation, predetermine all the appropriate responses, and have full control over frontline operators—who (should) consistently follow the corresponding prescriptions for success. Such a vision is challenged by past and contemporary research on organizational reliability (e.g., Charles Perrow's (1999) normal accident theory [NAT]; the "High Reliability Organization" [HRO] movement—La Porte, 1996; Roberts, 1990; and more recently, the resilience engineering [RE] movement]. Our (modest) experience is that in most current industrial processes, strict adherence to preestablished action guidelines is unattainable, incompatible with the real efficiency targets, and insufficient to control abnormal situations. There is nothing new here, the irreducible difference between prescribed work and real work, or "tasks" and "activities," or work as imagined and work as done, has been studied and depicted since the 1950s, for example, by the French Language Ergonomics movement (Ombredane and Faverge, 1955). Nevertheless, on reflection, many requests for assistance from our clients derive from their difficulty in reconciling this "old truth" with the inflation of applicable aviation safety standards and the compliance expectations of SMS frameworks. While neither the HRO nor RE movements have yet been able to produce really workable grids to assess and improve an SMS (see an attempt with the Resilience Assessment Grid in Hollnagel et al., 2011), they do provide a conceptual framework that we have found very useful. More specifically, we believe that the inevitable limitations of anticipation and predetermination lead to a need for organizational skills to manage the unexpected. These skills imply a specific organizational design and management style providing the requisite flexibility (through reasonable autonomy, variety of competencies, local responsibilities, and imagination), a specific design and principles for management of the operational processes, allowing a proper perception and management of margins of maneuver, as well as a specific kind of learning. We have found that these skills were underaddressed by most "official" SMS and safety culture frameworks, and we believe it is the role of HF/E practitioners to emphasize them.

HF/E PRACTICE IN AVIATION SAFETY OCCURRENCE INVESTIGATION

The investigation of aviation safety occurrences has also advanced significantly over the past 25 years. As stated, until the mid-1970s the blame for aircraft incidents and accidents was typically allocated either to operational staff, or to the aircraft itself, in case of critical failure. Accidents were seen as either due to a lack of human reliability, or a lack of technical reliability. While that can still occur, there has been a paradigm shift in many sectors in thinking about "accident causation." Aviation industry regulators, safety agencies, and operators (airlines, air navigation service providers, etc.) have begun to hunt more widely in their search for contributing factors to safety events. Without doubt a significant influence on this evolution have been researchers such as Rasmussen (1982), Reason (1990, 1997, 2008), Amalberti (2001), and Hollnagel (2004, 2008). Investigation methodologies such as the Incident Cause Analysis Method (BHP Corporate Safety, 2000), AcciMap (Rasmussen, 1997), the Human Factors Analysis and Classification System (Wiegmann and Shappell, 2003), and the Systemic Occurrence Analysis Methodology (SOAM; EUROCONTROL, 2005; Licu et al., 2007) have all evolved from the contributions of the aforementioned authors, and have been deployed in aviation.

This evolution of thinking is commonly referred to via the notions of "systemic accidents" or "organizational accidents." While these terms are often used interchangeably, we believe they should be differentiated, as they do not raise exactly the same questions, and do not lead to identical perspectives on accident causality. The systemic approach entails understanding how/why accidents happen within a "system." System properties result, at least partially, from the interactions between components, so that a system is "more than the sum of its parts." So when we say that the perspective on accidents has become "systemic," what is actually meant is that we shifted the focus to *interactions between system components*, so that naïve isolated causality terms such as "pilot error" are no longer accepted as a valid explanation of accidents. As humans are the main source of complex interactions within a system, reinforcing this perspective is a key role for HF/E practitioners within safety investigations.

The organizational approach is a human-centered perspective on a system including people. An organization is both *a process* (of organizing and self-organizing) and *its result* (a set of structured interactions operating to achieve shared goals). The notion of "organizational accidents," as contemplated by Reason (1997), introduced the idea that understanding a failure, such as an accident, occurring at the real-time, production level, implied understanding the weaknesses of the organization's backstage processes. In other words, an organizational perspective on accidents focuses on the past genesis of the weaknesses in the organization's functioning, leading to the accident. For example, in the 2009 AF447 aircraft accident [Bureau d'Enquêtes et d'Analyses (BEA), 2012], understanding what the crew did or did not do, in terms of human–machine interaction, situational awareness, crew cooperation, and flight context, reflects a systemic perspective that is familiar to HF/E practitioners (cognitive ergonomics). Understanding why the aircraft design, certification, and safety assurance processes failed to anticipate, then recognize, the possibility for such human–machine interaction failure, and how airline training failed to protect the system against that possibility, refers to an organizational perspective (macroergonomics).

HF/E practitioners should, in our view, help to analyze aviation accidents from both systemic and organizational perspectives. They need to go beyond contributing their knowledge about human performance characteristics and limitations, and encourage and educate fellow investigators to think both in terms of systemic interactions and organization.

However, this is probably easier said than done. There are at least two challenges to meet, or two main questions to answer:

- How do we establish an objective causality link between a specific human behavior (e.g., a "failure") and its background context, for example, system features (such as HMI design)? The notion of causality is obviously less clear-cut than in the case of technical system failures. There is a high risk of assuming causality on the basis of the mere coexistence of phenomena, such as fatigue or stress, and error.
- The same question arises for the organization perspective, with even more blurred causality links, as the cause–effect relationship may extend over many years and follow several layers of transformation. Here again, the mere existence of some perceived "organizational flaw" doesn't necessarily establish causality.

All this eventually boils down to the usual question: "Is our event more the fault of the individual, or the system?" A simple and well-known "tool" to answer this question is the "substitution test." The idea is to virtually substitute a number of "average" operators for the operator(s) involved in the failure, and refer to expert judgment, or an actual experiment, to assess whether the course of events may have reasonably/probably been the same. If no, there may be a deduction of individual accountability. If yes, it strongly suggests that systemic issues are the underlying determinants of the shared behavior. Identifying the systemic flaws will then allow confirmation of the hypothesis, if available HF knowledge provides an acceptable explanation.

Additional and more sophisticated reasoning methodologies have been developed, particularly by the Australian Transport Safety Bureau (ATSB, 2007), to rationalize HF analysis. The most interesting contribution of this document may well be in this note: "The term 'accident causation' is not used in this report due to semantic difficulties associated with terms such as 'cause' and 'causation'. The term 'accident development' also reflects the fact that the factors involved in many accidents develop over a period of time prior to the accident." Indeed, a common weakness of most safety investigation and analysis methodologies is that they are based on the attribution of causality: They seek to reveal a continuum from "direct" to "root" causes, with each successive level supposed to better explain the "real" or "deep" geneses of the event, to which corrective actions should then be applied. The notion of "cause," however, is too linear to account for what happens in a complex system. Because the benefit of hindsight removes the uncertainties that prevailed in real time, the "right way" and associated "correct" behaviors seem obvious after the event. Deviations from these behaviors are evidently perceived as direct causes, and the organizational features that allowed these deviations as root causes.

Analyzing the AF447 accident in the clear light of day, it seems obvious that following the "Unreliable Airspeed" procedure was the simple and only solution, and that a failure to do so meant a serious flaw in knowledge and the training process. Attentive HF/E practitioners should then raise their hand and ask, "Have you tried to imagine yourself in that situation? Do you know how an operator confronted in that situation with something both not understood and seemingly demanding immediate action would 'normally' react? Have you heard of the startle effect?" Then, the relevant HF/E question becomes: "Was it reasonable to put trust in a procedure like this, and the associated training, to maintain the safety of a flight in all imaginable (and unimaginable) similar situations?"

As far as root causes are concerned, things are even direr. While there is no scientifically established and commonly accepted model of a "safe organization" (but rather a set of partially antagonistic models: HROs/NAT/RE/SMS and a set of different safety culture frameworks, etc.), many safety investigators/analysts are primed to reveal the "organizational pathologies" which, in their minds, "obviously" led to the direct causes that explain the event. Reason (1990, 1997) developed his organizational accident model(s) in reaction to what he called the "proximity irony": As they are located closer to the "sharp end," frontline operators are destined to bear most of the causality, while being the less autonomous components of the system. This was indisputably a very positive shift. However several authors, including Reason himself (e.g., Reason et al., 2006), have argued that, since then, in many interpretations "the pendulum has swung too far," shifting the blame from frontline operators to managers, and obscuring the fact that a complex system: (i) could not be safe without some autonomy, hence responsibility, at the frontline and (ii) always includes uncertainty, incomplete information, conflicting goals, and constraints, as well as limited resources, hence calling for satisficing trade-offs within "bounded rationality" rather than straightforward and purely rational decisions. The role of HF/E practitioners is again critical in this perspective: They must analyze and highlight this complexity, and defend the investigation against the hindsight bias that makes "good" decisions retrospectively obvious. This requires a thorough and honest inventory of demands, options, available information, uncertainties, contradictions, complexities, pressures, and the like, as available at the time of the event (and not after the event). Interviews of managers and decision-makers should be conducted with the same methodological precaution and listening empathy as for frontline operators.

In an attempt to help investigators to overcome these difficulties in practice, the authors of this chapter have developed two safety event analysis methodologies. The first, SOAM (EUROCONTROL, 2005; Hayward and Lowe, 2004; Licu et al., 2007), evolved from the Reason model and guides the analyst in tracing back contributing factors from human involvement, through contextual conditions to organizational factors, and on to absent or failed barriers. The second, method of investigation organizational and systemic (MINOS) (Pariès and Rome, 2012), is the result of work undertaken with Airbus and EUROCONTROL (2003) in the early 2000s to overcome the limitations of causal approaches. The main steps are: Identification of the system of interest; identification of risks to be managed in the situation, construction of the corresponding safety model (what was supposed to protect operations from the loss of control and the accident?); analysis of the actual behavior of this safety model; deduction of lessons learned on the robustness of the safety model; and the definition of recommendations. The main distinction is that information extracted from the event is used to assess robustness of the safety principles rather than the reliability of the operators.

CONCLUSION

Writing about the practice of HF/E in aviation may sound rather commonplace and straightforward, but in the experience of your hosts for this chapter, it is not. One reason for this may be that we originate from quite distant backgrounds and cultures. One from the "old world": A French aeronautical engineer, and one from the new: An Australian aviation psychologist. However, our professional collaborations over the past 20 years do demonstrate contiguity in the practice of human factors and safety consulting. So the challenge must have been elsewhere. Having perhaps overthought this, we believe it rests precisely in the ambition to explain that proximity. What is straightforward about trying to make sense of hundreds of interventions spread over more than 20 years and scattered through organizations all over the world, ranging from CRM training to accident investigation, from human–machine interaction to safety culture, from human reliability to resilience engineering? What are the links between what we have done (or imagine we have done) and the most important theoretical contributions to the field? How does it connect to the historical evolution of a domain that is now more than a century old? This chapter was our best attempt at an answer.

REFERENCES

Amalberti, R. 2001. The paradoxes of almost totally safe transportation systems. *Safety Science*. 37(**2–3**), 109–126.

Anderson, J. 1985. *Development of Expertise in Cognitive Psychology and its Implications.* New York, NY: Freeman.

Australian Transport Safety Bureau (ATSB). 2007. *Analysis, Causality and Proof in Safety Investigations. Aviation Research and Analysis Report – AR-2007-053.* Canberra: Author.

Bainbridge, L. 1987. Ironies of automation. In: Rasmussen, J., Duncan, J., and Leplat, J. (eds.). *New Technology and Human Errors.* New York, NY: Wiley, pp. 271–286.

BHP Corporate Safety. 2000. *Incident Cause Analysis Method Investigation Guide, Issue 1, March 2000.* Melbourne, Australia: Author.

Billings, C. 1997. *Aviation Automation: The Search for a Human-Centred Approach.* Mahwah, NJ: LEA.

Bureau d'Enquêtes et d'Analyses (BEA). 2012. *Final Report: On the Accident on 1st June 2009 to the Airbus A330-203 Registered F-GZCP Operated by Air France Flight AF 447 Rio de Janeiro – Paris (English Version).* Paris: BEA.

Cooper, G.E., White, M.D. and Lauber, J.K. 1980. (Eds.). *Resource Management on the Flightdeck.* Proceedings of a NASA/Industry Workshop Held at San Francisco, California, June 26–28, 1979. (NASA CP-2120). Moffett Field, CA: NASA-Ames Research Center. pp. 3–16.

Damos, D.L. 2014. Editor's preface to the special issue on pilot selection. *International Journal of Aviation Psychology*. 24(**1**), 1–5.

Dockeray, F.C. and Isaacs, S. 1921. Psychological research in aviation in Italy, France, England, and the American Expeditionary Forces. *Journal of Comparative Psychology*. 1(**2**), 115–148.

Edkins, G.D. 2002. A review of the benefits of aviation human factors training. *Human Factors and Aerospace Safety*. 2(**3**), 201–216.

Eißfeldt, H. 2014. Commentary on the article by King: Select in/select out – What aviation psychology offers for pilot selection. *International Journal of Aviation Psychology*. 24(**1**), 78–81.

EUROCONTROL. 2003. *The Development of a Safety Management Tool Within ATM (HERA-SMART), Edition 1.0.* Brussels: Author.

EUROCONTROL. 2005. *EAM2/GUI8: Systemic Occurrence Analysis Methodology (SOAM), Edition 1.0.* Brussels: Author.

Goeters, K.-M., Maschke, P., and Eißfeldt, H. 2004. Ability requirements in core aviation professions: Job analyses of airline pilots and air traffic controllers. In: Goeters, K.-M. (ed.). *Aviation Psychology: Practice and Research.* Aldershot, UK: Ashgate, pp. 99–119.

Hayward, B.J. and Lowe, A.R. 2004. Safety investigation: Systemic occurrence analysis methods. In: Goeters, K.-M. (ed.). *Aviation Psychology: Practice and Research.* Aldershot, UK: Ashgate.

Hayward, B.J. and Lowe, A.R. 2010. The migration of crew resource management training. In: Kanki, B.G., Helmreich, R.L. and Anca, J. (eds.). *Crew Resource Management*, Second Edition. San Diego, CA: Academic Press.

Helmreich, R.L. 1993. Fifteen years of the CRM wars: A report from the trenches. In: Hayward, B.J. and Lowe, A.R. (eds.) *Proceedings of the Australian Aviation Psychology Symposium.* Sydney, Australia: Australian Aviation Psychology Association, pp. 73–87.

Helmreich, R.L., Merritt, A.C., and Wilhelm, J.A. 1999. The evolution of crew resource management training in commercial aviation. *International Journal of Aviation Psychology*. 9(**1**), 19–32.

Hoermann, H-J. and Goerke, P. 2014. Assessment of social competence for pilot selection. *International Journal of Aviation Psychology.* 24(1), 6–28.

Hollnagel, E. 1993. *Human Reliability Analysis, Context and Control.* London: Academic Press.

Hollnagel, E. 2004. *Barriers and Accident Prevention.* Aldershot, UK: Ashgate.

Hollnagel, E. 2008. Risk + barriers = safety? *Safety Science.* 46, 221–229.

Hollnagel, E. 2014. *Safety-I and Safety-II: The Past and Future of Safety Management.* Farnham, UK: Ashgate.

Hollnagel, E., Pariès, J., Woods, D.D., and Wreathall, J. (eds.). 2011. *Resilience Engineering Perspectives Volume 3: Resilience Engineering in Practice.* Farnham, UK: Ashgate.

International Civil Aviation Organization (ICAO). 2012. *Safety Management Manual (SMM), Third Edition. DOC 9859–AN/474.* Montreal, Canada: Author.

International Civil Aviation Organization (ICAO). 2013. *Annex 19 to the Convention on International Civil Aviation: Safety Management.* Montreal, Canada: Author.

King, R.E. 2014. Personality (and psychopathology) assessment in the selection of pilots. *International Journal of Aviation Psychology.* 24(1), 61–73.

Kirwan, B. and Shorrock, S.T. 2014. A view from elsewhere: Safety culture in European air traffic management. In: Waterson, P. (ed.). *Patient Safety Culture: Theory, Methods and Application.* Aldershot, UK: Ashgate.

La Porte, T.R. 1996. High reliability organizations: Unlikely, demanding and at risk. *Journal of Contingencies and Crisis Management.* 4(2), 60–71.

Lauber, J.K. 1980. Resource management on the flight deck: Background and statement of the problem. In: Cooper, G.E., White, M.D. and Lauber, J.K. (Eds.), *Resource Management on the Flightdeck.* Proceedings of a NASA/Industry Workshop Held at San Francisco, California, June 26–28, 1979. (NASA CP-2120). Moffett Field, CA: NASA-Ames Research Center. pp. 3–16.

Leplat, J. 1985. *Erreur humaine, fiabilite humaine dans le travail.* Paris: A. Colin.

Licu, T., Cioran, F., Hayward, B., and Lowe, A. 2007. EUROCONTROL – Systemic occurrence analysis methodology (SOAM) – A "reason"-based organisational methodology for analysing incidents and accidents. *Reliability Engineering & System Safety.* 92(9), 1162–1169.

Norman, D. 1981. Categorization of action slips. *Psychological Review.* 88(1), 1–15.

Ombredane, A. and Faverge, J.M. 1955. *L'analyse du travail.* Paris: Presses Universitaires de France.

Orlady, H.W. and Foushee, H.C. 1987. *Cockpit Resource Management Training.* (NASA CP-2455). Moffett Field, CA: NASA-Ames Research Center.

Pariès, J. 1996. Human factors training initiative that first emerged in the 1970s has reached maturity. *ICAO Journal.* 51(8), 19–20.

Pariès, J. and Rome, F. 2012. MINOS: Method of investigation organisational and systemic. In: *Poster Presented at the 30th Conference of EAAP.* Sardinia, September, 2012.

Perrow, C. 1999. *Normal Accidents: Living with High-Risk Technologies.* Princeton, NJ: Princeton University Press.

Rasmussen, J. 1982. Human errors: A taxonomy for describing human malfunction in industrial installations. *Journal of Occupational Accidents.* 4(2–4), 311–333.

Rasmussen, J. 1997. Risk management in a dynamic society: A modeling problem. *Safety Science.* 27(2–3), 183–213.

Reason, J. 1990. *Human Error.* New York, NY: Cambridge University Press.

Reason, J. 1997. *Managing the Risks of Organizational Accidents.* Aldershot, UK: Ashgate.

Reason, J. 2008. *The Human Contribution: Unsafe Acts, Accidents and Heroic Recoveries.* Farnham, UK: Ashgate.

Reason, J., Hollnagel, E., and Paries, J. 2006. *Revisiting the "Swiss Cheese" Model of Accidents, EEC Note No. 13/06.* Brétigny, France: EUROCONTROL.

Rippon, T.S. and Mannel, E.G. 1918. The essential characteristics of successful and unsuccessful aviators. *Lancet*. 192, 411–415.

Roberts, K.H. 1990. Some characteristics of high-reliability organizations. *Organization Science*. 1, 160–177.

Salas, E., Wilson, K.A., Burke, C.S., and Wightman, D.C. 2006. Does crew resource management training work? An update, an extension, and some critical needs. *Human Factors*. 48(2), 392–412.

Sarter, N.B. and Woods, D.D. 1992. Strong, silent and 'out-of-the-loop': Properties of advanced automation and their impact on human automation interaction. *CSEL Report 95-TR-01*. Colombus, OH: Ohio State University.

Schamel, J. 2012. How the pilot's checklist came about. FSS History. Accessed February 1, 2015. Available at: http://www.atchistory.org/History/checklst.htm.

Smith, H.P.R. 1979. *A Simulator Study of the Interaction of Pilot Workload with Errors, Vigilance, and Decisions. (NASA TM 78482)*. Moffett Field, CA: NASA Ames Research Center.

Weissmuller, J.J. and Damos, D.L. 2014. Improving the pilot selection process: Statistical approaches and selection processes. *International Journal of Aviation Psychology*. 24(2), 99–118.

Westrum, R. 2004. A typology of organisational cultures. *Quality & Safety in Health Care*. 13(**Suppl. II**), ii22–ii27.

Wiegmann, D.A. and Shappell, S.A. 2003. *A Human Error Approach to Aviation Accident Analysis: The Human Factors Analysis and Classification System*. Aldershot, UK: Ashgate.

Wiggins, M.W. and Griffin, B. 2014. Commentary on the article by Turner: Cultural complexity in pilot selection. *International Journal of Aviation Psychology*. 24(2), 96–98.

Wilmer, W.H. 1918. Plane News. Newsletter of Issoudun Army Air Field, AEF, France. 19 October 1918. A.1 (col. 1).

16 Human Factors Practice in Military Aviation
On Time and On Target

Ben Cook and Ryan Cooper

CONTENTS

PRACTITIONER SUMMARY

The tactical application of human factors (HF) in a military aviation context presents practitioners with an array of challenges and opportunities to enhance operational effectiveness and safety outcomes. Military commanders require pragmatic, culturally appropriate, and solution-focused support that is both on time and on target. Importantly, HF support that is either off time or off target more often than not misses the opportunity to influence command decision-making and inform action. Success as an HF practitioner requires more than competence. It stems from the capacity to build trust and a working alliance with commanders that is based on mutual respect, frank and forthright dialogue, compromise, and a client-centered agenda that recognizes the commander and the commander's team as the experts who hold the keys to any solution.

INTRODUCTION: MILITARY HF—WORKING IN A SPICE SHOP

If variety is the spice of life, then working as an applied HF practitioner in the military context is like working in a spice shop. The operational environment of our military forces is arguably one of the most rewarding and challenging domains for applied HF. The opportunity to work with advanced technology and complex systems, in combination with an intelligent, motivated, disciplined, and highly trained workforce, is an attractive recipe for any budding HF practitioner. Our day-to-day work is exciting, perplexing, sometimes frustrating but always varied and purposeful ... and always short for time.

Although there are a large number of high-performance workplaces across the Australian Defence Force (ADF), this chapter focuses on some recent bright spots for HF interventions in defense aviation units, in particular, tactical fatigue-management support and a safety culture program, which are enabling commanders to make more informed, risk-based decisions to enhance the well-being and performance of their war fighters (personnel). Aside from showcasing two successful HF interventions, the programs and lessons are considered relevant to any HF practitioner or organization (civil or military) seeking to implement risk-based interventions for greater efficiency and effectiveness.

PRAGMATIC INTERVENTIONS THAT BUILD TRUST

Few would argue with the premise that the profession of HF requires its practitioners to identify organizational challenges and to offer and act upon pragmatic solutions. However, the general, and sometimes vague principles, theories and tools of HF must be translated in order to deliver tailored solutions that are aligned to the culture, objectives, and resource constraints of the organization. This is, of course, the scientist–practitioner model. However, while the scientist–practitioner model forms the basis for HF interventions, in our experience, it is the capacity of the practitioner to build and maintain trust that can make or break even the best of HF interventions.

Just as this chapter begins with a practitioner summary, it is common for articles for a military audience to start with a "BLUF" (Bottom Line Up Front). This is a well-worn military acronym. The purpose is to ensure that individuals get to the point by placing the most crucial piece of information at the start of any communication. BLUF enables commanders to process large amounts of information in very short periods of time, a characteristic of decision-making within the military context, by providing the option to skip the details. It also hints at the importance of trust in the relationships between commanders and their advisors. Without trust, a commander is likely to ignore advice or become lost in the details and not deal with the issue at the appropriate level.

Trust is a key theme for this chapter and refers to how we (as practitioners) build and sustain high levels of trust, both through the delivery of results as well as by placing the local commanders at the center of HF solutions. Recognizing commanders as the experts who hold the keys to any solution, fosters trust both up and down the chain of command.

So, why is trust so important? Covey (2006), in his book *The Speed of Trust: The One Thing That Changes Everything*, provides us with two answers—increased speed and reduced cost. If we link the importance of trust to effective communication, Covey says: "In a high-trust relationship, you can say the wrong thing, and people will still get your meaning. In a low-trust relationship, you can be very measured, even precise, and they'll still misinterpret you" (p. 6). It follows that for low-trust relationships, people are constantly second guessing each other and management decisions are always questioned. The consequences are an increase in time, less effective processes and a significant and unnecessary cost to the organization.

In an unforgiving military setting where effective communication can mean the difference between mission success and failure, trust is not a luxury; rather, it is a force multiplier and a critical component of operational capability. You may struggle to find a better example of trust than the fighter combat instructor (FCI) or "top gun" course. The FCI course demands meticulous attention to detail from the life-support team, aircraft maintainers, and aircrew, all of which result in the ability to authorize a trainee FCI, with 600 hours on a single-pilot fighter aircraft, to lead a 14-ship formation into an air-to-air and air-to-ground mission. The rest of this chapter expands on HF/E practice issues in a military aviation context with two examples: tactical fatigue management and safety culture as successfully applied within the dynamic, fast-moving, and high-performance environment of air force fighter squadrons.

BRIDGING THE SCIENTIST–PRACTITIONER GAP—TACTICAL FATIGUE MANAGEMENT

After receiving a request to provide tactical fatigue-management support for a complex fighter aircraft training exercise, the obvious response was "When does the course start?" The reply "It has already started, we need you now" came as no real surprise. In the ideal world, a 2- to 3-month window to prepare for the intervention would have been optimal—but optimal is often neither practical nor achievable. So it is a matter of doing what you can with the resources at hand to deliver a 60% solution on time and, in doing so, establishing a solid foundation for further enhancements over the longer term. To achieve the commander's intent, our objective was twofold: first, to gather information on fatigue-related risk to support command decision-making while minimizing the footprint (i.e., don't get in the way); and second, to support the tactical management of fatigue at the coalface throughout a dynamic, high-stress, high-workload training exercise.

Our solution was simple yet robust and involved the use of wearable personal informatics devices and software tools, such as the well-known Actiwatch (Weiss et al., 2010) to enable the collection of accurate and objective activity, sleep, wake, and light-exposure data. The Actiwatches were worn for two 14-day periods as an opportunity to collect baseline data and for the entire phase (4 weeks) of the final exercise. A shift-scheduling software system, the fatigue avoidance scheduling tool (FAST), a biomathematical model of alertness, was also used to support predictions regarding the likely impact of fatigue throughout the activity. In the defense aviation system, the default FAST settings are considered adequate for routine fixed-wing transport operations. However, for specific operational environments the FAST settings should be adjusted as necessary. For example, operators of fighter aircraft are frequently exposed to an extremely high physical and cognitive workload and FAST thresholds were increased appropriately.

Importantly, this intervention was not research; rather, it was a risk-based intervention within the safety management system (SMS) that evaluated the effectiveness of existing risk controls (e.g., complexity of mission, experience, crew-scheduling). Throughout the intervention, individuals were regularly provided with one-on-one feedback on their results and tailored advice on techniques to manage fatigue before, during, and after duty periods.

The intervention provided many benefits in terms of information quality, improved speed of information collection and analysis, and speed of feedback. First and foremost, the client-centered solution delivered results through its ability to provide on time and on target information and, in doing so, informed command decision-making and action. Furthermore, the one-on-one engagement delivered benefits both in terms of buy-in and its ability to influence behavior.

> Prior to completing my first sleep study I was not aware that I had any major problems with my sleep. Simply having the ability to compare data with others creates a great awareness in someone who has poor sleep. It was extremely helpful, and now to this day I have sought mechanisms to both monitor and assist in my sleep.
>
> **(Fighter Aircraft Instructor)**

Notwithstanding the importance of longer term academic or scientific research programs, a 99% solution developed and delivered over months and years may not remain relevant. Ultimately, the levels of trust required to bridge the scientist–practitioner gap is engendered by delivering solutions to meet the day-to-day challenges encountered by commanders and their personnel. The success of the fatigue intervention has led to an enduring requirement for HF support to the training exercise. Moreover, the resulting high-trust relationship unlocked opportunities for further risk-based HF interventions that are driving cultural reform.

WORKING WITH SAFETY CULTURE WITHIN A SAFETY MANAGEMENT SYSTEM

The International Civil Aviation Organization (ICAO) recently released an updated Safety Management Manual (ICAO, 2013) that is based on the lessons learnt over many years across civil aviation. The ICAO guidance is similar to the SMS practices adopted by defense aviation, including the importance placed on developing and maintaining a robust safety culture. Safety culture remains a critical area for growth from an applied HF perspective. So, why all the fuss about safety culture?

Stated simply, the existence of an SMS is not sufficient to guarantee performance and safety outcomes. A strong safety culture ensures the SMS works in practice—it embodies the commitment to supporting the SMS (EUROCONTROL, 2008).

"The way things are done around here" is a very simple (albeit partial) definition of safety culture and, in practice, it has the capacity to render the formal SMS useless. This description emphasizes that safety culture is more than what people say about safety—it is concerned about the realities of safety. While all organizations value safety, the strength of their safety culture influences the relative importance of safety against the need to achieve other organizational objectives. For example, the want to complete the mission can, at times, be in conflict with the need to minimize exposure to risk or potential hazards.

Our experience has found time and time again that "culture eats management systems for breakfast." The statement implies that regardless of well-documented safety management procedures designed to provide higher margins of safety, with certain stressors (e.g., time pressure and fatigue) there can be a drive to simplify the process and bypass some formal procedures. This is often heavily influenced by supervision, the environment, and the broader organization. When it happens it can regularly lead to serious incidents or accidents within the workplace.

The term "culture eats management systems for breakfast" comes from a working paper written by Gunningham and Sinclair (2011). The authors found that the variation in the effectiveness of an SMS between similar organizations was directly linked to the strength of their safety cultures. Indeed, management style and motivation are considered more important than the management system (policy and procedures) in shaping performance (Gunningham et al., 2003). Gunningham and Sinclair (2011) emphasized that without organizational trust an SMS does not even have a chance of success. Similarly, in our experience, an absence of trust within an organization can quickly lead to a large gap between the standards/regulations (what is expected or demanded) and actual workplace practices (what is done).

But culture is a powerful force and changing an organization's safety culture, if change is required, can be a slow and difficult process. Defense aviation is heavily influenced by its "can-do" performance culture. While the benefits of encouraging a can-do culture are numerous, it must be acknowledged that some safety-management strategies, such as speaking up when things aren't right, can be impeded because of a strong sense of not wanting to let the team down. Having the moral courage to speak up when things aren't right is an admirable and essential trait when safety is being compromised. So enhancing safety culture is an essential enabler for establishing military workplaces that take care of the long-term well-being of their personnel (their most critical and valuable asset), and ultimately for creating a sustainable operational capability.

The implications for HF/E practice are many. It's not unusual for existing safety assurance processes (e.g., audits of an SMS) to fail to identify the misalignment of SMS policy and procedures against actual practices. In short, many well-intended shortcuts and deficient workplace practices are routinely not detected during audits. The outcomes of this can be an increasing gap between work-as-imagined and work-as-actually-done, and major system failures may be associated with this gap. With the benefit of hindsight, these issues will routinely emerge during any ensuing investigation—"we knew that was going to happen." This process can quickly destroy the trust relationship and degrade performance for many years. It remains critical to provide commanders with adequate information to understand where to focus and support enhancements and to ensure conversations are encouraged across all levels of the organization to identify and proactively address safety issues.

The recently developed defense aviation safety culture program has been designed to support commanders to foster a mindset of continuous improvement, through an open, just, and fair examination of safety-related issues. The safety culture program comprises the snapshot survey and safety culture workshop (SCW).

THE SNAPSHOT SURVEY

The snapshot survey is a tool to support local commanders in the management of safety climate within their units. Building on the work of Murphy and Fogarty

(2010), the annual survey was introduced in 2013 in an effort to strengthen the organization's surveillance and management of safety climate. The snapshot survey is based on the job demands–resources (JD-R) model that was first proposed by Demerouti et al. (2001). The snapshot survey version of the JD-R (Cooper and Fogarty, 2015) is shown in Figure 16.1. On the left-hand side of the model, there are two broad constructs (job demands and job resources) which are measured by a set of 12 scales. In the middle of the model, two scales are used to capture the strain construct. Separate individual scales have also been developed to measure compliance and job satisfaction as well as the four safety-related outcome constructs. The constructs are all linked by arrows that indicate the direction of influence and whether the influence is positive (+ve) or negative (–ve). The use of an underlying model is important because it allows particular aspects of the organization and the individual to be identified that may have an impact on safety performance. The model also shows how changes in variables such as job demands and job resources interact to affect the motivation and well-being of personnel, which then influence the safety outcomes (such as work-related errors). The survey has excellent psychometric properties with all scales demonstrating good reliability and validity (Cooper and Fogarty, 2015).

The success of the snapshot survey is underpinned by the use of innovative analysis and reporting methods ensuring a small team can provide a short turnaround between administration and the delivery of results (more than 160 unit reports within 4 weeks). The analysis uses the previous year's results to track a unit's progress. It addition, the results from all units are also used to identify those that are performing clearly above or clearly below the rest of the group. The unit reports are deliberately

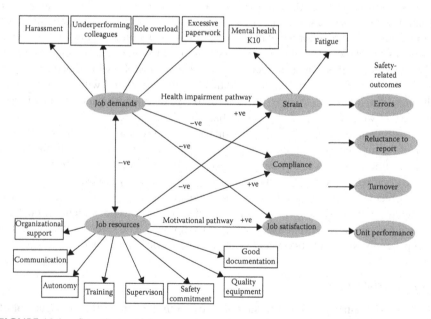

FIGURE 16.1   Snapshot model.

short to focus on the core issues and present information in a concise and direct manner. This includes a dashboard providing an overview of the unit's performance on all safety indicators on the first page of the report (the BLUF).

Though reports are produced at both the unit and organizational levels, unit-level reporting is prioritized. Placing local commanders at the center of the intervention provides the greatest return on investment as it generates local ownership and action. It also ensures that survey results are contextualized and reviewed in conjunction with other information that may be relevant at the time of administration.

The snapshot survey is a true example of an HF intervention that seeks to bridge the scientist–practitioner gap. The use of an underlying explanatory model and JD-R theory connects the snapshot survey to a rich literature and evidence base. Nevertheless, the ultimate success of the snapshot survey remains with the HF practitioner, including their capacity to engender trust. Failures to alleviate the privacy concerns of participants, uphold the principles of confidentiality, and/or protect information against misuse will thwart the success of any survey program. For the snapshot survey, trust is further promoted by ensuring a short turnaround from administration to the reporting of results (on time) and presenting relevant information in a concise manner to support command decision-making (on target). It is the role of the commander to take ownership of the results and to strengthen trust by engaging the workforce in an open discussion of safety, well-being, and performance-related issues.

SAFETY CULTURE WORKSHOP

The SCW is a structured face-to-face program, designed to complement the snapshot survey. The SCW has been adapted from the successful United States Naval Safety Center program. The program contends that safety performance and operational excellence is built on a foundation of trust, integrity, and leadership, created and sustained through effective communication.

The SCW is typically conducted over 2–3 days and involves facilitated group discussions targeting specific occupational groups and rank structures. This group information is supplemented by one-on-one interviews with key personnel. Participant response cards are utilized as a part of the focus groups to gain live, de-identified responses on issues related to safety culture. The preliminary results for the SCW are collated and provided to the commander prior to exiting the unit and a more formal report completed within 10 business days.

Again, trust is a central issue and it has a powerful impact on the outcomes of the SCW. The ability to conduct frank and forthright discussions targeting sensitive safety issues requires a high degree of trust between the HF practitioner(s) and the commander as well as between the HF practitioner(s) and the workforce. Balancing the sensitivities inherent in this dual relationship, and ensuring transparency in the motivation behind the intervention, is fundamental to the formation of trust and the overall success of the SCW. The feedback to date from commanders has been very complimentary of the program and its ability to provide further context, in addition to the snapshot survey, on the causes of unit issues as well as those elements that are performing well.

CONCLUSION: THE COMPLETE SOLUTION COMING SOON

Thankfully, when it comes to the tactical application of HF there is no one-size-fits-all solution. Instead, as HF practitioners, we work in an environment littered with imperfect solutions that are implemented against competing (and sometimes conflicting) organizational demands. In our experience, supporting the HF need of commanders in the rapidly changing military aviation context requires pragmatic, culturally appropriate, and solution-focused support that is both on time and on target. There are no clear-cut answers, only workable solutions grounded in evidence.

Concerning the fatigue management and safety culture programs discussed in this chapter, these interventions deliver a risk-based approach to the tactical application of HF. They are designed to empower a commander to enhance the well-being of personnel for sustained combat effectiveness and positive cultural reform. Ultimately, the capacity of the HF practitioner to build trust and a working alliance with commanders is critical to the overall success of applied HF interventions.

REFERENCES

Cooper, R.L. and Fogarty, G.J. 2015. The snapshot survey: An x-ray view. *Aviation Safety Spotlight*. 032015(**03**), 35–39.

Covey, M. 2006. *The Speed of Trust: The One Thing That Changes Everything*. New York, NY: Free Press.

Demerouti, E., Bakker, A.B., Nachreiner, F., and Schaufeli, W.B. 2001. The job demands-resources model of burnout. *Journal of Applied Psychology*. 86(**3**), 499–512.

EUROCONTROL. 2008. Safety culture in air traffic management (a white paper). EUROCONTROL/FAA Action Plan 15 Safety, December 2008. Available at: https://www.eurocontrol.int/sites/default/files/article/content/documents/nm/safety/safety-atm-whitepaper-final-low.pdf.

Gunningham, N., Kagan, R.A., and Thornton, D. 2003. *Shades of Green: Business, Regulation, and Environment*. Redwood City, CA: Stanford University Press.

Gunningham, N. and Sinclair, D. 2011. *Working Paper 83 – Culture Eats Systems for Breakfast: On The Limits of Management-Based Regulation*. Canberra: National Research Centre for OHS Regulation.

International Civil Aviation Organization (ICAO). 2013. *Safety Management Manual*. ICAO Doc 9859-AN/474, Third Edition. Quebec, Canada: ICAO.

Murphy, P. and Fogarty, G. 2010. Safety culture in defence explosive ordnance: A preliminary study to develop a safety climate measure. In: International Military Testing Association Conference, Lucerne, Switzerland, 27 September to 1 October 2010. Available at: https://eprints.usq.edu.au/8887/1/Murphy_Fogarty_AV.pdf.

Weiss, A.R., Johnson, N.L., Berger, N.A., and Redline, S. 2010. Validity of activity-based devices to estimate sleep. *Journal of Clinical Sleep Medicine*. 6(**4**), 336–342.

17 Human Factors and Ergonomics Practice in the Oil and Gas Industry

Contributions to Design and Operations

Rob W. Miles and Ian Randle

CONTENTS

PRACTITIONER SUMMARY

Oil and gas installations are particularly hazardous both for personnel and the environment. The oil and gas industry has traditionally relied upon generic human factors and ergonomics (HF/E) standards for equipment design, overlooking the other aspects of the HF/E systems approach, including the characteristics of the users, tasks, working environment, and organizational factors. Relatively recently, however, following high-profile accidents, the industry has started to consider human contributions to major accident risks. Recent initiatives by industry bodies, regulators, and major operating companies are now starting to get HF/E taken seriously. To fully realize the potential of the formal HF/E analyses, they should be carried out into the work planning and operational procedures. This chapter looks at HF/E in oil and gas from the viewpoints of major capital projects and assesses HF in the operating asset.

A CHALLENGING ENVIRONMENT

Before discussing HF/E issues in oil and gas, it is worthwhile mentioning some of the issues and challenges that characterize the upstream part of this industry. These include the following:

- A high-hazard environment (large quantities of hydrocarbon, often under high pressure or temperature and sometimes containing toxic substances such as hydrogen sulfide).
- Drilling and production installations are often located in remote and harsh environments, requiring work in extreme temperatures (from deserts to arctic conditions), sometimes with limited support due to the remoteness of the location.
- Long working hours: Typically 12-hour shifts (7.00 AM to 7.00 PM, day and night shifts), and 2- to 4-week tours of duty with no days off during the tour.
- A reliance on skilled human intervention to keep drilling operations or process plant operations running efficiently and safely.

These factors present significant HF/E challenges, which need to be accounted for in the design and operation of facilities. However, the industry has a historical pioneering culture, and prides itself in being able to build and operate assets in the most extreme conditions. This may have contributed to a tacit acceptance of facilities with problematic operability and habitability in the past. HF/E practitioners working in the industry may witness hundreds of photographs of "poor ergonomics" in the design of oil and gas installations around the world.

There are a number of different players: Operating companies (private and state-owned), engineering companies, large vendors/subcontracted operators, etc., with complicated interactions. The design and engineering of new assets is usually contracted out by the operating companies (e.g., Shell Global, BP Global, ConocoPhillips) to specialist engineering and construction contractors. These contractors may not receive direct feedback on the success of, or problems with, their previous designs in the field, and most engineers designing the asset will not have worked on or even visited an operating installation. This is a barrier to the continuous improvement in design that has been seen in some other industries.

Another challenge to integrating HF/E principles in the design and operation of oil and gas assets comes from the relatively short timescale for design and engineering. Oil and gas exploration is very expensive and once a viable field has been discovered, the race is on to start producing as soon as possible to recoup the exploration and capital costs. This may partly explain why the number of HF/E studies that are undertaken in a typical oil/gas project is significantly fewer than in a comparable nuclear or defense project.

DRIVERS FOR HF/E IN OIL AND GAS

A key driver for increased focus on HF/E in design and operations has been health and safety regulation. The United Kingdom has the most extensive regulatory environment, both onshore and offshore. The Control of Major Accident Hazards (COMAH) Regulations and the Offshore Installations (Offshore Safety Directive) Safety Case Regulations require operators to demonstrate that they have considered the human contribution to major accident risks.

In the United States, the Safety and Environmental Management System (SEMS) II rule, introduced in the wake of the Macondo blowout in the Gulf of Mexico, requires operators to "incorporate Human Factors into the design, operation, and maintenance of their offshore facilities." Norway has a number of working environment and technical and operational safety regulations that contain prescriptive workplace ergonomics requirements, which are beginning to consider human contributions to major accident risks. In the other major oil and gas producing regions of the world, the regulatory environment with respect to HF/E is less extensive, and in some cases nonexistent or limited to a few internal company standards.

HUMAN FACTORS IN OIL AND GAS CAPITAL PROJECTS

HF/E in oil and gas capital projects started as the application of HF standards. Only a handful of the larger international operating companies have introduced these, and only one of these has made HF/E a mandatory requirement, at the time of writing. These were integrated into the wider engineering specifications, which drive and govern large engineering projects. These standards were largely generic guidelines on accessibility, equipment layout, and interface design, with the premise that if you apply the standards then the design of the asset will be acceptable from HF/E perspective.

This crude "one-size-fits-all" approach found favor in its simplicity and coherence with the contemporary engineering design approaches. However, it lacked fundamental aspects of the HF/E systems approach to design. That is, it did not take into account the specific target audience, the organizational factors, the particular tasks being undertaken, or the range of working environments encountered for the installation being designed.

The practice of human factors integration (HFI) in the oil and gas industry has a relatively short and patchy history. It is only within the past 10 years or so that the major oil/gas operating companies have started to embrace the systems approach to HFI, something which has been embedded in some other sectors for much longer. HFI extends beyond the use of standards, both in approach and scope. The scope of HFI is much wider and includes HF/E issues also, for example, this applies to the following:

- Physical requirements in design (standards and specifications)
- Interface design and alarm handling
- Equipment testing and review (e.g., factory acceptance test [FAT])
- Mental and physical workload

- Environmental factors and personal protective equipment (PPE)
- Staffing levels
- Workflow and functional linkages
- Safety critical task (SCT) identification and analysis
- Safeguard design
- Procedure development and review
- Training and competency requirements
- Performance measurement

HF/E can be applied to all key phases of major projects: From early concept design through front-end engineering and design (FEED) to the detailed engineering, procurement, and construction (EPC) and commissioning phases.

The International Association of Oil & Gas Producers (IOGP) published a seminal guide in 2011 on Human Factors Engineering in Projects (IOGP, 2011). This sets out a road map for HFI and identifies the key HF/E activities in the various phases of a project. Importantly, this includes both workplace design requirements and the human contributions to major accident risks, which sets it apart from previous industry guides that focused mainly on the former. The IOGP guide also sets out roles and responsibilities and competency requirements for HF/E assistance on a project. This guide is becoming the standard upon which the major operating companies are basing their internal HF/E processes.

FUTURE DIRECTIONS FOR HF/E INTEGRATION IN PROJECTS

One future direction for HF/E integration in projects relates to the continuous drive within the industry to reduce manning for safety and cost reasons. Replacing people with technology is an aspiration for many new projects, but the implications are not always properly considered. For example, a skilled process technician who walks on an inspection round in the plant daily may develop tacit knowledge on how to keep the plant running smoothly, learning the idiosyncrasies of the plant based on their "on the ground" experience. If the process technician is removed and the daily walk-around is replaced by increased instrumentation and a more complex control system in a remote control room, there is a risk that the skilled functions of the technician are not adequately replaced by the technology, and the same level of efficiency and plant availability will not be achieved.

There is a role played by HF/E here, to analyze the skilled behaviors and information acquired by the technicians on the ground, and help to identify how these can be effectively replicated by the control and instrumentation system. However, this level of HF/E influence on a project is exceptional. The challenge for HF/E is to become sufficiently embedded in the project design process to allow this type of input to become routine.

There are many other areas in which HF/E can further contribute to increased efficiency and safety in the design of oil and gas installations. One of the key future developments is in the integration of HF/E within process safety studies. Hazard and operability (HAZOP) studies, layer of protection analysis (LOPA) and bow-tie analysis are established tools for assessing and managing major accident risks in the oil and gas industry, and are routinely undertaken as part of the project risk analysis. In these analyses, the human can be considered both as a contribution to process deviations or upsets and as part of the safeguards against them. However, the extent to which HF/E knowledge is utilized within these approaches is often superficial. "Human error" is often listed as a generic error mode without qualification and "operating procedures" are often listed as a safeguard in the assumption that these will always be fully and correctly executed (and will be effective). HF/E tools and methods that seek to refine these safety studies to take a more realistic account of the human element are beginning to be used, but at present they are far from being a routine part of these studies.

Other areas to which HF/E practitioners contribute include emergency response, behavioral safety, incident investigation, and safety culture. In general, within the past few years, HF/E in oil and gas has started to consider human contributions to major accident risks more routinely. This is despite it being a "high-hazard" industry that has a history of high-profile, major accidents with their roots in HF. The next section therefore gives some insight into the process of assessing HF in oil and gas operations, and the issues that the HF/E practitioner may encounter. This gives a good insight into what it is like to be an HF/E practitioner in the oil and gas sector in the operating asset itself.

ASSESSING HUMAN FACTORS IN OIL AND GAS OPERATIONS

HF/E practitioners embedded in regulatory agencies, insurers, independent auditors, potential clients or joint venture partners, or the operating companies will sooner or later be tasked with making an assessment of HF/E on-site. This indeed could be a task that would take a team several weeks; however, the reality is that it will be one HF/E practitioner and a colleague and it will take a week. Site visits should involve at least two people. There is a lot of information to cover and it is much more effective to interview in pairs and compare notes. It also helps to discuss observations as the site visit progresses to identify priority areas. It can be very effective to put together a multidisciplinary team of three, combining HF/E with processes control and (say) mechanical engineering.

All offshore installations in operation in UK North Sea waters will have a safety case (sometimes referred to as an "Operational Safety Case") and this case will be "accepted" by the relevant regulatory authority. In many regions beyond the North Sea, operators choose to adopt a voluntary safety case to maintain parity with their North Sea operations. From 2015 the United Kingdom offshore safety case includes major accident prevention, pollution prevention, and regulatory compliance, and will be assessed and accepted by the UK "Competent Authority," a joint activity of the Health and Safety Executive (HSE), and the Department of Energy and Climate Change (DECC). If the safety case has not been accepted, the installation will not have permission to operate. If the installation is found to be in significant breach of the safety case, the permission to operate can be withdrawn. This sanction is reserved for the most serious breaches but it gives the safety case regime a powerful lever over the operator.

In the operational safety case the company legally responsible (the duty holder) sets out all the things that can go wrong (the hazards), the measures to prevent these from happening [the safety critical elements (SCEs)], the management system in place to ensure these measures are effective [the safety management system (SMS)], and the arrangements for auditing and verifying compliance with the safety case. Safety cases always include a description of the assets and activities as this underpins the basis for establishing the hazards. There will also be a set of major accident scenarios that set out foreseeable major events and how these are planned for and prevented.

It is a legal requirement that the safety case includes a consideration of HF, both as a contribution to, and mitigation of, potential accidents. This is set out more in detail in the assessment principles for offshore safety cases (APOSC) (HSE, 2006). While the UK legislation has been updated with the introduction of the revised safety case regulations in 2015, APOSC remains unchanged at the time of writing. There is no prescribed method so it is necessary to work through the major accident scenarios and SCE looking for the human elements. This will be explained in more detail shortly. The following sections give some practical advice to HF/E specialists from experience, again with an emphasis on UK installations.

Before You Go

There is no point in visiting any major hazard site, and especially not an offshore installation or onshore refinery, without preparation. These are complex sites that undertake hazardous activities. You are in greater danger as you are unfamiliar with the worksite and less likely to read the signs that all is not going as per plan. To visit a UK offshore installation you will have to have at least the following:

- Basic Offshore Safety Induction and Emergency Training (BOSIET), which will equip you to escape a capsized helicopter, fight fires, and right a life raft (and more)
- Minimum Industry Safety Training (MIST) course, an industry induction that explains the risks, behaviors, and standard practices offshore
- Any company- and installation-specific inductions
- An Oil and Gas UK offshore medical certificate

With these you can apply for a Vantage Card, the UK North Sea workplace "passport" which enables you to travel offshore [the vantage system also records annual hours offshore per employee so remember this when you are looking at a fatigue risk management system (FRMS)]. At some point you will be measured to make sure you fit through a standard helicopter window.

Understanding the Safety Case

It is important to read the operational safety case. Safety cases in the United Kingdom are commercially confidential and so are only obtainable from the operators (duty holders) and regulatory authorities, with permission (e.g., from the duty holder's office). A typical safety case is 600 to 900 pages. Typically, it is possible to take notes but not copies or printouts.

There is no set format for safety cases but most conform to the following:

An introductory statement with the corporate policy: Since 2015, this should include the corporate major accident prevention policy (CMAPP). This sets out the high-level policy and may include HF. It is a legally binding commitment by the duty holder's executive board.

A description of the activities on installation: This gives a high-level overview of what work goes on (e.g., "drilling for oil").

A description of the equipment and facilities: This is often a large list of equipment plus some general arrangement drawings. It is useful to look for equipment that is known to be hazardous and to find the locations where work takes place. This may not be obvious from other sections of the safety case. Examples would be machinery spaces with equipment that requires periodic maintenance. Some of these can be deep down below sea level.

A description of the hazard: This will include HAZOPs, quantitative risk assessment (QRA), and other formal risk techniques. The key HF/E concern is how human activity is treated in these analyses (qualitatively and quantitatively).

The risk controls: Here we are looking for the SCEs and how human agency is linked to these SCEs. In reality, all SCEs have a human element, be it design, maintenance, or operation. It is important to consider how they are maintained and operated, so procedures, access, competence, and task demands are relevant. For the central control room (CCR), operational procedures, alarm handling, and other CCR tasks are relevant. If the installation has interfaces with other installations, small satellite platforms or unloading vessels, then decision-making and communications in both normal and abnormal situations are relevant.

The SMS: There will usually be a chapter describing SMS, often combined with the operational management system (OMS). The SMS should consider procedures (how they are checked and reviewed), competence (how this is assured and linked to the SCT and elements), and work control [primarily, supervision and the permit-to-work (PTW) system]. It is important to pay particular attention to tasks that do not require a permit; how is it determined whether tasks require a PTW or not? In practice, it is usually unrealistic for all tasks to require a PTW. Another point is the treatment of routine tasks; are these covered by standard procedures? The SMS section should also include a section on contractor and supply chain management. Contractor competence is a key issue, both at contract placement and when the work takes place.

Occupational health: This may have a section to itself or it may be included in the SMS. Relevant to HF/E are working hours and fatigue.

Human factors: There may or may not be a section on "HF." If there is, it tends to be generic with a high-level summary of corporate policy on HF, task design, and human performance. This can be useful to refer to on-site as a comparison between corporate policy and actual practice.

Incident investigation: Investigation, near miss reporting, learning, and continuous improvement may be in the SMS section or in a stand-alone section about handling incidents, investigation, and learning. Learning from incidents (LFI) is a key issue in improving safety and there are many HF/E issues in LFI. The primary focus for the HF/E practitioner is the inclusion of HF/E in this process of investigation, learning, and action, including action tracking.

Verification and audit: The safety case will contain details of the independent verification scheme. This is a legal requirement and the independent verifier can provide their report. This may alert you to HF/E issues to target when you get to the installation. If the installation is also a vessel (which floats and has propulsion), then it may be operated "under class" and will have regular independent SMS audits by the classification society (Lloyds Register Group Limited, DNV GL, American Bureau of Shipping, for example). The most recent audit report must be held for inspection on the vessel and can be useful to read once you get offshore. If the rig is under class then the majority of the crew will be mariners. This will have a major impact on how competence is managed as mariners' qualifications are highly regulated, prescribed, and externally examined. You may find a "them and us" divide between the mariners and the rest of the crew.

Employee involvement: In the United Kingdom, there are legal duties in relation to safety representatives (SI 971) and employee consultation. This includes employees' involvement with the safety case. It is important to be familiar with the company arrangements to see how they are implemented. If you find a disengaged workforce with antagonism between managers and workers, it can be useful to revisit these company commitments to engage employees and see whether these processes are working as intended. Unfortunately, the workplace is still a "them" and "us" culture between workers and management. Even the fact that we use terms like "workforce engagement" shows that we assume a disengaged group. As an HF/E practitioner, you are neither "them" nor "us"; you are independent and an honest broker for what you see and hear. This is a very fine line to tread, but if you deviate from it you undermine your credibility and potentially the credibility of HF/E as a profession.

You may ask why so much emphasis on the safety case. What about injuries, slips, and falls? Falls remain a very significant cause of injury and death offshore and HF/E issues are often associated, but focusing on the human element in major accident prevention is where you can make the most positive impact. "HF" is often wrongly interpreted as mean behavioral safety. It is important to redress this imbalance and demonstrate the considerable value HF/E brings to major risk reduction. If you find problems with PPE and handrail compliance, it is useful to address these at the leadership and management system level, not as an individual behavioral issue.

OTHER AVAILABLE SOURCES

The company annual reports and other promotional material: Annual reports will provide information on plans to decommission or expand operations on an installation or at a site that is not presented in the safety and operational material. This can be important when interpreting other decisions observed in this field.

Company websites: Technical and financial information on websites can provide valuable context to operational decisions observed at the site. Videos can be particularly informative and not necessarily in the way the company intended. A recent promotional video on a corporate website showing operations on the company's new drillship showed workers in the exclusion zone while the rig was operational and moving. The same company's safety case stated that workers would never be in the exclusion zone during operations. Injuries to drill floor workers caused by contact with moving machinery are a significant issue on these more automated drill rigs.

HF/E Topics

The sources above will help to identify a number of HF/E issues, including the following. If there are any problems with any of the following issues, the solutions will be strongly biased towards HF/E. They are as follows:

Control of work and the PTW system: How is the risk assessment for PTW completed and reviewed? How are the PTW conditions established and verified on-site? How are PTWs reviewed and approved?

Supervision: How is the worksite supervised? What are the principal roles of the supervisors in relation to risk controls? Once offshore, you can observe how they divide their time between worksite, their desk, planning the job, or reading and responding to e-mails.

Competence: How is the necessary competence for a task or role determined? How is competence assured for core crew (i.e., operator/duty holder employees) and for contractors? How would lack of competency be identified and what would happen next?

Human performance: How is human performance assured by (for example) FRMS including the monitoring of working hours and overtime?

The CCR: How are the CCR operators trained for normal and emergency operations? Have there been alarm rationalization studies to ensure the alarms are manageable? What is the intended level of automation? Failed automation is a significant cause of work overload.

Incident investigation: Have there been any recent incidents (including near miss situations) with a subsequent investigation? Was HF/E considered and if so, what was the outcome?

Other specific activities: The rig may be drilling or undertaking well intervention (workover) activities. These provide an opportunity to look at specific hazards such as the drill floor equipment, the monitoring and control of the well, and the interfaces between the operator and the specialist contractors.

It is useful to review these issues and select a few topics and provide these to the site along with a few sentences about what you will want to observe. Two activities are worthy of special note: Control room activity and maintenance.

As an HF/E practitioner offshore, it is important to spend some time in the CCR and, if appropriate, the drillers' control room (cabin or doghouse). Pay particular attention to displays, information flow, and work environment; it is important that people know what is going on. Observe how alarms and other upsets are handled. Are there alarm floods, standing alarms (alarms left inactive for long periods), or many alarms being cancelled without action? Are there a lot of distractions? There will be CCTV and for the drill floor there will be windows. Are the lines of sight effective and can the operators see what they need to see? There will be a lot of communications with technicians out on the plant or drill floor. Are these effective and clear or is there scope for misunderstanding or ambiguity? Many operational elements, for example valves, have similar names and some very serious incidents have occurred as a result of simple misunderstandings.

Maintenance is a major concern offshore and it is important to spend some time observing maintenance activity. The best way to do this is to attend the daily permit review meeting and select a relevant job planned for the next shift, such as any job requiring isolation and isolation certificates or any complex or seldom used procedures. Once you have identified the job to be observed, it will be necessary to notify the relevant manager and then meet the supervisor to arrange to be at the start of the job for the "toolbox talk." Permission and PPE will be needed to observe the worksite and arrange to go there once work has begun. Read the PTW and observe how the team complies with the required conditions for safe working. If the job extends over a shift change you can observe the handover as well.

Useful Things to Take Offshore

You are limited in what you can take offshore, but a few things are noteworthy. Photographs are valuable for capturing equipment controls for usability assessments and work environment features such as access and light levels. It is advisable to use the installation camera, which is intrinsically safe. For some areas, you may need a PTW and this will need to be raised before the permit review meeting.

Portable PCs may be restricted, along with Internet access, and only intrinsically safe electrical equipment will be allowed. There will be shared PCs into which you can put a memory stick and card reader. It is a good idea to type notes onto the memory stick each day. If the weather deteriorates then nonessential personnel (including you) may be "down manned." You will only be able to take your keys, passport, onward ticket, and a memory stick. The same rules may apply if the weather causes weight imitations on the helicopter flights; for the core crew this simply means leaving their luggage until they return. For you it could mean leaving all your work.

It is also useful to take electronic copies of the following:

- *HSE's "Reducing error and influencing behaviour (HSG 48)"* (HSE, 1999). This is widely used in the offshore sector and includes a number of assessment questions and a three-factor framework of individual, job, and organization, which can be useful in discussions.
- *Step Change in Safety's "Human factors guide"* (Step Change in Safety, No date). This report uses James Reason's safety barriers approach to explain

HF in a series of offshore case studies and can be used for structured interviews.

- *IOGP report 368 "Human factors"* (IOGP, 2005). This is helpful for introducing HF/E to the workforce and as a primer in interviews.
- *HSE's "Guidance on permit to work systems (HSG 250)"* (HSE, 2005). This is useful when observing a PTW and when interviewing the PTW coordinator and supervisors.

PEOPLE TO MEET

It is necessary to inform the site or installation who you will want to meet. These will typically include the following:

- The offshore installation manager (OIM) or site manager
- The safety representatives
- The offshore/rig (on-site) medic
- The principle discipline managers—i.e., production superintendent, maintenance engineer, construction superintendent, chief tool pusher
- The permit controller or their equivalent
- A maintenance supervisor and their team

It is worth mentioning the first three here. It is important to meet the OIM early on to explain the purpose of the visit and the deliverable. You should aim to make your assessment, determine any actions, at least in draft form, and commit to feed these initial results back to the OIM and relevant crew members before you leave for the shore.

The OIM will introduce you to the management team. Having a clear and achievable plan with objectives and a commitment to provide feedback will ensure that you are given access to the staff and tasks that you need. You have to make the case that interrupting the worksite for your HF/E intervention or assessment is worthwhile. Many areas on an offshore installation are hazardous so you may need to be accompanied and your access may have to be planned (e.g., for hearing protection, emergency recovery, emergency breathing equipment).

Safety representatives play an important role in representing workers' safety concerns. Meet them in private and ask open questions about any safety concerns they may have, any recent incidents or investigations and how they were handled, and any jobs that are seen as dangerous. Ask what would help. You can be a conduit for their concerns and use HF/E techniques to help find solutions. If they refer to a dangerous task, ask to see how and where it is done, and see if a task redesign is possible. This can be a very positive outcome to a site visit.

The *offshore medic* will have information about working hours and any sleep or fatigue problems, for example, fatigue and disturbed sleep as result of noise in the accommodation or motion sickness. The HF/E focus is on how these human performance risks are monitored and mitigated to ensure operational safety.

PROFESSIONALISM OFFSHORE

In all of this activity, it is important to position yourself. You are familiar with your subject but may be unfamiliar with the offshore environment. Be honest about this. You are neither management nor workforce but you will be aligned with management simply by virtue of being "sent" to the installation. When talking to the workforce, your role is to elicit information, interpret that in an HF/E context, and translate it into meaningful actions for the management that will make the work environment better and tasks safer and easier. This is a powerful selling point and you can use it to engage people in discussion. There will be many issues that have been raised before with no progress, often because the message has been lost when it reached onshore management or there was perceived to be no practical solution. Make the message clear, and link it to the risk controls in the safety case so that it is in management terms, then provide a possible solution or at least a path to a solution.

You must also obey the rules. This especially applies to working hours. This includes stopping work at the end of the 12-hour shift. If you need to work longer then ask the OIM for permission and abide by the answer. You will normally work day shifts but if you need to observe night work then plan how you will rest and control your overall hours, agree this with local management, and explain it to the workforce.

FORMING CONCLUSIONS AND GIVING FEEDBACK

By about day 3 on site, you should have a clear understanding of any problems and some suggestions for improvements. Testing these ideas with the on-site management and supervisors will help you refine them into a small number of improvement actions. Link these suggested improvements back to the safety case and risk control measures to ensure that safety is at least maintained and if possible improved. An estimate of the time and cost benefits will make the case for changes. Note that improvements in getting the job done will ensure that the changes are sustained: Changes that make a job safer but harder will fall out of use unless closely managed, while changes that make the job easier and safer will be self-sustaining.

It is a good idea to select the most beneficial three or four improvements and feed these back to the installation management before you leave. Listen to any suggestions and criticism and act on these. The recommendations you take back to onshore management must be effective and achievable. You may face a difficult decision if you observe something that, from experience, you believe could lead to an imminent danger. In this case, you must raise the issue with local management before you leave the installation or site. The local management must have the opportunity to correct the situation before you take your concern to the senior management onshore. Even if the situation is resolved satisfactorily, you may conclude that local management oversight is weak as the dangerous situation or working practice was allowed to develop and this may be part of your onshore feedback. The importance of raising issues immediately cannot be overstated; an incident may occur between your observation and your report, raising questions over why you felt able to leave the worksite without acting, if you were so concerned. Getting this wrong may undermine your professional standing and the reputation of HF/E.

Often the task specification will only require verbal feedback, usually accompanied by a presentation, but these do not provide a permanent record and they are not usually filed or archived. Your output should become part of the organizational memory that can contribute to how the organization learns. It should also be stand-alone so that the organization can circulate it to other similar operations. It is wise to deliver your report in hard copy as well as electronically so that it is properly dealt with and filed, not lost in e-mail in-boxes. Make a commitment to revisit your improvement suggestions after one year so that both you and the organization can learn from your intervention.

CONCLUSIONS

HF/E in the oil and gas industry has been a late starter, but is now catching up. Stronger regulation and emerging industry standards have been a recent driving force for improvement, both in the ergonomics of the workplace and in the management of human contributions to major accident risks. However, there is still a long way to go before HF/E becomes a routine part of the global industry and becomes a standard and accepted professional discipline within this industry.

Looking forward, we specifically need better translation of HF/E outcomes into ongoing operational practice. HF/E analysis usually identifies many opportunities for human error and identifies appropriate means of error reduction, but the persistence of such error reduction during the operation life of the plant is not assured. In operation, the focus shifts to operational controls such as the PTW and the quality of procedures, and the link to design intentions can be lost. To fully realize the potential of the formal HF/E analyses, we need to become more effective at injecting the outcomes into work planning and operational procedures.

REFERENCES

Health and Safety Executive (HSE). 1999. Reducing error and influencing behaviour (HSG 48). Richmond, UK: HSE. Accessed March 18, 2016. Available at: http://www.hse.gov. uk/pubns/books/hsg48.htm.
Health and Safety Executive (HSE). 2005. Guidance on permit to work systems (HSG 250). Richmond, UK: HSE. Accessed March 18, 2016. Available at: http://www.hse.gov.uk/ pubns/books/hsg250.htm.
Health and Safety Executive (HSE). 2006. Assessment principles for offshore safety cases (APOSC). Aberdeen, UK: HSE. Accessed March 18, 2016. Available at: http://www. hse.gov.uk/offshore/aposc190306.pdf.
International Association of Oil and Gas Producers (IOGP). 2011. *Human Factors Engineering in Projects. Report no. 454.* London, UK: IOGP.
International Association of Oil and Gas Producers (IOGP). 2005. *Human Factors – a Means of Improving HSE Performance. Report No. 368.* London, UK: IOGP. Accessed March 18, 2016. Available from: http://www.ogp.org.uk/pubs/368.pdf.
Step Change in Safety. No date. Human factors: How to take the first steps.... Accessed March 18, 2016. Available from: https://www.stepchangeinsafety.net/safety-resources/ publications/human-factors-first-steps.

18 Human Factors and Ergonomics Practice in the Nuclear Industry

Helping to Deliver Safety in a High-Hazard Industry

Clare Pollard

CONTENTS

PRACTITIONER SUMMARY

The application of human factors/ergonomics (HF/E) in the nuclear industry provides practitioners with opportunities to design systems and processes, while ensuring the safety of workers and the public. There are multiple features that make the industry unique such as the regulatory regime, political aspects, public perception, timescales, level of detail, and the hazards. The safety case for a nuclear facility requires qualitative and quantitative elements. It is a challenging industry with many different types of projects from building new facilities to decommissioning aging plants.

INTRODUCTION: NUCLEAR HUMAN FACTORS SPECIALIST— JACK-OF-ALL-TRADES, MASTER OF INTEGRATION

I have been an HF/E specialist in the United Kingdom's nuclear industry since I graduated in 2000. I wanted an interesting job and that is exactly what I got. Through the years I have worked on a number of different projects, across many of the different nuclear sites in the United Kingdom, including the following:

• Designing and building new facilities
• Modifying existing plants with new equipment or processes
• Decommissioning aging facilities
• Periodically reviewing the safety of existing plants against modern standards

In the nuclear industry, we address a wide variety of HF/E topics, including training, procedures, equipment design, emergency actions, alarm philosophies, work planning, and resourcing. We carry out both qualitative and quantitative assessment for complex systems and processes to ensure the safety of our operators and the public in this high-hazard industry, with an understanding of both safety and engineering. As such, there is a tendency for HF/E practitioners to be regarded as a jack-of-all-trades, with an understanding of a wide range of HF/E areas and an ability to integrate many aspects of HF/E into projects, rather than as an expert in one area or method of application.

WHAT MAKES THE NUCLEAR INDUSTRY UNIQUE?

VARIETY

The variety of the nuclear industry is incredible, and includes power stations, research facilities, control rooms, cranes, remote operations, building new plants, defense sites, submarines, weapons, nuclear medicine, medical diagnostic equipment, waste processing and storage facilities, and former research and reactor sites ready for decommissioning or being decommissioned.

They all have their unique aspects, such as design and use of equipment, space constraints, different end goals, etc. Nuclear submarines bring their own challenges, not least equipment design and environmental conditions, as do nuclear weapons. Another aspect, the healthcare industry, pushes boundaries in terms of technology that can diagnose and treat the most serious of illnesses. As such, the industry offers a great many opportunities for HF/E specialists.

HAZARDS

The consequences of a nuclear accident are unacceptable, as seen at Three Mile Island, Chernobyl, and Fukushima, and a huge amount of work continually goes into making accidents highly unlikely across all types of sites. There are the same industrial hazards as in other heavy industries but we also have to protect ourselves against radiological hazards from ionizing radiation and contamination, which can cause significant health hazards. It is this hazard that drives the level of detail required in HF/E assessments.

Operator-initiated fault sequences, operator-dependent safeguards and mitigators, and recovery operations all have to be considered to a level of detail proportionate to the hazard and the potential consequences of an error to both the workforce and the public. Design of equipment to protect against the hazard has to be usable.

REGULATORY REGIME

To assure the safety of nuclear installations in the United Kingdom, the Office for Nuclear Regulation (ONR) independently regulates nuclear safety and security at all nuclear licensed sites in the United Kingdom. They also regulate the transport and safeguarding of nuclear and radioactive materials. There is a robust licensing process by which a corporate body is granted a license to use a site for specified activities. The Nuclear Site License is a legal document, issued for the full life of the facility. The ONR sets out site license conditions that each licensee must comply with in order to operate in the concerned area (see ONR, 2016). There are internal reviews of all projects, as well as scrutiny from the regulator who produce guidance in the form of safety assessment principles (SAPs) (ONR, 2014). Here "safety" means nuclear safety, with conventional health and safety hazards dealt with in a different manner. The SAPs provide ONR inspectors with a framework for making consistent regulatory judgments on nuclear safety cases and provide nuclear site duty holders with information on the regulatory principles against which their safety provisions will be judged. However, they are not intended or sufficient to be used as design or operational standards, due to the nonprescriptive nature of the United Kingdom's nuclear regulatory system. There are HF SAPs covering all aspects of assessment to ensure appropriate HF analysis is carried out for each project (the term "ergonomics" is not used). The current SAPs cover a wide range of issues from HF integration and task screening to training, procedures and environment. The section on "HF" begins as follows:

A nuclear facility is a complex socio-technological system that comprises both engineered and human components. The human contribution to safety can be positive or negative, and may be made during facility design, construction, commissioning, operation, maintenance, or decommissioning. A systematic approach to understanding the factors that affect human performance, and minimising the potential for human error to contribute to or escalate faults, therefore needs to be applied throughout the entire facility lifecycle. Assessments of the way in which individual, team, and organisational performance can impact upon safety should influence the design of the facility, plant, equipment, and administrative controls including emergency arrangements. The allocation of safety actions to human or engineered components should take account of their differing capabilities and limitations. Safety cases need to demonstrate that interactions between human and engineered components are fully understood, and that human actions that might impact on safety are clearly identified and adequately supported.... (p. 102)

POLITICS

The nuclear industry is highly political, with public perception being very important to ensure continued governmental support. Over the past few years alone we have had the following:

- Consideration of the role that nuclear power has to play in the future of energy production (e.g., see Department of Trade and Industry, 2006)
- UK Public consultations to discuss the nuclear new build program and the proposal for long-term underground radioactive waste storage
- An accident that occurred at the Fukushima Daiichi nuclear power plant following the major earthquake and tsunami
- Development of nuclear programs with other countries through international agreements with France, the United States, etc.

The nuclear industry is very important economically. Generating plants are the size of small towns, supplying power to millions of people. The global market is worth billions.

TIMESCALES

In the nuclear industry, everything takes a long time, which can be seen as unreasonable to those new to the industry. It is a lengthy process to design and build a new facility. From "optioneering" and concept design stages to operation, clear justification has to be made to ensure it is the safest option and all design decisions have been substantiated in terms of the potential fault sequences, and their consequences. The emphasis is on thoroughness, to ensure robust safety arguments supported by evidence such as task analysis assessments of operator actions. The timescales also link back to political aspects with lengthy delays in the progress of the much-needed new nuclear power generation facilities due to funding concerns, being of the many reasons.

DETAIL

We deliver a safety case to present the arguments and evidence that a facility is safe to carry out planned activities. It is not enough to say that something is safe; we have to demonstrate this is the safest way to perform the task. The term "safety case" encompasses the totality of a licensee's (or duty holder's) documentation to demonstrate high standards of nuclear safety and radioactive waste management (ONR, 2016). All nuclear facilities have an operational safety case that defines the boundaries of the plant and its safety systems. Every plant change and every minor modification has to be reviewed in terms of what the potential impact of that change could be. Real data is hard to come by. The nuclear industry is not all control rooms and complex software, logging all keystrokes. It is also manual work on plant, handling and moving items, storing waste, etc. As such, exact data on individual errors, if they are recovered from immediately and therefore do not result in an incident, may not be captured. Also, other tasks may not be undertaken on a regular basis, such as lifting/handling of equipment around the facility. As such, fault sequences require detailed assessment well beyond a reliance on historical data.

WHO WORKS IN THE INDUSTRY?

The nuclear industry is global and comprises many diverse participants ranging from individuals to companies, industry associations to intergovernmental bodies, as well as appointed bodies responsible to national governments.

A large number of people are responsible for the design, building, commissioning, running, and eventual decommissioning of our nuclear facilities. I have worked alongside many of these people and found that the majority of their work is of interest to HF/E specialists as it affects people, processes, and procedures, and influences how we work. They include the following:

- HF/E specialists. When I started, these were likely to be people who had many years of nuclear experience and who mainly stayed in the industry. Given the increase in work over the past decade, this is no longer the case. There is much more movement of personnel with many HF/E specialists working in multiple sectors, particularly across the high-hazard sectors.
- Designers and engineers from many backgrounds, including civil, mechanical, electrical, control and instrumentation, process, building services, etc.
- Safety specialists, including safety case, fire safety, criticality, environmental, radiation protection, shielding, etc.
- Plant personnel, including operators, maintainers, planners, emergency response teams, trainers, human resources, document management, quality, health physics, etc.
- Others, including security and classification officers, IT services, medical services, dosimetry, etc.

The operating teams are impressive, with a reliance on highly skilled operators and maintainers ensuring continued safe operation.

Interacting with designers to ensure human capabilities and limitations, such as size, strength, capacity, etc., are taken into account as the design develops is part of the standard work of an ergonomist, but other areas may surprise you. One aspect that fascinated me was maintenance planning. Consider a power-generating reactor that has to be "live" for a certain number of days a year in order to meet power supply needs and their targeted financial budget. That site may have a planning regime that plans each individual maintenance activity on each system component up to a year in advance. This requires considerable planning which culminates in detailed "outages" requiring many trades and professions to safely work closely together on the plant under intense time constraints to ensure the system can be brought back online when required. This requires incredible people-centered management skills.

WHAT IS IT LIKE IN AN OPERATING FACILITY?

What is most visible in nuclear facilities is the range of heavy equipment, including tanks, pipes, and valves out on plant, and cranes and other handling, construction, and demolition equipment elsewhere. The facilities, power-generating reactors, and equipment needed are huge, with heavy machinery working alongside highly pressurized vessels with delicate instrumentation. There may be ponds for storing fuel, rows of electrical cabinets, vaults and glove boxes with glove ports, enclosed "cells" with master–slave manipulators, fuel casks, and waste drums. They may have control rooms where staff can monitor or carry out operations remotely. These can be old facilities or new, full of "state-of-the-art" equipment, depending on the facility and its age or refurbishment program.

The nuclear industry is not the male-dominated world you might expect; there is an increasing number of female operational staff as well as specialist support and management personnel. There is a group called Women in Nuclear United Kingdom (WiN UK), a nonprofit organization whose mission is to address the industry's gender balance, improve the representation of women in leadership, and engage with the public on nuclear issues. WiN United Kingdom is supported by both the Nuclear Industry Association and Nuclear Institute (NI).

A great deal of work is ongoing to recruit the next generation to ensure a highly skilled workforce is maintained. So while some niche operational areas may be staffed by older workers, in the last 10–15 years there has been a change in the age profile of the workforce. In many parts of the industry there has been a big increase in the graduate population to ensure sustaining our workforce and specialist skills. There is a Young Generation Network group, as part of the NI, created to offer the younger members of the Institute the opportunity to further their knowledge and facilitate networking between generations.

For many plants, you wear personal protective equipment. If there is no need for you to be in a particular area you would not enter, but instead view remotely or undertake tabletop reviews to understand processes and systems. All plants operate on the rule of "essential personnel only," so if you are not vital to the task, you will not be allowed access to these areas.

DEMAND FOR HF/E

There is a huge demand for HF/E specialists across the nuclear industry. There are many sites and lots of ongoing projects, most of which will require HF/E input on some level for all types of activity like the following:

- Normal operations
- Plant transients
- Fault diagnosis and recovery
- Maintenance
- Emergency response

Many site licensees have their own HF/E team, but even so, most of these teams are small and still need ongoing support for the many projects on a single site at any one time. For this they turn to consultants, either employed by specialist safety consultancies or self-employed HF/E practitioners.

Many different things can trigger the need or call for an HF/E assessment, including, but not limited to the following:

- A change in task, equipment, or key aspects of plant
- If the consequences of an error in a particular task could give rise to a significant dose to the operator or to the public
- Revised operating or emergency procedures
- In response to an incident, e.g., following investigation into a problem area
- Decommissioning of part or all of the facility

There are some differences between countries in terms of what activities the HF/E specialists undertake. These differences include the varying level of detail required for qualitative and quantitative analysis, the requirement for safety case support and the prescriptive nature of the safety case, and related documentation. Some countries focus almost exclusively on the design of control systems in control rooms, whereas, in the United Kingdom we get more involved in the consideration of the entire socio-technical system.

So, who decides if HF/E is required for a project or not? Usually the requirement is stated in their safety manuals/procedures for specific HF/E tasks to be undertaken that are linked to the hazard level of the plant and the associated safety case and, therefore, the project manager requests support. However, if aspects of HF/E have not been included, but are thought to be needed by the reviewers, specialist support may be requested at the review stage of the delivered safety case by the external regulator or internal Nuclear Safety Committee. In the United Kingdom, there are plans for new facilities and there is a large program of decommissioning of old plants. As such, there is high demand and many different types of assessment work, e.g., concerning decommissioning (Pollard, 2014).

Nuclear ergonomics can be complex, with training, mentoring, and coaching provided for graduates by specialist teams, supplemented by nuclear-specific classroom-based training modules that help to increase sector knowledge.

In terms of what is demanded of HF/E practitioners, HF/E analysis in the safety case needs to employ a combination of approaches. It needs to inform the fault schedule and probabilistic safety assessment aspects of the safety case, but not be driven by it solely. The HF/E approach makes use of plant requirements, Hazard and Operability (HAZOP) studies, and aspects of safety assessment and engineering to inform screening and depth of analysis. It also uses task analysis for detailed error identification and analysis of significant events to inform the fault schedule, to input probabilistic and deterministic arguments, and to ensure all conceivable routes to faults have been identified. The HF/E analysis needs to be reconciled with the wider safety case to ensure all foreseeable significant error modes and claims and assumptions on operator actions and managerial systems are valid, reliable, and substantiated.

Other aspects involve equipment design. This can range from a small plant modification, such as a new stand-alone display, to an entirely new facility introducing a whole set of control systems using the latest technology. The design development stages may involve many aspects such as job design, alarm reviews, and training needs analysis. HF/E specialists get involved in many aspects such as running workshops to support identification of operational requirements, assessing and supporting the layout of controls, and leading mock-up trials at many different stages.

WHAT IS IT LIKE TO BE A PRACTITIONER?

Nuclear projects are so different, with constraints and requirements so unique, that you are constantly learning new systems and processes as you move from one project to another. This learning takes place on the job, developed during operator interviews, plant walk-downs, project briefings, etc. The majority of projects are one-offs and so you have to learn each time about specific project constraints, operating requirements, plant history, etc. These varied projects are what make the industry interesting. The start of every project is exciting. We have to learn whole new systems with history, constraints, unique challenges, different radiological consequences, and hazards.

HF/E specialists liaise closely with both designers (helping to support the design through its development to ensure it delivers a usable, intuitive system) and with safety case personnel (putting in place the evidence to demonstrate human error levels are minimized by the design and the safety management system).

The industry is mostly staffed with highly motivated people who work hard to make a difference. The majority are happy to be inclusive of all disciplines but most HF/E practitioners have had to fight their corner to get themselves heard at one time or another (though this is certainly not unique to the nuclear industry). It helps if you understand the wider arguments, the safety concerns, and the plant constraints so that you can talk in a shared language and justify your conclusions. Justification of decisions is the most important aspect, and must be based on assessments (usually involving task analysis)—evidence that can be presented and discussed.

I spend a lot of time with operators and facility management. This may be on plant walk-downs, running workshops on culture or procedure writing, delivering training in elements of HF, supporting design reviews and hazard identification studies, forming working groups to tackle problem areas, presenting to safety committees. Aside from all of this, I spend a lot of time carrying out task and error analysis, reading procedures, reviewing training plans and writing reports, and presenting the reasoned argument to justify an action.

The solutions can be complex. There has to be a focus on adding value and ensuring nuclear safety as well as applying ergonomic principles. Project constraints on aspects such as the environment or aging plant interface design have to be traded-off against the safety of the operator and the public. As such trade-offs and compromises sometimes have to be made to ensure the most appropriate solution for all stakeholders too.

CONCLUSION

The nuclear industry is a challenging industry with many different types of projects from building new facilities to decommissioning aging plants. The industry provides practitioners with the requirement to work towards ensuring the safety of workers and the public. By working alongside designers while embedded in the safety case process, HF/E specialists consider operator-related fault sequences and carry out both qualitative and quantitative assessments.

There are multiple features that make the industry unique such as the variety of facilities and applications, regulatory regime, political aspects, the timescales involved in projects, the level of detailed assessment required, and the nature of the hazards inherent in the industry.

REFERENCES

Department of Trade and Industry. 2006. *The Energy Challenge Energy Review Report 2006*. Norwich, UK: HMSO.

Office for Nuclear Regulation (ONR). 2016. *Licence Condition Handbook*. Bootle, UK: Office for Nuclear Regulation.

Office for Nuclear Regulation (ONR). 2014. *Safety Assessment Principles for Nuclear Facilities*. Revision 0. Bootle, UK: Office for Nuclear Regulation.

Pollard, C. 2014. Human factors aspects of decommissioning: The transition from operation to demolition. In: Sharples, S. and Shorrock, S.T. (eds.). *Contemporary Ergonomics and Human Factors 2014*. London, UK: Taylor & Francis, pp. 390–394.

19 Human Factors and Ergonomics Practice in Manufacturing

Integration into the Engineering Design Processes

Caroline Sayce and Fiona Bird

CONTENTS

PRACTITIONER SUMMARY

Human factors and ergonomics (HF/E) in manufacturing is employed across many industries. While the techniques and methods we apply are consistent, the unique requirements and constraints of each industry drive the manner in which HF/E is applied. Within both the rail and defense sectors, HF/E is part of a multidisciplinary design process within a requirements-driven engineering environment, involved with the design and integration of numerous subsystems in the development of the end product. Detailed processes are followed to enable effective HF/E input through each design phase of a project, requiring both an organized and flexible approach by the HF/E practitioner. While safety and operability are the key considerations for HF/E in the development of a submarine, the development of rolling stock also considers comfort and minimizing the risk of injury for a wide range of end users. The challenges experienced across the industries can be both common and unique, ranging from managing scope and incorporating legacy design to understanding complex nuclear systems, or managing the political influence of unions.

INTRODUCTION

This chapter considers the role of the HF/E practitioners in the engineering arena of the manufacturing domain. It explores their role as part of a multidisciplinary team, the integration of subsystems into larger complex systems, the factors that influence the HF engineering approach that is applied, and the management of compromise to optimize the final design.

We have drawn from our respective experience at Rolls-Royce in the development of the Nuclear Steam Raising Plant (NSRP), which powers the United Kingdom's Royal Navy Submarine fleet, and at Bombardier Transportation in the development of passenger rolling stock. Rolls-Royce's submarine business operates from six sites across the United Kingdom with the headquarters at Raynesway, Derby. Rolls-Royce has been the sole technical authority for the NSRP for more than 50 years and is responsible for plant design, project management, manufacturing and procurement, safety assessment, and commissioning of the Royal Navy's submarine fleet as well as the management and operation of the Royal Navy's land-based reactor, the Vulcan Nuclear Reactor Shore Test facility. The submarine business manufactures and supplies pressure vessels, fuel cores, propulsors, flexible couplings, and turbogenerators and is currently involved with developing the successor submarine, which incorporates the company's PWR3 reactor technology.

Bombardier Transportation owns the Litchurch Lane Works in Derby, which has been manufacturing rolling stock since the mid-nineteenth century. The most recent rolling stock to be designed, engineered, and manufactured at the site have been the ELECTROSTAR™ and TURBOSTAR™ electric and diesel multiple units that operate on UK mainline infrastructure and the Victoria Line 09TS and Subsurface Line S-Stock for London Underground. Bombardier is currently developing the AVENTRA™ multiple units for London Rail to operate on the Crossrail and Overground networks and further developing the AVENTRA™ platform for operational use across the United Kingdom.

THE ROLE OF A MULTISYSTEMS INTEGRATOR

Throughout the design for manufacture of a complex product or system, HF/E input is applied by a number of organizations throughout the supply chain, each with differing requirements and constraints. For example, HF/E may be required during the following situations:

1. The design and manufacture of a component part
2. The integration of components to create a subsystem
3. The integration of a series of subsystems in preparation for the manufacture of a larger system

Rolls-Royce and Bombardier are responsible for managing all aspects of the design, safety, manufacture, performance, and through-life support of their respective systems. This involves the integration of multiple subsystems, with HF/E playing an important role in all aspects where human interaction with any of the subsystems is necessary. While Bombardier is responsible for manufacture of the train as a whole system, and manages the integration of subsystems from multiple suppliers to create the end product, the integration of the NSRP within the submarine as a whole is undertaken by BAE Systems and therefore Rolls-Royce is both an integrator and a supplier in the manufacture of the end product.

The primary role of HF/E within Rolls-Royce is to support the safety case for the NSRP which provides evidence that the strict safety requirements have been met, including those relating to safe operation and human reliability. HF/E in the rail domain has a wider application, as the operational environment is very different. Here, while HF/E is an input to the safety case, HF/E is used primarily to minimize costly operational delays associated with human performance, reduce the risk of musculoskeletal injury, and create an environment that is comfortable for users.

THE ROLE OF HF/E IN A MULTIDISCIPLINARY TEAM FOR MANUFACTURING

In the manufacture of any piece of equipment, HF/E is one discipline among many that contributes to the overall design. While human-centered design is a noble aspiration, in highly automated systems, humans are less involved in the operation of the system. As reliability targets and penalties for failing to achieve them are set ever higher, there tends to be a greater reliance on automated systems. Therefore, HF/E is recognized as a critical discipline to be integrated within the design process, though it receives no positive discrimination over other disciplines.

The HF teams work with a large range of internal specialists. These could typically include engineering (mechanical, electrical, system, software, etc.), industrial design, safety, acoustics, materials, environment, maintenance, testing, manufacturing, and commercial. There are also links to the equivalent disciplines within the customer's organization as well as other organizations linked to legislative and standards compliance (e.g., notified body [rail], Office for Nuclear Regulation). The requirements and feedback placed on or by all of these actors have to be incorporated into the HF/E activities.

HF/E IN THE ENGINEERING PROCESS

Both Bombardier and Rolls-Royce have established engineering and design processes that are applied to every project. The HF/E processes are designed to integrate seamlessly so that the HF/E activities add value at the right points in the project. The activities are managed by a human factors integration (HFI) plan, an integrated HF/E program and HF/E actions, and a decisions log, all of which are live and frequently updated throughout the project.

In this chapter it is not possible to describe in detail the HF/E methods that are utilized during the engineering program due to the sheer amount of HF/E work that is carried out. For example, a brand new train will utilize anywhere between 3,000 and 10,000 hours of HF/E depending on the amount of novel design that is required and a new submarine will utilize approximately 65,000 hours. The HF/E methods utilized by Bombardier and Rolls-Royce have evolved over many years to take into account research-based HF/E good practice, rail industry good practice, the experience of qualified HF/E specialists who work in-house for our customers, and the realities of implementation in engineering projects. This has led to a set of HF/E methods that are specific to the application and needs of the industry in the present times.

HF Integration at Bombardier

At Bombardier, HFI is a key input to the design of the cab, saloon, exterior, and maintenance of the whole vehicle. During the concept design phase, the HF/E requirements are defined and the HF/E program of activities is harmonized with that of the other disciplines. Draft task analysis, style guides, and specifications are produced (e.g., for human–machine interfaces and audible alarms); HF/E good practice is applied to the concept design of the cab, saloon, and exterior; and support is provided to other analysis such as "day in the life of the train" conducted by the operations team and hazard and operability studies (HAZOPs) conducted by the safety team.

All of these activities draw on internal HF/E processes and employ design solutions that have been proven on previous builds, as well as novel items that require more detailed input. During preliminary and detailed phases, the HF/E team works with the system engineers and suppliers to develop the subsystems, and the industrial designers, vehicle engineers, and the design for maintenance team in the integration of the subsystems into the vehicle. Computer-aided design (CAD) assessments using mannequins are carried out to ensure anthropometric fit for the target audience and compliance with sight line requirements. During these phases, the activities conducted during the concept phase are further refined, human error and workload assessments are undertaken, and HF/E testing is carried out with end users using test rigs and mock-ups to de-risk the design as much as possible before the first train is built.

The first trains off the production line are subject to months of intensive testing by all disciplines to prove the design and as such it is critical for the HF/E team to identify their testing requirements and write test specifications (to verify the design against the requirements) to secure access to the test facilities. The tests for alarms, operability, and likewise are the final level of HF risk reduction and are carried out with support from the system and test engineers. By the end of each project, the HF/E activities are characterized by a suite of HF/E analysis documents, style guides, test specifications and reports, and design assurance documents demonstrating compliance against requirements.

HF INTEGRATION AT ROLLS-ROYCE

Rolls-Royce undertake a specified set of activities during each project phase, beginning with requirements capture and early HF analysis during the concept phase, followed by task analysis and development of design guidance and style guides during the preliminary stages of design. During later design phases, the HF team integrates with system designers, component owners, operations and installations teams, and electrical control and instrumentation engineers to provide design guidance and HF analysis to ensure compliance with all HF requirements. The HF team also works closely with the safety engineers throughout the project to support the safety case, for example, through attending HAZOPs and providing input to the categorization and classification of safety systems.

The key HF concern is the safe operability of the NSRP under all operational modes, covering normal, abnormal, and emergency scenarios. Operability assessments are carried out via task analysis, manning, and workload assessments to gain an understanding as to whether the system designs are operable with the time allowances and manning levels proposed. Where issues are identified, HF recommendations are made for system design change, in order to improve operability. Human reliability assessments are then undertaken to provide evidence that the operability of the systems and procedures meet the required reliability targets, and training needs analysis (TNA) is performed.

Individual component assessments are carried out, but during the earlier design phases this is limited to assessment of the physical design of the component, rather than the way it is used in conjunction with other components, or the installation of the component within the submarine as a whole, which is confirmed at a later stage in the design process. Therefore, components that meet the specified HF design requirements when considered as a stand-alone item may be noncompliant when a CAD mannequin assessment is carried out on the proposed installation of the component. This may mean modification is required—either in the component part or in the way in which the parts are integrated to form a wider system.

A series of HF verification and validation activities is also undertaken, although, due to the complexity and scale of the end product, prototypes to aid design iteration are costly. As such, these are generally limited to activities such as interface development, which can be run through a simulation of the plant behavior, or small-scale component trials. Therefore, the majority of HF carried out by Rolls-Royce is theoretical analysis based on the proposed component and system designs, and associated operating procedures using traditional HF techniques. However, during the realization phase of the project, a detailed program of on-boat testing is undertaken during build and commissioning, supported by a shore integration facility.

Like Bombardier, the HF activities are characterized by a suite of HF documents that are formally issued to the customer or released internally at key stages of the project. These demonstrate compliance against requirements.

INFLUENCING FACTORS

There are many factors that influence the organization, for example, political, media, and budget. These increase project complexity and affect the HF/E activities. This section identifies those that have a direct impact on the HF/E activities of each organization.

USER POPULATION

The end user population varies according to industry. In a submarine, the end users are predominantly military personnel who are highly trained and disciplined to work within a strict hierarchy of responsibilities. In the rail industry the users are civilian and trained (but not with the same level of discipline) but the culture is more open to engage in debate, for example, regarding design or management decisions as well as the political influence of the unions and passenger groups is visible. Historically, both the rail and defense environments predominantly employed male staff. However, the rail industry has been designing to include women since the 1990s, whereas, in the defense industry this is a more recent consideration, particularly for a submarine environment.

The rail industry designs for comfort as well as reduction of injury, with an extensive anthropometric dataset being applied throughout the design process. The defense industry, however, is primarily concerned with safety and reliability. Due to the spatial and environmental limitations on board a submarine, the application of anthropometric data for comfort and injury reduction is more limited and operators will be expected to work in less comfortable environments than expected of drivers of a passenger train.

REQUIREMENTS AND STANDARDS

Requirements have a significant impact on design. For example, in rail, these are primarily from the Technical Specifications for Interoperability, EuroNorm Standards, Railway Group Standards, and Customer Standards and Specifications, and for the NSRP these are from Defense Standards, the Ministry of Defense, and industry good practice. These are managed according to company processes through a database such as the Dynamic Object Oriented Requirements Management System (DOORS).

On occasion, requirements (ranging from operational targets to functional engineering requirements) or existing supplier products can conflict with HF requirements and careful assessment is needed to understand the risk and identify a compromise that reduces risks to as low as reasonably practicable (ALARP). Compliance is demonstrated through all of the HF/E activities and documented in the suite of HF/E reports, with satisfaction arguments recorded in the requirements management database.

LEGACY SYSTEMS

Legacy systems (these are generally older systems that remain in operation until a more modern replacement is implemented) are present on both submarines and rolling stock, and they present unique challenges. For example, new submarine designs must accommodate women as crew members, whose requirements have not been present on previous designs, and therefore legacy components that were previously acceptable in terms of manual handling or force requirements may no longer be acceptable due to the differing capabilities of women operators. Although it is preferable to make design changes where the component is noncompliant with HF/E requirements, in some circumstances design change is not feasible and procedural modifications will be considered.

In other cases, systems cannot be changed as they are familiar to the users, for example, legacy alarms and indicators in a rolling stock cab with which the driving population is familiar and therefore a change would increase the risk of human error. In such cases, new systems have to be designed around legacy systems, and carefully integrated where functional relationships play a role in the management of human error. The use of proven technology and subsystems is also sometimes preferable to both the customer and the manufacturer as it can create efficiencies in terms of time frame and cost of manufacture, and increase reliability, therefore reducing commercial risk for all parties.

COLLABORATIVE WORKING

A key consideration for Rolls-Royce during the design process is the integration of the NSRP alongside secondary systems within the wider submarine, which is manufactured by BAE Systems. This requires collaborative working between the two organizations in order to produce a final design solution that is consistent and compliant with HF requirements as well as the requirements of the wider engineering environment.

For this to be effective, a detailed set of requirements is essential to ensure that both organizations are working towards the same goal. However, a clear process is also required as a method for dealing with noncompliance with the requirements, to ensure both organizations are consistent in assessing any mitigating factors. For example, a restriction in floor to ceiling height may be acceptable in an area that is infrequently manned, but unacceptable in a main thoroughfare.

Greater challenges are encountered where more detailed design is required, for example, in the development of a human machine interface through which both Rolls-Royce and BAE Systems subsystems are operated. Although requirements can be met through a variety of design solutions, the interface as a whole must be consistent across all systems to minimize the opportunity for error and inevitably compromise will be essential.

SPATIAL ARRANGEMENT

HF/E input during the installation planning is essential for reviewing the location of components that are manually operated. Although in an ideal world all components would be optimally positioned for operation, the space limitations dictate that this is not always possible. Therefore, items that must be accessed quickly in an emergency, are critical to operations, or are operated regularly, need to be optimally positioned, whereas items that are accessed less frequently (e.g., for irregular maintenance tasks) may be located in a less optimal position.

In the design of rolling stock, additional complexities come in the form of external sight line requirements and the space envelope requirements for the components and wiring that accompany the controls on the desk. All of these lead to some level of compromise in the layout.

CHALLENGES AND REWARDS

In the following sections, we reflect on some challenges and rewards from our experiences as practitioners in these manufacturing environments.

WORKING AS AN HF/E PRACTITIONER AT BOMBARDIER

Bombardier has been employing HF/E on their trains for over 20 years. HF/E is an accepted part of the engineering process. While the scope of work proposed will always be challenged to ensure that it is adding value to the final product, as with any project or customer, the value of HF/E is readily acknowledged.

The work in Bombardier is varied insomuch that every project has a different scope of work and requires specialists who can adapt existing HF/E methods to suit the scope and be flexible to changing requirements and timescales. Flexibility is also essential in the application of HF/E good practice and the findings of HF/E assessments to the design to ensure they are applicable to the operational railway; onerous HF/E requirements can be accommodated but only if they can be justified to benefit (e.g., operational efficiency, improved safety) within the scope of the project. The role also relies on a range of nontechnical skills including teamwork, project management, communication and negotiation, self-motivation, ability to follow procedures, and report writing. Over 15 years, I have developed a wealth of knowledge of the rail industry, although much of this has been obtained from the many experienced railway colleagues that I have worked with who are always willing to share their knowledge for the benefit of the industry.

The role of an HF/E specialist at Bombardier is challenging, but it is rewarding to see the efforts of the team realized in the final product as it rolls off the production line and into service on the operational railway.

WORKING AS AN HF/E PRACTITIONER AT ROLLS-ROYCE

At Rolls-Royce, an organized and pragmatic approach is required to ensure work programs across disciplines are aligned throughout projects that can run for many years, and that HF is incorporated at an appropriate time during the project.

One of the key difficulties I have experienced as an HF engineer within the submarine domain is lack of exposure to the end product. We can all relate to design issues that we can observe on everyday items such as mobile phones or cars, but understanding the problems experienced within a submarine environment is challenging when opportunities to access the boats are few and far between, and security constraints severely limit the level of information available within the public domain. I have visited two different classes of submarine, which has helped to a full appreciation of the space constraints and other limitations on board. I also benefit from having access to current and ex-Royal Navy operators who have a wealth of operating experience, vital in providing appropriate HF guidance and support.

The other fundamental challenge in the application of HF to the design of the NSRP is the importance of having an understanding of the operating philosophy for the reactor. This is where a specialization in physics alongside ergonomics would come in handy, but it is amazing how much you manage to pick up as you go along!

Despite the challenges I have experienced since I began to work in this industry three years ago, I very much enjoy the variety that my job provides, and I am proud to be involved with the continuous at sea deterrent for the United Kingdom.

CONCLUSION: IT IS ABOUT COMPROMISE

HF/E plays an important role in the manufacture of complex systems such as submarines and rolling stock. It is improbable that HF/E good practice, in its totality, could be applied to the design and manufacture of any product as the costs and timescale would outweigh the value added.

There are many challenges inherent in the integration and management of HF in the manufacture of engineered systems, and HF/E practitioners tailor their approach and methods applied to ensure that HF/E adds value while ensuring requirements compliance and the mitigation of risks that are relevant to that industry.

FURTHER READINGS

Commission Regulation (EU) No. 1300/2014 of 18 November 2014 on the technical specifications for interoperability relating to accessibility of the Union's rail system for persons with disabilities and persons with reduced mobility. *Official Journal of the European Union*, L 356/228, 12 December 2014. (Also associated EuroNorm Standards.)

Commission Regulation (EU) No. 1302/2014 of 18 November 2014 concerning a technical specification for interoperability relating to the 'rolling stock—Locomotives and passenger rolling stock' subsystem of the rail system in the European Union. *Official Journal of the European Union*, L 356/228, 12 December 2014. (Also associated EuroNorm Standards.)

Dadashi, N., Scott, A., Wilson, J.R. and Mills, A. 2013. *Rail Human Factors: Supporting Reliability, Safety and Cost Reduction*. London: CRC Press.

EEMUA. 2007. *Alarm Systems: A Guide to Design, Management and Procurement. Publication no. 191*. Second Edition. London: EEMUA.

Ministry of Defence. 2007. *Defence Standard 00-56. Safety Management Requirements for Systems: Part 1 and Part 2, Issue 4*. June 2007. Glasgow, UK: Ministry of Defence.

Ministry of Defence. 2008. *Defence Manual of Training Management, Joint Services Publication (JSP) 822. Issue 1.0*. September 2007. Glasgow, UK: Ministry of Defence.

Ministry of Defence. 2008. *Defence Standard 00-250. Human Factors for Designers of Systems (all parts)*. Glasgow, UK: Ministry of Defence

Railtrack. 1995. GM/RT2161. *Railway Group Standard, Requirements for Driving Cabs of Railway Vehicles, Issue 1*. August 1995. London: Railtrack.

Wilson, J. and Norris, B. 2005. *Rail Human Factors: Supporting the Integrated Railway*. Aldershot, UK: Ashgate.

Wilson, J., Norris, B., Clarke, T. and Mills, A. 2007. *People and Rail Systems: Human Factors at the Heart of the Railway*. Aldershot, UK: Ashgate.

20 Human and Organizational Factors in Regulation

Views from a Former Regulator

John Wilkinson

CONTENTS

PRACTITIONER SUMMARY

Human and organizational factors (HOF*) specialists help regulators and the regulated with good legislation to protect health and safety. Experience in the UK Health and Safety Executive (HSE) suggests that developing the right HOF capability with a diverse and independent team of specialists is critical. A small and credible HOF team with a clear and simple strategy can do a great deal in a relatively short time to improve HOF uptake and activity, internally and externally. This includes developing freely available, practical guidance tools, developing guidance for safety report criteria, providing effective regulator training, and performing routine HOF inspection and support for enforcement. While it helps to understand HOF first by "looking in the mirror" (by investigation), regulators also need to focus on design and human factors integration (HFI)—getting it right first time. This integration also needs to be embedded in the safety management system and the organizational culture, otherwise there will be barriers to effective implementation and possible unintended consequences. Those working on regulated sites also need to be aware of the time required to implement HOF effectively, and how best to manage complexity and the rate of change.

INTRODUCTION: A PERSONAL JOURNEY

This chapter considers some critical issues for a regulator specializing in human factors in hazardous installations (e.g., the process industries such as chemical manufacturing factories) in the UK. I was one of the founder members of the original UK HSE HOF Team when it was set up in 1999, and led the team from 2003–2011 as Principal Specialist Inspector (Human Factors). A strong team ethos was adopted at the start and the team was always much more than the sum of its parts, which, given its limited numbers, was and is helpful. It is fair to say both that I stood on the shoulders of giants and also trod heavily on many of their toes while I and the team grappled with HOF.

I came into HSE by a circuitous route. I spent eight years in logistics as a driver, warehouseman, supervisor, logistics planner, and manager before leaving to study for a dual degree in Psychology and Philosophy. As a 33-year-old graduate, on the basis of seeing a short recruitment video by the HSE, I chose to become an inspector thinking vaguely that I might be able to do some good (the possibilities for harm only become clear later…).

My complete lack of factory and manufacturing background made it a challenging start but I adapted slowly. Inspecting the Welsh valleys and North and West Wales was another form of education. Relative naivety confers some advantages to the psychologically and philosopically minded and I became increasingly conscious that something was missing from our inspection approach. The heavy focus on machinery guarding and enforcement did not seem to be enough even though my colleagues were generally very experienced and pragmatic in practice.

After a spell on the foundries and steelworks national interest group I was even more aware that we were missing out on something. Investigating the death of a

---

* HOF is used for the HSE topic-focused approach; HF/E is used where the wider HF/E community is being referred to.

young electrician working on a 440V system at a steelworks really challenged me. I searched in vain for an explanation for his apparent working on live equipment despite isolation and permit arrangements being in place. I spent weeks at the site looking for other explanations. While this may seem naïve, my puzzlement was widely shared at the time by my colleagues and the site. It only dawned on me much later that he had most likely reversed the isolation for some purpose—perhaps testing—and then forgot to reset it. He was also working at the end of a long and intensive shutdown that was running late so he probably felt there was time pressure or other pressures on him. At that time I and my colleagues simply lacked the training, support, or the tools to understand this properly.

Soon after this, I joined the (then) Chemicals and Hazardous Installations Directorate in 1996 (later the Hazardous Installations Directorate, HID). Even as a new recruit I could see that the Control of Industrial Major Accident Hazards (CIMAH) Regulations (1984) and the associated inspection and assessment regime were at best a poor fit for the major hazard sector. Inspection and assessment was variable and often dependent on the incidental background industry experience of the inspectors themselves. Safety reports had the classic disjointed approach—lots of words on plant, equipment, and processes, but nothing on people except that they were, as ever, well trained and competent, and working according to procedures. The occupational health and safety management system was invariably just added alongside this in the report but with no meaningful major hazard links. Without an effective major accident focus, both inspection and assessment were problematic. By 1999, I had tried various inspection approaches and was even more convinced that we were missing out on something.

In this chapter I give an experiential account of how HOF was implemented in the UK major hazard inspection and assessment regime, and highlight the key areas we focused on: team forming and strategy, team training and topic development, and inspector and specialist training. While this is a reflective and UK-based account, it may useful to others who work as or with regulators because the main issues are likely to be similar for them. I also discuss our main influences and attempt some evaluation of what we did, including industry take-up, barriers, achievements, and what I think can be done further in this field. I begin with the drivers and demand for HOF in UK regulation.

DRIVERS AND DEMAND FOR HOF IN UK REGULATION

REGULATION

There is no doubt that the big drive for the HSE on HOF was the coming of the European Seveso II Directive (Council Directive 96/82/EC) and the implementation of this in the United Kingdom through the Control of Major Accident Hazards (COMAH) Regulation in 1999. The Seveso Directive was triggered by a series of industrial accidents including a dioxin release in 1976 from a small chemical manufacturing plant close to the village of Seveso, near Milan in Italy. The intent of the directive was and is simple—to avoid such catastrophes in future or at least to effectively mitigate the consequences of any hazardous release. In advance of the UK regulations, much work was done by a core group of centrally led inspectors and others in what later became the HSE's HID, who helped to determine what the organization required to support effective implementation. In truth, the Seveso requirements provided few real clues and the COMAH regulations themselves were, to say the least, succinct. Staying true to their European Community (EC) stereotype and resource level, the United Kingdom built a sizeable inspection and assessment regime onto this, and of course remained true to the clear intent of the Directive: to prevent major accidents and where that failed, to mitigate their consequences (Seveso II, 2003).

An effective human factors capability was necessary to underpin COMAH and the specific HOF requirements that this implied for the regulator. A consultant-led internal and external review concluded that this needed to be a hands-on capability, i.e., a small team of HOF specialist inspectors working with and directly supporting and enabling field inspectors and specialists. This meant developing inspector and specialist capabilities, safety report assessment criteria, and plans and implementation for inspection and verification. Demand developed quickly once the new support was publicized and early adopters among the inspectorate soon publicized this and the outcomes further.

There had been earlier demands on the HSE to implement the so-called 'Six Pack' of EC directives, implemented in the United Kingdom, e.g., as the Display Screen Equipment (DSE) Regulations 1992 (implementing Council Directive 90/270/EEC 1990), which specifically called for ergonomic assessment and expertise. HF/E was implied or explicit in several of these regulations but not as a coherent HF/E approach and certainly not with a major hazard focus. But the COMAH regulations were of a very different nature and required a different approach.

The HSE had produced HOF guidance as early as 1989 (HSG48; HSE, 1989—revised as *Reducing Error and Influencing Behavior*—HSE, 1999). While this was introduced during inspector training as a core document, it was not used much in practice, mainly due to a lack of direct support or further and more specific HOF training. There was no direct expertise or support available then anyway for major hazard work except for more research-based specialists or on the quite separate nuclear side. That support was already fully committed to nuclear inspection and the HOF requirements were already considered to be "mature" and well-embedded by the nuclear industry.

MAJOR ACCIDENTS

Aside from regulations, another key early influence was Andrew Hopkins' book *Lessons from Longford* (Hopkins, 2000). This came to the team's notice at the time when they were supporting the investigation team at the BP Oil Refinery in Grangemouth, Scotland, after a series of incidents which had taken place there in 2000 (HSE, 2003a). The book was the first really accessible source to address the overfocus on occupational health and safety at major hazard sites rather than a more balanced focus between this and process (major accident) safety. It was clear that the Esso Gas Plant at Longford was, at that time, focused on the wrong indicators and issues to allow it to manage and keep track of its major hazard control. BP had similar issues at Grangemouth and of course later in the even better-known subsequent Texas City Refinery disaster in

2005. "Lessons from Longford" could be read and assimilated easily, and the team borrowed from it and publicized it widely. We invited Andrew Hopkins to visit and speak to the HSE shortly after this on one of his regular UK trips. Industry was beginning to take an interest in his findings and views at the same time.

The Buncefield explosion and fire occurred on December 11, 2005, when petrol vapor from an overfilled tank ignited causing an explosion measuring 2.4 on the Richter Scale. This was Britain's most costly industrial disaster (see HSE, n.d.[a]). The investigating inspectors had been on the HOF training course and had made regular use of the team for inspection. They asked for early HOF support, including direct input to the early investigation . This was a painstaking and very slow forensic task with witness interviews much delayed by legal interventions. However, the delay allowed significant preparation time. I helped design the basic question sets for the interviews, assisted in retrieving and evaluating documents and procedures from the damaged control suite, and used Health and Safety Laboratory deep topic specialist support to reconstruct the central control room (CCR) layout and activities, the human–computer interface, and to assess fatigue and shift-work issues.

Eventually, the investigation and a prosecution were completed and the main lessons were widely publicized. The HOF learnings for the team led to the development of several new topics and sub-topics for the HSE web pages, e.g., safety critical communication (driven by the significant shift handover issues) and design (reflecting the CCR and human–computer interface issues). The topics briefly expanded to 14 before they were wrestled back into a "top ten" framework (covered later). The thorough and detailed investigation work on these issues produced a much clearer understanding for the team on how and what to inspect on these issues.

Immediately prior to Buncefield, further learning opportunities had emerged from the Texas City disaster. The U.S. Chemical Safety Board (CSB) contacted the HSE—including the HOF team—to help with background information on BP and to help them develop their approach with regard to some of the key issues that their investigation was uncovering, for instance, on the organizational culture. The CSB's quick, smart, and professional approach to investigation prompted some envy in the team.

DEVELOPING THE RIGHT HOF CAPABILITY

For many regulators, developing the right HOF capability is a challenge, and there are trade-offs between developing internal capability and bringing in external HOF expertise. The HSE initially chose to develop its own capability to underpin the COMAH regulations; inspectors already working on major hazard inspection were recruited from within. This was preferred to using technical consultants, third-party arrangements or a more research-based, or academic approach. The intention was to bring a pragmatic focus to implementation, some immediate "face" credibility, and to allow the new specialists to "sell" HOF in the field.

I was selected in 1999 along with another inspector, a former safety, health, and environment (SHE) manager from a global business, who brought a different and valuable perspective. Both of us were challenging the status quo, competitive, ready to do something different, and willing to innovate. This mind-set was essential for getting started as HOF specialists, especially given our lack of immediate formal expertise and qualification in HOF. Importantly, though, our previous inspection experience helped inform this process and kept it credible in the field. Our personal training was carried out through the structure of the qualifying arrangements set by the (then) Ergonomics Society alongside the day job.

Our very experienced team leader was a well-known, highly qualified, and experienced HOF practitioner with wide-ranging major hazard and railway experience and gave us further immediate credibility and professional support. We made good progress in developing HOF experience through support requests and the early safety report post-assessment site visits. Expertise developed rapidly from applying new skills and knowledge to difficult problem areas at the COMAH sites arising from, incidents, the assessment process, or ongoing inspection. We had to think on our feet and apply our developing knowledge and skills pragmatically, and often from first principles.

We were soon joined by two human factors specialists who brought consultancy, research, and other industry experience to the team, and later by an additional small cohort of inspectors who had undertaken an MSc in safety and human factors. Some of these came from the outside the major hazard sector and from offshore with a view to exporting the discipline more widely across HSE. This helped to roll out HOF support to the field by recruiting existing inspectors and leaving them located where they were. Overall, having having such diverse perspectives on HOF and backgrounds is a highly effective approach to developing a team.

DEVELOPING A STRATEGY

The team developed a simple strategy, covering inspection, assessment, guidance, awareness, training and education, and research. This was regularly reviewed but proved robust and provided a solid basis for progress. It benefited from keeping close links to the real issues arising from inspection and assessment, and in keeping things as simple as possible. Taking a more strategic view beyond the immediate HOF team strategy was also important, especially in focusing on internal stakeholders. Looking upwards and "selling" HOF benefits to HSE managers and leaders could also be difficult, e.g., because of individual changes and increasing organizational change.

KEY ACTIVITIES

The HOF "Top Ten" and the Inspector's Tool Kit

A key part of the HSE strategy involved the identification and promotion of the "top ten" HOF topics (see Box 20.1; HSE, n.d.[b]). This approach has become well known in the major hazard industries. It "operationalized'" HF/E by focusing on the key HF/E topics most relevant to the onshore nonnuclear industrial major hazard sector, though we consulted with offshore and found broad alignment there as well. The selection process was based on a combination of expert and inspector opinion (the

BOX 20.1   UK HSE "TOP TEN" KEY TOPICS

1. Managing human failures (including human factors in risk assessment, incident investigation)
2. Procedures
3. Training and competence
4. Staffing (including staffing levels, workload, supervision, contractors)
5. Organizational change
6. Safety critical communications (including shift handover, PTW [permit to work])
7. Human factors in design (including control rooms, human–computer interfaces (HCI), alarm management, lighting, thermal comfort, noise, and vibration)
8. Fatigue and shiftwork
9. Organizational culture (including behavioral safety, learning organizations)
10. Maintenance, inspection, and testing (including maintenance error, intelligent customers)

latter confirmed more formally by internal survey) on investigated incident history (UK and worldwide via existing HSE research and reports) and industry liaison. This basic approach has stood the test of time and has been widely mirrored. It was deliberately "theory-lite" to allow industry to focus on key HOF topics and areas rather than engage in sterile debates about what HF/E was or was not—we simply identified the topics and asked them to get on with it.

The management of organizational change provides a good example of early topic development. After providing HOF support at several "problem" sites, either just after or during significant organizational change, we developed outline guidance and tools from experience, and reports and recommendations, along with relevant research. This was tested with various stakeholders before being turned into a formal draft. Other major hazard directorates—including nuclear and railways—were then directly engaged and consulted, and a final draft was produced, agreed upon, and adopted (HSE, 2003b).

In line with good HOF principles, this was kept as a freely accessible simple web document. Initially, it was placed with similar documents on the HSE web pages but later a suite of HOF web pages was designed and launched by the team, each with a particular topic focus (see HSE, n.d.[b]). Alongside this, an "Inspectors' Tool kit" was developed based on existing guidance, team experience, and inspector input (HSE, 2005). The web pages were added to the existing one as more experience and information became available and is now quite fully developed for most topics. It represents the largest single part of the HSE web page and is the most frequently visited (source: HSE web team) site.

The "key topics" approach was later evaluated independently. The Energy Institute also stayed with the key topic approach after their own stakeholder review in 2011 (Energy Institute, 2008), and the offshore Step Change in Safety initiative still uses a very similar approach. The top ten approach has proved robust and successful. Success in this case means that HOF became embedded in general inspection and assessment and that specialist HOF support was available, and regularly used for "deserving cases" and for inspector training and development. It also means that generally the onshore (and offshore) industry uptake of and activity in HOF steadily increased from a 1999 baseline of near zero to a significantly improved position by 2009 (HSE, 2009; Section 1.2.4). The latter research report explores the remaining barriers to HOF uptake but acknowledges the progress made in all the major hazard industries.

There has been good industry and other stakeholder take-up on HOF for the on-and offshore major hazard sector, e.g., through the UK Energy Institute, the Institution of Chemical Engineers (IChemE), the Oil and Gas Producers (OGP), the Step Change for Safety initiative. As an example, the Energy Institute was an early leader in engaging the team in taking up HOF and developed their own suite of resources based on the key topics. In fact, they were the first to develop guidance in their Briefing Note Series, including simple topic self-assessment tools. The UK influence—from HSE and the industry—can also be seen in post-Texas City activities of non-UK bodies such as the Center for Chemical Process Safety (CCPS) in the US.

GUIDANCE FOR SAFETY REPORT CRITERIA

Another early project was to develop guidance for those safety report assessment criteria that had HOF relevance. The COMAH criteria were already developed at this early stage so the team had to work with what was available. This is a key constraint for HOF in terms of regulation. To better focus the relevant criteria on HOF, some specific short guidance was developed for each criterion and was made widely available (later via HSE, 2004). This was again kept simple and was structured on the key topics. This was essential for the team when they were asked to assess safety reports on HOF issues. It was also designed to "face outwards" so that inspectors and specialists could use or consult them as a "first pass," and also so that safety report writers could use it to help them develop their reports in the first place.

With a small team, there were practical limits on how many reports could be assessed, so the guidance helped safety report assessors to discern whether or not a full HOF assessment would be beneficial. For example, in many cases there was simply insufficient detail in the reports so it made more sense to "go and see" the site, and then provide guidance to the site operator for a revised submission. Many early safety reports did not mention people at all; they were "ghost reports" that just described plant and processes. The early assessments allowed the team to quickly develop presentations and workshops for COMAH organizations and a series of "HOF in Safety Report" road shows were conducted, and a wide range of allied speaking and educational/promotional opportunities were also pursued. We talked to everybody that we could initially (conferences, workshops, seminars, individual company events, and so on), and then gradually refined this over time to target opportunities more strategically. Though we "talked the talk" we had to make sure we "walked the walk" on site and in assessment too. Otherwise, our impact would have been much smaller.

INSPECTOR AND SPECIALIST TRAINING

As part of the strategy, a lot of promotion and industry training work was done as well as more basic inspector support e.g., for enforcement in new areas such as HOF risk assessment (safety critical task and error analysis) and organizational change.

A full HOF course was designed and piloted for inspectors and specialists in the non-nuclear, major hazard sectors, on- and offshore, in 2000–2001. The pilot course gained a very positive response with some suggestions for improvement. The course was then rolled out as a core part of inspector and specialist training and is still functioning (as at 2016). The "Inspector's Tool-kit" (HSE, 2005) provided both a structure for the course and a takeaway "try this at home" pack. Again, the focus on key topics created an instant and easily understood framework and the priority topics for inspection were highlighted.

The training course positioned HOF as part of the mainstream expectation for inspectors and specialists and allowed them to meet the HOF team face-to-face. A follow-up and consolidation visit was offered to each delegate and they were encouraged to use the Inspectors' Toolkit questions for appropriate problem areas. The course helped to establish some boundaries. For instance, for each topic, clear operational guidance was given on how to address the topics, how to recognize where it was appropriate to ask for specialist support, and where and how to enforce the same.

Industry training for HOF "champions" was also pursued, and the IChemE set up a Human Factors professional development course for its members and others. This is now onto its fifth round. The UK Energy Institute has also developed a training package.

HOF INSPECTION

The primary driver for improvement in HOF within HSE and in industry was the use of direct HOF inspection informed by normal inspection experience and safety report assessment. The team also worked out the best way to provide clear recommendations and benchmarks to inspectors and sites. A couple of early requests for support came to me following a fatality and a major accident at top-tier COMAH sites. In both cases, the inspectors concerned felt strongly that HOF and process safety issues were the main contributors but the links were hard to pin down directly. The fatality followed an over-ambitious organizational change—too much change, too fast, without adequately considering the major hazard status of the site. I gathered evidence and we enforced on training and staffing shortfalls. As described above, what we learnt from this and similar interventions led to us developing HSE guidance on organizational change.

The major accident involved a refinery unit explosion. In this case, I set up an intensive five-day organizational culture audit of the whole site to assess their focus on process safety. Unsurprisingly, this was lacking and we enforced a thorough reappraisal.

Offshore inspectors and specialists attended the training course as well. Some raised HOF issues directly as part of their team inspections, and also asked for occasional direct support otherwise. However, it was not until 2008–2009 that they recruited their own HOF specialist inspectors. This had a significant impact on the offshore operators both for inspection and assessment, and greatly increased HOF activity and interest. Getting HOF specialist inspectors onto offshore assets was (and remains) the key, just as it was for onshore.

As part of inspection, we learnt that a careful look is needed for those slippery organizational factors—such as resourcing, workload, planning, and coordination of work—which can topple any company's defenses (Wilkinson and Rycraft, 2014). These are the "wallpaper"—just the way things *are* round here, and because of that they are not always noticed, let alone identified and challenged. So they can and do recur as contributors in later incidents.

SUPPORT FOR ENFORCEMENT

For the HSE, any investigation has to be balanced by the need to consider enforcement, including prosecution. This second focus can conflict with the need to gain all of the lessons quickly and systematically. In the United States, the CSB can act quickly and independently and get its reports out within a year (usually). They can focus wholly on the key lessons and what would prevent similar events, and communicate these quickly. My experience in the United Kingdom has been that once the evidence necessary for a prosecution has been secured there is then less appetite to persevere while looking for the lessons and recommendations to prevent a recurrence. And in looking for evidence to support legal action, this can skew the final investigation picture that emerges. With a few exceptions—generally for major accidents—the investigation reports also remain unpublished.

For example, it took five years to progress the Buncefield case to prosecution (in July 2010, five companies were fined a total of £9.5 million). For the small remaining investigation team left after the main investigation was complete, this time-line actually represented something of a triumph. The "story of the night"—a simple account of the lead-up and the key contributory factors, including many of the human and organizational factors—finally emerged in 2011 (HSE, 2011), once the prosecution was complete. This was after a plethora of more technically focused reports from 2006 onwards, through the Major Incident Investigation Board (MIIB), the Buncefield Standards Task Group (BSTG) and, later, the Process Safety Leadership Group (PSLG). But to make real sense of what happened from an HOF viewpoint you also need to know that those on duty that night were simply not expecting a full tank (and to know why they did not), so that for them the fact that there was no cause for alarm. The "story of the night" is, in my view, still incomplete in its treatment of the basic events of the night and the HOF contributory factors. As with the industry, there also was a tendency by the regulator and other key stakeholders after the accident to focus on the higher-level issues such as culture and process safety leadership without an equal focus on the HOF basics. This was very frustrating for me and the team.

PRISM

A European Union project called PRISM provided funding to develop a network focusing on human factors approaches to major accident safety (The European Process Safety Center [EPSC], 2001–2005). This provided an early opportunity to share the United Kingdom's developing approach through attendance and presentations at workshops and conferences. This led to further exchanges through, for example, the Major Accident Hazards Bureau (MAHB) work, such as Seveso inspectors' Mutual Joint Visits ([MJV]; information and experience exchanges on specific topics including HOF in Portugal in 2007). This resulted—in 2006—in an EPSC award to the team for its contribution within the EC on HOF. The different EC inspection models—and generally lower levels of regulator resource—allowed only limited take-up of the U.K. approach but the exchanges were useful for all parties. And of course many Seveso operators are multinationals working in the United Kingdom, European Union, and elsewhere, and so can use the UK or other approaches to set higher HOF standards elsewhere.

SAFETY CRITICAL TASK ANALYSIS

The idea behind Safety Critical Task Analysis (SCTA) is that the focus should be on what people do (their tasks and activities) and specifically on those that are most hazardous, i.e., critical from a major accident viewpoint. The potential for error is then examined very thoroughly and ideally as part of the design process, not later when tasks are already being done. This approach came out of the nuclear industry originally and is very well established there and in some other high hazard industries (HSE, 1999).

One mistake I made was standing off from full engagement with SCTA and HOF in design in the early years. Despite enforcing SCTA, it was many years before I was fully persuaded of the benefits and I tended to take a very pragmatic procedures-led approach for existing critical tasks (not that this was wrong in itself but it needed more balance in terms of HOF process). Lacking project-based design experience meant that I paid less attention to the detail of good HF/E and HFI and took a more pragmatic route focusing on, for example, control room and HCI design standards and guidance rather than a more structured HF/E project approach. SCTA and HF/E in design have still been slow to take off though there has been significant uptake offshore in recent years.

BARRIERS TO IMPLEMENTATION

There are several internal and external barriers in HOF regulation. The following are just a few key barriers from UK experience.

SENIOR MANAGEMENT SUPPORT

As in all things, fashion trends in safety measures change and, while HOF was warmly welcomed and embraced in HSE over the first three to four years, some of the central and senior management support dropped off later as key individuals moved on, and other organizational priorities intervened. There was a perception that HOF was now "done" and embedded and so it was "business as usual." The problem with HOF is that it never gets "done" because things constantly change and get more complex. It also requires a focus, not just on the "sexier" high level issues such as process safety leadership and culture, but on the nuts and bolts of HOF in design, projects, hazard, and risk analysis and in basic ongoing operating and maintenance arrangements. For me this is necessary HOF "homework" before addressing the higher-level issues like culture. I am not sure that this message has got through though I know the HOF specialists are still working on it and trying to implement the same.

MAGIC BULLETS

Many operators put significant emphasis on culture and behavior initiatives (specifically behavioral modification programs such as behavior based safety [BBS]) and many mistakenly equated these with HOF. BBS was often seen as a "magic bullet" for all HOF and cultural ills (Anderson, 2014). There is still too much focus on the noncompliance side of behavior (and so BBS) rather than on error. There is also too much emphasis on "empty" process safety leadership without the necessary underpinning of the HOF detail in the major accident measures. Without a more integrative HOF approach, such initiatives can never do enough.

For example, there still is insufficient focus on what people do—the safety critical tasks and activities—and the suboptimal and unexamined performance shaping factors that represent a much larger contributor risk for the major hazard sector. There is also a shortfall in integrating HOF effectively into investigation; most investigation methods do not do this adequately (EI, 2008). Even for apparently well-established issues, such as shift handover, there is still work to do. Simple procedures and standards for handover are often still missing or inadequate, even offshore ones—in my experience—in the offshore sector.

ATTITUDES TOWARD HOF

Other barriers included attitudes of "it's all common-sense" and "we are doing these things already, we just call them something else." Generally these views were wrong, judging by results from incidents and inspection. Common-sense is famously poorly distributed and our very human set of cognitive biases skew our individual approach to this. However, there was—and is—truth in that a more structured HOF approach can piggyback on existing arrangements and initiatives. But it is not just a matter of rearrangement; there are invariably some new things to do as well. And although operators were pressed by the HSE to pursue workforce involvement vigorously, they did not then necessarily see that this needed to be active and informed engagement in specific issues such as design projects and hazard/risk analyses. They could also fail to take opportunities of simple user improvements to critical procedures, which could also improve operations in a practical and focused way.

Unsuccessful Initial Applications

A particular risk for the uptake of HOF lies in early applications that are less than fully successful. For example, SCTA—whether done on "mature" tasks or as part of HFI approach in projects—can have mixed results for a few reasons. These include poor inputs to the process, late scheduling within projects, lack of informed and active frontline representation, and overtight project management that is at odds with the desired longer-term operational outcomes, i.e., projects being specified and closed out to the satisfaction of the project team and senior managers but delivering significant long-term user issues (operability and maintainability). If initial efforts to bring HFI into design and projects are less than fully successful, project managers can have their anti-HOF beliefs confirmed. But poor outputs simply reflect the poor inputs to the HFI process. This is not all the fault of project managers; they simply align with their perceptions of the organization's priorities, and their lack of HOF and HFI training. The same is true of wider HFI work.

UNINTENDED CONSEQUENCES

The experience in the United Kingdom is that developing HOF expertise for direct inspection and assessment makes for a pragmatic and reasonably successful approach. However, those HOF specialists need to have good domain experience, i.e., experience of inspecting the industries they are working with. One unintended consequence of recruiting and developing more HOF specialists in the United Kingdom was that for a brief period the center saw their specialists as a "pool" capable of deployment anywhere. The specialists themselves were quickly uncomfortable when deployed to sites and processes where they lacked experience and the experiment quietly failed. They needed time to develop that inspection and industry experience before they could apply their HOF skills and knowledge effectively.

ADVICE TO THOSE WORKING WITH REGULATORS

For those working with regulators, my advice is to get ahead of the curve—in other words use the HOF toolkit or similar resources to assess your own HOF progress and gaps. The regulator is more worried by complacency or failure to be proactive than by identified gaps where an action plan is in place and progressing. It is very uncomfortable for a site to find itself driven by someone else's agenda—the regulator's—and in timescales not of its own choosing. Clearly, this implies some skilling-up on HOF internally and perhaps some pump-priming by external parties. My specific advice is to develop those skills through investigation initially before deploying them more proactively in your auditing, monitoring, and review arrangements, i.e., prospectively before something happens. It also makes sense to integrate HOF into your design and project arrangements, and into your hazard and risk analyses. That way you avoid designing error traps and avoid the need to "work around" rules and procedures to get things done.

If an organization or site is "stuck" then initial enforcement can get things moving and is often welcomed at site-level—but not always. I have experienced sustained outrage: "We have experienced, well-trained, and motivated people, and lots of procedures in place," usually qualified by "if only they would do what they are told...." Regulator persistence has usually turned this around later but this is not productive for either party. For ongoing success though, the relationship with the regulator needs to become a dialogue based on the site's own positive engagement with HOF and the recognition that this runs hand in hand with business success. For example, processes that run smoothly and with minimal unplanned interruptions are generally safer and more profitable. Sites that are in denial can turn the relationship into a wearisome and fruitless game and may even claim to be "managing the regulator." In my experience, if they are, they are often not managing much else. Sites that engage in HOF early in the process find themselves in demand to offer advice to others on how to do this and are usually very positive about the benefits to the business.

CONCLUSION

There has been substantial progress in HOF regulation for major hazards in the UK. This is not perfect, and has not always been as tidy or strategic as it could have been, but there is an ongoing healthy dialogue within and outside HSE, and there is more HOF activity. The HOF discipline is well-established too. There is now a wider focus to include non-major hazard activities and HOF champions or HOF advisers are widely used in industry. Of course industry and processes are constantly changing so the regulator cannot stand still; there are always new challenges to be met. This means engaging with industry and with inspectors and specialists to make sure these developments are picked up and assessed early.

The key, in my view, is taking a user-focused approach, engaging staff in an active and informed way in HOF and safety activities, and focusing on what matters—the control of major accidents and key occupational health, safety, and environmental hazards.

For those working with regulators, the most comfortable relationship is one of ongoing dialogue and partnership on HOF issues (though the regulator must keep some distance to avoid regulatory "capture"). Find out your gaps and prepare a plan to address these. This makes good business sense too; your human "capital" is then free yet within proper boundaries to contribute their innovation and creativity safely to the business.

REFERENCES

Anderson, M. 2014. Behavioural safety and major accident hazards: Magic bullet or shot in the dark? [Online]. HSE. Accessed August 22, 2016. Available at: http://www.hse.gov. uk/humanfactors/topics/magicbullet.pdf.
Energy Institute. 2008. Guidance on investigating and analysing human and organisational factors aspects of incidents and accidents. [Online]. London: Energy Institute. Accessed August 22, 2016. Available at: http://publishing.energyinst.org/topics/ human-and-organisational-factors/guidance-on-investigating-and-analysing-human-and-organisational-factors-aspects-of-incidents-and-accidents.
Health and Safety Executive (HSE). (n.d.[a]). Prosecution resulting from Buncefield explosion. [Online]. Accessed December 15, 2014. Available at: http://www.hse.gov.uk/news/ buncefield/.

Health and Safety Executive (HSE). (n.d.[b]). Human factors top ten. [Online]. Accessed December 15 2014. Available at: http://www.hse.gov.uk/humanfactors/top-ten.htm.

Health and Safety Executive (HSE). 1989. *Human Factors in Industrial Safety*. London: Her Majesty's Stationery Office.

Health and Safety Executive (HSE). 1999. *Reducing Error and Influencing Behaviour*. London: Her Majesty's Stationery Office.

Health and Safety Executive (HSE). 2003a. Major incident investigation report: BP Grangemouth Scotland: 29th May–10th June 2000, Health and Safety Executive. [Online]. Accessed December 15, 2014. Available at: http://www.hse.gov.uk/Comah/bpgrange/index.htm.

Health and Safety Executive (HSE). 2003b. Organisational change and major accident hazards. HSE information sheet, Chemical Information Sheet No CHIS7. [Online]. Sudbury, UK: Health and Safety Executive. HSE Books. Accessed December 15, 2014. Available at: http://www.hse.gov.uk/pubns/chis7.pdf.

Health and Safety Executive (HSE). 2004. Safety Report Assessment Guide: Human Factors via Human factors internet pages. [Online]. Accessed December 15, 2014. Available at: http://www.hse.gov.uk/humanfactors/resources/safety-report-assessment-guide.pdf.

Health and Safety Executive (HSE). 2005. Inspectors human factors toolkit: Human factors in the management of major accident hazards. [Online]. Health and Safety Executive. Accessed August 22, 2016. Available at: http://www.hse.gov.uk/humanfactors/toolkit. htm.

Health and Safety Executive (HSE). 2009. Repositioning human factors. Research Report 758. [Online]. Health and Safety Executive, Norwich, UK: HSE Books. Accessed August 22, 2016. Available at: http://www.hse.gov.uk/research/rrpdf/rr758.pdf.

Health and Safety Executive (HSE). 2011. Buncefield: Why did it happen? The underlying causes of the explosion and fire at the Buncefield oil storage depot, Hemel Hempstead, Hertfordshire on 11 December 2005. [Online]. The Competent Authority. Accessed August 22, 2016. Available at: http://www.hse.gov.uk/comah/buncefield/buncefield-report.pdf.

Hopkins, A. 2000. *Lessons from Longford: The Esso Gas Plant Explosion*. Sydney: CCH Australia.

Seveso II. 2003. Directive 2003/105/EC of the European Parliament and of the Council of 16 December 2003, amending Council Directive 96/82/EC on the control of major-accident hazards involving dangerous substances. *Official Journal of the European Union*. L 345, 31/12/2003, pp. 0097–0105.

Wilkinson, J. and Rycraft, H. 2014. Improving organisational learning: Why don't we learn effectively from incidents and other sources? Paper presented at the *IChemE Hazards 24 Conference*. Edinburgh, Scotland, May 2014.

21 Human Factors and Ergonomics Practice for Consumer Product Design

Differentiating Products by Better Design

Dan Jenkins

CONTENTS

PRACTITIONER SUMMARY

The category of consumer products is incredibly diverse; it ranges from relatively simple products, such as toothbrushes, to complex electronic devices, such as smartphones. Unlike the domains in which human factors/ergonomics (HF/E) was born (i.e., aviation and industrial settings), there are few standards or regulations that mandate the involvement of HF/E. Accordingly, the place for the HF/E practitioner on the development team is far from guaranteed. To gain this place, the additional up-front costs associated with HF/E involvement must be demonstrably offset by an improvement in the quality of the final design and, in turn, the product's market potential. This typically involves defining a measure of performance and using a range of tools and techniques to demonstrate how product features or attributes can influence these values.

WHAT ARE CONSUMER PRODUCTS?

Many of the artefacts that we interact with on a daily basis (smartphones, white goods, cooking equipment, televisions, entertainment systems, cosmetics, etc.) fall under the definition of a consumer product. The term "consumer product" is typically reserved for artefacts that are purchased by individuals or households for use or consumption. Just like any other physical artefact, the design of these products will have a marked impact on their acceptance, the way they are used, and the broader system performance.

WHAT MAKES CONSUMER PRODUCTS DIFFERENT?

High profile events such as the Three Mile Island reactor incident, the Gulf of Mexico oil spill, or more recently the Santiago de Compostela derailment, all appeared as headline news and grabbed the public's attention. Accordingly, the case for HF work is often very clear in controlling risk toward human life and catastrophic damage to the environment. More critically, it is often enshrined in regulations and standards. Consumer products are also associated with hundreds of thousands of harmful cases every year; the Royal Society for the Prevention of Accidents (RoSPA) provides statistics on the many accidents that occur every year from their use. For instance, in 1999, more than 10,000 people were injured by vacuum cleaners and 1,500 by tin openers [Department of Trade and Industry (DTI), 1999]. However, product recalls aside, these injuries alone rarely provide a clear motivation for design change.

The case for including HF/E in consumer product design is typically made based on greater ease of use, customer experience and, ultimately, increased sales revenue. However, the association between usability and sales is far from straightforward (it would be a lot easier to sell HF/E services if it were!). There are many commercially successful products with poor usability that retain market share until user expectations are redefined by a new entrant in the market. Increasingly, these introduced products do not necessarily have to be direct competitors in order to shift user expectations. For example, the mass adoption of smartphones and tablets created a user expectation for more intuitive, visually appealing devices that was capitalized upon by products like the Nest Learning Thermostat (see Figure 21.1a and b). The great news for HF/E practitioners is that arguably, now more than ever, product usability is viewed as a way of differentiating products and, ultimately, increasing sales. This is particularly relevant for consumer products as they are typically purchased by their end users, providing a great opportunity to sway purchase decisions.

Another aspect specific to consumer products is that, unlike many products used in workplace settings, there is less control or no control over the artefact, or wider system, once the product is in the hands of the user. This is fundamental to the work of HF/E practitioners other than expensive product recalls for safety issues; it is not possible to modify designs once a product has been launched. While the design of a panel in a control room can be changed if the layout is confusing, the design of an interface for a food mixer remains fixed.

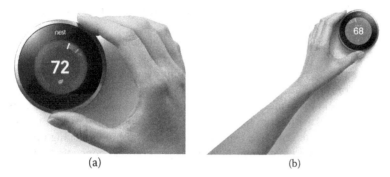

(a)   (b)

FIGURE 21.1   (a) and (b) Nest Learning Thermostat—user experience and usability is frequently cited as the key to the success of this product.

Consumer products can also be differentiated from products used in workplace settings by training and education. While in a workplace setting, it is possible to define a minimum education or training level, the users of consumer products will have a large range of competencies. There is also limited opportunity to modify user behavior through supervision or mandatory training. Furthermore, it would be naïve to think that any instruction manual will be read by a large proportion of users. This, necessarily, shapes how the artefact is designed and developed.

The last key difference relates to the scale and scope of projects. Due to the different scales of projects within a domain, it makes it difficult to make direct comparisons; yet, it is probably fair to say that, when compared to many other domains discussed in this book, HF/E interventions for consumer products need to be relatively low cost. Perhaps, more importantly, these interventions also need to be highly pragmatic. As a result, the focus tends to be on methods that have a very clear and well-established link between analysis activities and positive design changes.

To summarize, consumer products tend to be different from the other products and systems in this book in the following ways:

- There is rarely a regulatory case for HF/E involvement. As such HF/E must be "sold" on its ability to increase the perceived value of the product.
- Consumer products tend to be purchased by their end users. As such, user experience is likely to have a relatively greater sway over purchase choice.
- There is very limited control over the product once it is in use.
- There is no control over who will use it and no minimum training level.
- Interventions need to be focused and highly cost effective.

GETTING HF/E INTO CONSUMER PRODUCTS

As previously stated, unlike automobiles, power stations, and medical devices, there are no clear standards or regulations in place that force the explicit consideration of HF and usability in the consumer product development process. Accordingly, before HF/E specialists are invited to join the design team or give inputs in the design process, a clear business case must often be created to demonstrate how the added cost to the design process will add value.

Within the world of consumer products, the financial case is often the strongest. As such, a compelling argument can often be made by discussing how the performance of a product can be measured and improved upon. Products stand a great chance of commercial success if they better meet the user's primary need (efficacy), allow them to do more (flexibility), are easier to use and understand (usability), provide users with more time (efficiency), or reduce the level of risk (safety). Quantification of each of these aspects can be critical in identifying opportunities for improvement and measuring the success of proposed concepts (reducing risk) (see Figure 21.2).

For in-store assessments and even online reviews, usability can serve as a key differentiator. Consumer products typically exist in the home and are used by many that are excluded from the work environment. This may be because they are young, old, or have some form of sensory, physical, or cognitive impairment. As such, the importance of inclusive design is often clear. Accordingly, most designers and ergonomists are obsessed with creating or refining products so that they allow their users to live more comfortable, enjoyable, and independent lives. To be usable, products must consider the capabilities of their target audience. Focusing on the sensory, physical, and cognitive capabilities of users not only increases the target market, but it also provides a greater user experience, which can form a competitive advantage.

Breaking down each of these elements, as illustrated in Figure 21.3, it is possible to start to quantify performance. Acceptable forces, height, distances, and color combinations can all be measured, defined, and tested.

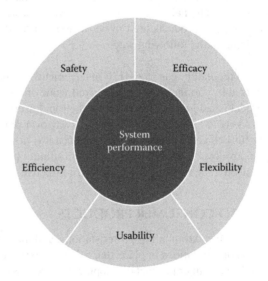

FIGURE 21.2 Example of measures of performance.

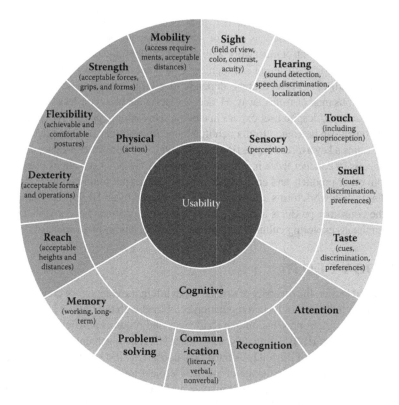

FIGURE 21.3   Factors influencing usability and inclusive design. (*Note*: This is not intended to be a complete list.)

USER BEHAVIOR WITH PRODUCTS IN THE WILD (THE HUMAN SIDE)

People use artefacts in different ways and find their own novel uses for them. This has always been true, of course (see "task–artefact cycle" in Carroll et al., 1991), but has interesting consequences for design of consumer products. For example, when products such as smartphones combine so many functions into a single device (a smartphone can act as an alarm clock, a notepad, an e-book, a light, a camera, an audio recorder, etc.), we cannot assume homogeneity of interaction behavior. This increasing level of complexity should also be considered in the light of short replacement life cycles. Sticking with the example of a mobile phone, some people will choose new models because of ease of interaction or familiarity of interface, while others will adapt to learning new interaction paradigms over and over again in a way that rarely happened before. Others may own a device without ever knowing how to use (or even being aware of) the majority of its possible functions. For HF/E practitioners, it is no longer acceptable to design with a particular persona in mind. To gain the widest appeal, products must be flexible enough to support different modes of operation.

To compound this challenge further, increasingly, consumer products (especially electronic products) are used as part of systems that go beyond the artefact itself. A television remote control must work with digital services and on-screen menus which might be provided by different companies. Computer and mobile device usability, particularly with cloud-hosted apps, depends to a large extent on software running on servers thousands of miles away. Hence, the designer has even less "control" over users' interactions, despite user experience being increasingly prized and valued.

While much can be gained from quantifiable measures of performance (MOPs), user testing remains a key aspect of assessing user behavior. For consumer products, it is often advantageous to conduct these in a home setting or at the point of sale as prior experience from other products and environments shape both expectation and mental models. Accordingly, fieldwork forms a key part of the role played by an HF/E practitioner working in the consumer products domain. Given that many products have a global market, this often means exploring cultural differences and expectations in a variety of countries.

HOW DO WE DO IT?

As discussed earlier, the diversity of artefacts that fall in the category of consumer products, along with the lack of formal regulation combined with the limited level of control over the wider system presents unique challenges and requires a different approach to the consideration of HF/E. Ultimately, the level of HF/E involvement in a design project and the thoroughness of the analysis must be balanced against project time and budget constraints. As such, the first stage of any new project involves planning to develop a suitable approach. Typically, this will involve building an understanding of the project constraints and prioritizing design developments that are expected to have the greatest impact on the usability, and in turn the desirability, of the product.

Figure 21.4 describes a simplified process for designing a product (for simplification, the numerous iterative loops are not shown). The first stage of the process (stage 1) is to define the overall purpose of the product and the wider system in which the product is required to play a role. This activity is normally conducted through engagement with a suitable range of stakeholders including representatives from end users, maintenance, marketing and sales, production, etc.

Once the high-level purpose is established, a range of HF/E tools can be applied in stage 2 to gain an understanding of the people that will be using the product, the context or environment that they will be working in, and the activities that they could be conducting. In turn, by exploring and defining these constraints, MOPs or measures of effectiveness (MOEs) can be set. These should relate to the defined purpose and will be specific to the project. Typically, these metrics would include the kind of measures illustrated in Figure 21.3, such as efficiency, efficacy, flexibility, usability, and safety. Once MOPs have been defined, it is possible to benchmark existing competitors and parallel products (stage 3). This will not only set testable acceptance criteria for the new product, it will allow desirable and undesirable attributes of the products to be identified.

Thus far, there has been little to differentiate the HF/E process, described in Figure 21.4, from many other domains discussed in this book. The more salient departure relates to the next steps (stages 4 and 5), which seek to take the analysis findings and use them to directly inform the design. Normally, some combination of creative

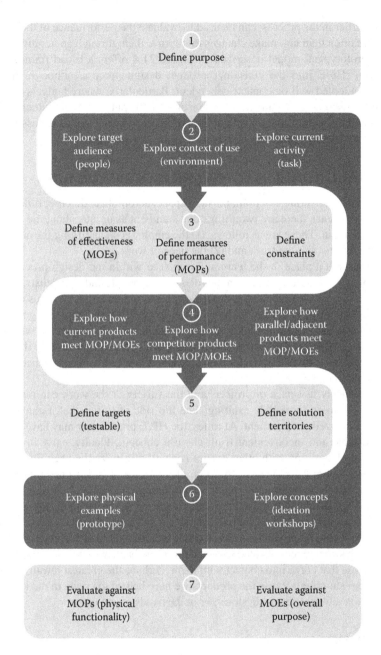

FIGURE 21.4 Example of design process.

workshops and prototyping activities (stage 6) will be employed involving representation from design and engineering disciplines, potentially, alongside end users and other stakeholders. Ideally, insights from the HF/E analysis will be presented back and used to inspire and focus the creative process. Finally, (stage 7) the MOPs and MOEs, defined earlier in the process, can be used to evaluate the performance of the concepts and, more often than not, make changes to improve them through an iterative loop.

The prototypical model described in Figure 21.4 offers a broad framework for integrating HF/E into the consumer product design process, although the exact approach adopted will be context-dependent. Particularly, where budgets and time-lines are constrained, it is imperative to ensure that the purpose of a product is clearly defined and the start of the project and the MOPs correlate, as far as is possible, with market and user acceptance.

HF/E PRACTITIONER IN CONSUMER PRODUCT DESIGN

Having worked in a range of domains (aerospace, automotive, defense, rail, medical, maritime, nuclear), there are two things that stand out as unique about the consumer products domain. These are as follows: (1) the emphasis on justifying the place of the HF/E practitioners in the team and (2) the pace of work.

To gain both a place in the team and influence within the design process, HF/E practitioners are required to take on the roles of salespeople and evangelists. At times, this requires a very methodical approach of presenting data that provides a compelling case. Other times, a case can be made by storytelling and explaining the potential impact on the lives of the end users and stakeholders. Regardless of whether the importance of HF/E is sold to the head or the heart, it helps tremendously if the HF/E practitioner has genuine passion for their role and is able to instill confidence that they will be able to make positive changes.

The relatively fast pace of projects and the variety of the work can make working in this domain incredibly exciting. The flip side of this is that it can also be a highly pressurized environment. At times, the HF/E practitioner may have to deliver uncomfortable and inconvenient truth about a concept. Ideally, early involvement and close collaboration with other disciplines will minimize the need for this, however, that is not always the case. As with all domains, it is not uncommon for the HF/E practitioner to be brought in a fair way through the lifecycle design.

Ultimately, though, for those who chose to work in this domain, the excitement resulting from the variety and the pace massively outweighs the challenges. While the financial success of a product remains a crude indicator of good design, most ergonomists and designers are motivated to create products that have a positive impact on the lives of those that come into contact with them. Perhaps the greatest reward, however, comes from stumbling across the products we have helped to create in the real world, on television advertisements, in stores, or in the hands of end users.

CONCLUSIONS

As discussed in this chapter, there are a number of unique challenges that must be addressed in order to design products that are more usable and desirable, and have the greatest chance of commercial success.

Products must be viewed in the context of their diverse environments of use. They must also be largely intuitive and designed to meet the needs of users with a wide range of ages and abilities (sensory, cognitive, and physical). It is only by focusing on stakeholder values through a semi-structured iterative and integrated development process that these complex challenges can be met.

REFERENCES

Carroll, J.M, Kellog, W., and Rosson M.B. 1991. The task artefact cycle. In: Carroll, J.M. (ed). *Designing Interaction: Psychology at the Human Computer Interface*. New York, NY: Cambridge University Press, pp. 74–102.
Department of Trade and Industry (DTI). 1999. Working for a safer world: 23rd annual report of the home and leisure accident surveillance systems. [Online]. London, UK: Department of Trade and Industry, Consumer Affairs Directorate. Accessed April 07, 2015. Available at: http://www.hassandlass.org.uk/reports/1999.pdf.

22 Human Factors and Ergonomics Practice in Inclusive Design

Making Accessibility Mainstream

Edward Chandler and Phil Day

CONTENTS

PRACTITIONER SUMMARY

The needs of disabled people in the information society have, in many respects, not been fully catered for, in line with technological advances. This means that disabled people can become disenfranchised and risk social isolation. While legislation exists to enable inclusion, this often leads to a minimum viable product scenario where the letter of the law is met but not the spirit. This can result in accessible products and services that are not really usable for disabled people. This chapter demonstrates that the needs of disabled people can be included in mainstream digital and technological solutions, avoiding a "race to the bottom" or building niche solutions for separate user groups. One example and one case study are discussed to provide insight for human factors and ergonomics (HF/E) practitioners to help demonstrate that disabled people can and should be considered within a user-centered design process and not as a group outside the norm or excluded.

INTRODUCTION

Accessibility is often thought of as being concerned with guidelines, standards, or laws. Although these are useful sources of information and can help in raising the profile of accessibility, it is not always a particularly useful emphasis. Instead, accessibility is about people with differing needs and capabilities. A helpful definition is given in ISO 9241 (parts 20 and 171, ISO/IEC 2006 a,b): "Accessibility is the usability of a product, service, environment or facility by people with the widest range of capabilities." The newly formed International Association of Accessibility Professionals (IAAP, 2014) also mentions accessibility as "providing access to information for everyone, regardless of age or ability, so that each individual can realize their full potential."

The World Health Organization estimates that 15% of the world's population lives with some form of disability, and states that rates of disability are increasing (WHO, 2011). This is particularly important as populations are changing. Europe, North America, and some parts of Southeast Asia have populations that are aging. For example, it is estimated that in the United Kingdom, 23% will be age 65 or more by 2034 (UKONS, 2010). A similar trend is predicted across the European Union with approximately 33% predicted to be 65 or above by 2050 (Zaidi, 2008).

By contrast, other regions like South Asia have a population where the majority is young. In India, every third person in urban cities is young (under the age of 34) and is predicted to have a median age of 29 by 2020 (UN-HABITAT, 2013). For those regions with aging populations, however, this poses a significant accessibility challenge. Those above the age of 65 are much more likely to have some form of impairment, with an estimated 20%–50% who are over 65 being disabled, and similarly 66% of disabled people are over 65 (UNESCO, 1995).

Clarkson and Coleman (2010) proposed that we should think in terms of "design for our future selves." This way of thinking helps when promoting the cause of accessibility by making it more personal.

Designing accessibility features from the beginning of the development of a product often results in less costly features that are better integrated and fit for purpose. In addition, some accessibility features can be of benefit to everyone. For example, predictive texting was originally designed as an accessibility aid but was widely adopted as it made sending text messages much easier. In a similar manner, text to speech engines are very useful for blind and partially sighted people, but the higher quality synthetic voices now mean that mobile phones can talk to us, which can sometimes be a benefit or a nuisance depending on the context.

Successfully designing these accessibility features requires HF/E practitioners to combine expert knowledge with the valuable insights offered by involving end users, particularly as participants in evaluations of any solution with possible accessibility benefits. It is also worth noting that we all have varying levels of capability, and therefore, it is often unhelpful to think in terms of designing for a distinct group of "disabled people" rather than thinking in terms of designing for different capability levels.

Below, we describe two case studies to illustrate issues for HF/E practice; the inclusion of accessibility features into mobile telephones enabling businesses to reach more people, and a novel input device for self-service use called the Universal Navigator (uNav for short).

MOBILE ACCESSIBILITY AND A METHOD FOR DELIVERING ACCESSIBLE SERVICES

The mobile phone is an example of ubiquitous technology improving the lives of disabled people, including those who are deaf or hard of hearing, those who are blind or partially sighted, and in some cases, those with cognitive impairments. This has been achieved by the inclusion of accessibility features into the mainstream product as the mobile industry has matured. An early example was the Short Message Service (SMS), which was not designed specifically for deaf people but enabled them to communicate with others (deaf or not) in the same way as nondisabled people, breaking down barriers and revolutionizing communication (Power and Power, 2004). In 2005, Nokia phones were made more accessible to blind and partially sighted people as third-party software could be installed on their high-end phones. However, this software is considered niche as it only served blind and partially sighted people, and had to be purchased and installed post-purchase (therefore needing intervention by a sighted individual).

Fast-forward to 2009 (two years after the initial release of the iPhone), and Apple released the iPhone 3GS, which included built-in accessibility for several user groups, including those who are visually challenged. This meant that the iPhone was more accessible "out of the box" for disabled people. All of these features were available from the settings menu regardless of whether you needed them or not—they were part of the core operating system (see Figure 22.1). This allowed people to interact with the device in a way that met their needs and capabilities. This has changed the market (and user expectations) so much that the other mobile phone software providers (like Google Android and Windows Phone) have also introduced built-in accessibility features.

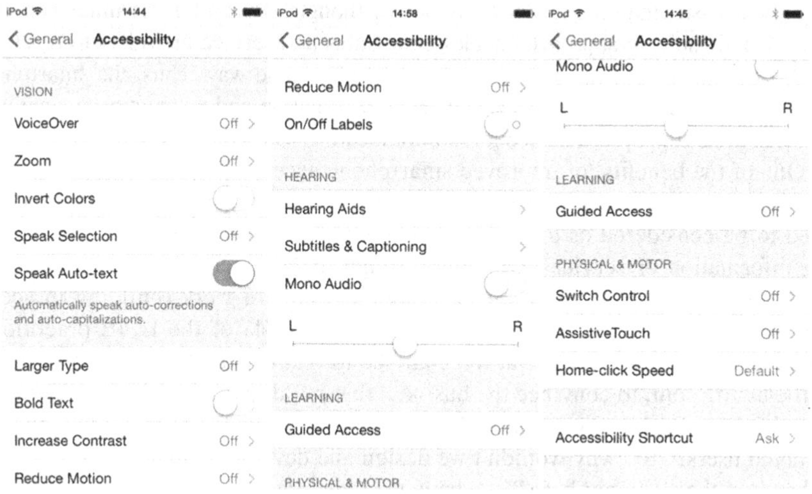

FIGURE 22.1   Screenshots of the accessibility features available on Apple devices.

By building accessibility into its products and software, Apple demonstrated that accessibility does not have to be an expensive add-on or require specialist devices to be built for different user groups. Both Apple and Google have produced guidance to make apps and services more accessible to enable developers to accommodate disabled people more easily (Apple, 2015; Google, 2015).

Disabled people can therefore take advantage of many of the third-party apps that are available in the same way as nondisabled people. Some of these, such as navigation, shopping, and banking apps, can offer particular benefit to blind and partially sighted people and, in some cases, to those with cognitive impairments, helping them to live an independent life. Through smartphones, service providers can more effectively engage with their customers (disabled and nondisabled) via apps that integrate into their services, rather than bespoke specialized services for specific groups that are often prohibitively expensive.

An example of this can be taken from the work that John Lewis (a British chain of stores) have done alongside RNIB to make their app more accessible for disabled people (RNIB, 2015a,b). One of the core themes of RNIB's 2009–2016 strategy was to create a more inclusive society by reaching out and working with businesses to showcase examples of improved inclusion. RNIB worked with John Lewis to show them how to make their app more accessible, to enable disabled people to shop more independently (RNIB, 2014). For this type of engagement, the HF/E practitioner needs an initial focus on understanding and showing the "state of the nation" (i.e., why is this not as accessible as it could be) and then needs to work alongside the users to provide solutions.. Within accessibility in the digital space, to avoid accessible and not-usable syndrome, the need to develop and write accessible codes goes hand in hand with the interaction and information that a user needs when they cannot see the screen. This relies on the HF/E practitioner to translate the visual interface and nonaccessible interaction into a meaningful and accessible interface and interaction. One of the biggest lessons learnt on this project was that accessibility could and should become business as usual and it was easier than the business previously thought (RNIB UK Channel YouTube, 2015b). This holistic approach enables companies and service providers to reach the largest customer market possible, in the most affordable way. Thus, the smartphone has become an important tool for gathering information and connecting people with services rather than just a device for communicating with other people.

One of the benefits (of improved smartphone accessibility enabling delivery of services and information) for the HF/E practitioner is that disabled people do not need to be considered as a niche group needing specialist solutions. When making information or services available through mobile technologies, the needs of disabled people can easily be incorporated as the hard work (building in accessibility) has already been done. At this point, the role of the HF/E practitioner turns from solution identifier at the coalface to that of an advocate and strategist in the boardroom, to convince the business that building in accessibility enhances their business and customer base. The discussion moves from "should we include disabled users?" to "why wouldn't we design and develop in an accessible manner to broaden the customer base?"—a truly user-centered design approach. Once past that point, the HF/E practitioner turns their attention back to details such as the design of the app or responsive website which still needs to include all elements of good design in terms of simple navigation, structure, information architecture, and interaction design in a user-centered design process. The HF/E practitioner should be doing this anyway, so disabled users become a user group within the design rather than separate and niche.

One of the unexpected outcomes of improved smartphone accessibility has been the shift in attitudes toward touch-screen accessibility for personal devices. Previously, this technology had been dismissed as largely inaccessible (especially for blind and partially sighted people) but the iPhone has changed these attitudes. The industry is now revising touch-screen accessibility in light of this innovation. This is an important milestone as it shows how innovation can change attitudes, and that HF/E practitioners and other stakeholders should not accept the status quo in building an equal experience. That being said, a universally recognized accessible experience for touch-screen self-service kiosks is yet to be realized, but this is being researched.

UNIVERSAL NAVIGATOR—AN ACCESSIBLE USER-CENTERED SOLUTION FOR SELF-SERVICE

The uNav (see Figure 22.2) is a device designed to offer an accessible complement to kiosks and automated teller machines (ATMs) equipped with touch screens. This provides benefits to those who are visually impaired and also those with limited reach (such as people in wheelchairs). There is a trend toward using touch screens more widely for these types of devices, and therefore, a need to find an accessible complement as touch screens are not accessible to blind people without some additional features. The intent was to balance the need for tactilely discernible, easy to reach controls, with the desire to offer an attractive, mass-market aesthetic, making this device something that anyone might wish to use.

The project started with a user-centered design ethos with a multidisciplinary team of in-house experts, including industrial designers, usability specialists, accessibility specialists, and interaction designers. The team created several early ideas with different input modalities (sliding, pushing, rotating, clicking) being considered. This led to a larger number of ideas, which were evaluated based on expert review and reduced to a smaller set of concepts. Low-fidelity functional prototypes were

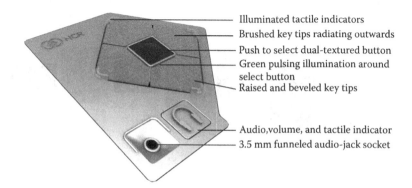

Illuminated tactile indicators
Brushed key tips radiating outwards
Push to select dual-textured button
Green pulsing illumination around select button
Raised and beveled key tips

Audio, volume, and tactile indicator
3.5 mm funneled audio-jack socket

FIGURE 22.2 The Universal Navigator showing key features.

then produced and tested with 25 sighted participants to identify the most promising type of input device (Day et al., 2012). The preferred concept from this testing was then refined and a higher fidelity prototype was produced.

At this stage, RNIB was approached to give input, and collaboration began focusing on the speech output that would accompany such a device. A larger evaluation with 48 blind and partially sighted people was then conducted in the United Kingdom (Day et al., 2013), with additional testing also being conducted in the United States where 20 participants with varying physical and sensory impairments, including reduced mobility, dexterity, and visual acuity (Day et al., 2014), were tested. This example illustrates how multiple stakeholders from industry can successfully collaborate with experts in charity and consultancy roles.

GUIDELINES AND LEGISLATION: MOVING TOWARD HARMONIZATION AND COMPLIANCE

Accessibility is covered in a number of standards, guidelines, and laws around the world. Although many of these documents have similar aims (such as making sure that those who cannot walk or see can use the product or service), the detail varies significantly between them.

For example, when considering a physical self-service product like an ATM, a number of standards around the world stipulate heights at which the interface should be placed. This is to ensure that the ATM is usable by those in wheelchairs, as well as those of small and large stature. Despite this common aim the minimum and maximum heights vary significantly between countries. Figure 22.3 shows heights of people in different countries, with the dark area representing minimum to maximum heights for the interface. Some countries also separate out the interface into core and noncore elements (for primary and secondary tasks); a lighter shaded area represents this. This diagram was derived by the authors from height data published in the standards or guidelines for the named countries.

This points to a need for closer harmonization between regions and countries. There are some signs of hope with the latest European and US standard makers attempting to align more closely, such as the latest update of the US Section 508 of the Rehabilitation Act (US Access Board, 2011), and the European standard EN 301

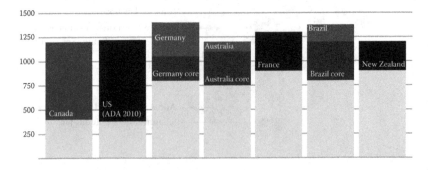

FIGURE 22.3   Average reach heights for various countries.

549, "Accessibility requirements suitable for public procurement of ICT products and services in Europe" (ETSI, 2014). Both standards concern accessibility requirements for public procurement of information and communications technology (ICT).

Some countries have very specific legislation that has brought accessibility into the spotlight. For instance, the Americans with Disabilities Act (ADA) (US Department of Justice, 2010) resulted in an increased awareness of accessibility matters, and in some cases business change as a result of the same. However, legislation on its own is not the answer. It can lead to the application of minimum standards on products and, therefore, a race to the bottom. HF/E practitioners are essential in giving input to new standards, in interpreting those standards and applying them to a specific context or problem area, and also to continue to research methods of going beyond what is currently mandated. The potential market for solutions that offer accessibility features is growing as screens. This provides benefits, and therefore, this presents a useful business opportunity along with a beneficial area of application for practitioners.

On this note, the idea of passing or failing an accessibility standard is often discussed, but this is not always helpful, or even possible. For example, a hardware product may have built-in accessibility features but if it is installed at the wrong height or above the stairs it is still inaccessible. Instead, the standard should be used as a way of demonstrating good practice. Also, making a solution accessible requires a more holistic approach than just taking into consideration requirements from a single document and working through them mechanistically.

A more recent development is the move toward wider adoption of the Web Content Accessibility Guidelines (WCAG), which was originally derived for websites (Caldwell et al., 2008). The requirements from this guideline are generic enough that they can be applied to software in general, and this has been the case in both EN 301 549 (ETSI, 2014) and US Section 508 (US Access Board, 2011). This may as well be a trend that is more widely applicable across the HF/E domain; that of attempting to harmonize guidelines and standards, and where appropriate, to incorporate de facto standards or working practice rather than reinventing new standards just for the sake of it.

BRINGING EVERYTHING TOGETHER: THE USER-CENTERED APPROACH

Through the two case studies we have tried to demonstrate what success looks like for disabled people when accessibility is embodied in part of the core process.

Apple's initial foray into accessibility was to maintain US Government contracts. They could have stopped at their computing solutions, rather than embedding accessibility into their entire range (iPods, iPads, iPhones, as well as the Macs). This highlights how seriously Apple takes accessibility. Companies can start with good intentions and processes, but this can go astray as processes, teams, and working practice changes. HF/E practitioners should promote the continuation of user-centered design processes, including testing with end users, even in companies where this has been achieved in the past.

There are dangers in the quantification of accessibility via standards and guidelines. While this is possible to a certain degree, it often cuts out the disabled end user and reduces accessibility in reaching a goal or accessing information instead of the usability of the method of getting there. The ADA, Section 508 (in the United States) and WCAG (from the Web Accessibility Initiative [WAI]) have all quantified what accessible is or is not, and have completely removed the user from this. Companies seek to comply (meet the minimum) on a technical level rather than deliver systems and services that people with disabilities can use with ease. To illustrate this, Figure 22.4 highlights an accessible route to the platform. Note that it is a very long way to the other side. The amount of effort required by a wheelchair user to cross the platform using both ramps has not been accounted for yet.

Experience suggests that increasingly accessible and usable products and services increase consumer expectations. With rising expectations amongst disabled users, the minimum requirement of legislation and guidelines will be insufficient to meet these expectations. Meeting these expectations will therefore require more involvement of disabled users in the design process, and this requirement will intensify with aging populations.

There is anecdotal evidence that business that offer good customer services in-store are more likely to receive repeat business by disabled people. Businesses that provide in-store support enable disabled people to have a less stressful shopping experience (RNIB, 2015a, b). These include providing assistance around and through the store, selecting items and other accessible features (step-free access, accessible toilets, and checkouts). Once an expectation for a good in-store experience (such as John Lewis and Marks & Spencer) has been established, disabled people will naturally take that expectation forward to their digital sales channels. From the business side, this means mobile apps and responsive web design in addition to "click and collect" services that are accessible, enabling consumers to interact in a way that they prefer through an "omnichannel" approach (in-store, online, and on mobile). Therefore, to maintain customer loyalty and increase customer numbers, businesses need to provide holistic customer-centric services that meet users' needs and expectations across a multitude of touch-points (i.e., any point in

FIGURE 22.4 An "accessible" route to the platform.

which a customer needs to interact with any part of the business). What this means for HF/E practitioners is that a systems thinking approach needs to be applied in order to design and deliver accessible and usable services and solutions across all of the customer touch-points. Service design as a discrete discipline embodies this holistic approach, but the systems thinking approach adopted by HF/E practitioners draws on multidisciplinary skills, including physiology, psychology, engineering, and design.

Designing successful solutions for disabled people is not just for accessibility specialists. A multidisciplinary team of industrial designers, interaction designers, and HF/E practitioners (including accessibility and user experience specialists) is needed to deliver inclusive solutions that meet the needs of as many users as possible (including those with or without disabilities).

Buxton (2007) advocated a design space that enables design concepts to "bake in" by being edited, commented on, and reviewed by the wider team. By doing this, any issues can be identified early and rectified before the solution is put in front of users. The National Capital Region (NCR) team uses a design space to allow experts to work on designs, and then conducts user testing to confirm that designs meet users' needs. The end result is that uNav shows how a user-centered design approach within a multidisciplinary team can produce a meaningful and engaging solution that is both accessible and usable.

CONCLUSIONS

Accessibility, and in particular designing to meet the needs of disabled people, is at the core of the HF/E discipline. When the question "who are our users/customers?" is asked, the user group will invariably include people with varying levels of physical and cognitive capabilities. People with disabilities should therefore be included in your user groups during any discovery and user needs gathering phases.

The growing body of literature, standards, and expert reviews is useful, but is no substitute for engaging with end users throughout the development process, including testing with participants who have a wide range of capabilities and needs. In doing so, we not only provide designs that cater to their needs, but we also cater to the possible future needs of ourselves and our changing society. Accessibility is essential for the future and strikes at the heart of meeting user needs.

REFERENCES

Apple. 2015. Accessibility for developers. [Online]. Accessed March 14, 2016. Available at: https://developer.apple.com/accessibility/ios/.

Buxton, W. 2007. *Sketching User Experiences Getting the Design Right and the Right Design*. Amsterdam, The Netherlands: Elsevier/Morgan Kaufmann.

Caldwell, B., Cooper, M., Reid, L.G., and Vanderheiden, G. 2008. Web content accessibility guidelines (WCAG) 2.0. [Online]. W3C World Wide Web Consortium December 2008. Accessed March 4, 2016. Available at: http://www.w3.org/TR/WCAG20/.

Clarkson, P.J. and Coleman, R. 2010. Inclusive design (editorial). *Journal of Engineering Design*. 21(**2–3**), 127–129.

Day, P.N., Chandler, E., Colley, A., Carlisle, M., Riley, C., Rohan, C., et al. 2012. The universal navigator: A proposed accessible alternative to touch screens for self-service. In: Anderson, M. (ed.). *Contemporary Ergonomics and Human Factors 2012*. London: Taylor and Francis, pp. 31–38.

Day, P.N., Carlisle, M., Chandler, E., and Ferguson, G. 2013. Evaluating the universal navigator with blind and partially sighted consumers. In: Anderson, M. (ed.). *Contemporary Ergonomics and Human Factors 2013*. London: Taylor and Francis, pp. 355–362.

Day, P.N., Johnson, J.P., Carlisle, M., and Ferguson, G. 2014. Evaluating the universal navigator with consumers with reduced mobility, dexterity & visual acuity. In: Sharples, S. and Shorrock, S. (Eds.),*Contemporary Ergonomics and Human Factors 2014*. London: Taylor and Francis, pp. 183–190.

ETSI. 2014. ETSI/CEN/CENELEC. EN 301 549 V1.1.1. Accessibility requirements suitable for public procurement of ICT products and services in Europe.

Google. 2015. Accessibility. [Online]. Accessed March 14, 2016. Available at: https://developer.android.com/guide/topics/ui/accessibility/index.html.

ISO/IEC. 2006a. *ISO/IEC 9241-20. Ergonomics of Human-System Interaction–Part 20: Accessibility Guidelines for Information/Communication Technology (ICT) Equipment and Services*. Geneva, Switzerland: ISO/IEC.

ISO/IEC. 2006b. *ISO/IEC 9241-171. Ergonomics of Human-System Interaction–Part 171: Guidance on Software Accessibility*. Geneva, Switzerland: ISO/IEC.

IAAP. 2014. About IAAP. [Online]. Accessed March 14, 2016. Available at: http://www.accessibilityassociation.org/content.asp?contentid=1.

Power, M.R. and Power, D. 2004. Everyone here speaks TXT: Deaf people using SMS in Australia and the rest of the world. *Journal of Deaf Studies and Deaf Education*. 9(3), 333–343.

RNIB. 2014. RNIB *Group Annual Report and Financial Statements 2013/14*. London: RNIB.

RNIB. 2015a. Shopping provider success stories. [Online]. Accessed March 4, 2016. Available at: http://www.rnib.org.uk/information-everyday-living-shopping/shopping-provider-success-stories.

RNIB. 2015b. John Lewis and RNIB Business create accessible technology. [Online]. Accessed March 4, 2016. Available at: https://www.youtube.com/watch?v=C498rfqB7X8.

UKONS. 2010. Older People's Day 2010 Statistical Bulletin. UK Office for National Statistics. [Online]. Accessed March 4, 2016. Available at: http://www.ons.gov.uk/ons/rel/mortality-ageing/focuson-older-people/older-people-s-day-2010/focus-on-older-people.pdf.

UNESCO. 1995. Overcoming obstacles to the integration of disabled people, (disability awareness in action). [Online]. Accessed March 14, 2016. Available at: http://www.daa.org.uk/uploads/pdf/Overcoming%20Obstacles.pdf.

UN-HABITAT. 2013. State of the urban youth, India 2012: employment, livelihoods, skills. Mumbai, India: UN-HABITAT Global Urban Youth Research Network/IRIS Knowledge Foundation. [Online]. Accessed March 4, 2016. Available at: http://www.esocialsciences.org/general/a201341118517_19.pdf.

US Access Board. 2011. *Information and communication technology (ICT) standards and guidelines*. Advance Notice of Proposed Rulemaking December 2011. Section 508 of the Rehabilitation Act of 1973, as amended (29 U.S.C. 794d).

US Department of Justice. 2010. 2010 ADA standards for accessible design. [Online]. Accessed March 4, 2016. Available at: http://www.ada.gov/regs2010/2010ADAStandards/2010ADAstandards.htm.

World Health Organization (WHO). 2011. World report on disability, 2011 edition. [Online]. World Health Organization. Accessed March 4, 2016. Available at: http://whqlibdoc. who.int/publications/2011/9789240685215_eng.pdf.

Zaidi, A. 2008. Features and challenges of population ageing. Policy Brief March (1) 2008. [Online]. Vienna, Austria: European Centre for Social Welfare Policy and Research. Accessed March 4, 2016]. Available at: http://www.euro.centre.org/ data/1204800003_27721.pdf.

23 Selling "Ergonomic" Products
Different Approaches for Different Applications

Guy Osmond

CONTENTS

PRACTITIONER SUMMARY

This chapter considers issues in the selling of "ergonomic" products, emphasizing different approaches for different applications: individual, departmental, and organizational. The supply of "ergonomic" products is most effective in individual cases to address disabilities or musculoskeletal problems. In dealing with new activity or specific departmental needs, "ergonomic" product vendors can be highly effective but are often introduced after problems have started to arise. In enterprise installations, ergonomics may be sidelined by aesthetic considerations. IT channels are now involved in the supply chain of "ergonomic" products but may have no idea about their function of purpose.

INTRODUCTION

Almost every supplier of office furniture, IT, industrial handling equipment, tools, clothing, vehicles, and, indeed, any product designed for human interaction, will offer "ergonomic" products or refer to the "ergonomic" features of their products. To limit the scope of this chapter, I shall be specifically discussing "ergonomic products" designed to enhance the comfort and/or productivity of workers using computers, laptops, tablets, and other IT devices at work, in their homes, or in mobile working situations such as hotels, airports, trains, and coffee bars. Geographically, my experience has been in the United Kingdom, though many of the issues can be generalized internationally.

In the context of office furniture, workstation accessories, and (especially) seating, the word "ergonomic" is widely used and widely misrepresented. Many products are marketed as "ergonomic" in much the same unregulated way that foods are labeled as "low-fat" and all sorts of products are labeled "eco" or "green." British ergonomists tend to think that no product can be "ergonomic" in and of itself: it must be given a context and an application before the situation can be considered ergonomic. In US English, this differentiation is often not made and products are routinely described as "ergonomic," frequently by "ergonomic consultants."

For the purposes of this chapter, I shall refer to "ergonomic" products (using inverted commas), applying the common parlance, for any product specifically designed with the intention of enhancing the comfort and/or productivity of workers.

"ERGONOMIC" PRODUCTS IN THE WORKPLACE

There are three typical scenarios in which ergonomics and "ergonomic" products become part of the workplace conversation: (1) individuals with disabilities or musculoskeletal problems (individual), (2) specific departmental or operational needs, such as hospital sonographers or office personnel being reassigned as homeworkers (departmental), and (3) building refits or enterprise relocations (organizational). These three scenarios are outlined in the following.

INDIVIDUAL

Compared to the number of suppliers of office furniture, the number of companies specializing in this sector is very small and the majority of these specialist companies are quite small (usually with fewer than 25 employees). However, the knowledge and experience within them is substantial (Williams and Haslam, 2006). This is also the sector in which the majority of money specifically focused on "ergonomic" products is spent.

Typically, the process starts with an assessment of the individual. Often, this is through the employer's own Display Screen Equipment (DSE) assessment programme, which results in an escalation because the needs of the individual cannot be met through in-house resources. A third-party escalated assessment will then be required. This might be through an occupational health contractor, government programmes (e.g., "Access to Work" [AtW] in the United Kingdom) or a Health and Fitness Education (HFE) or Health and Safety consultancy.

Another common option is for a vendor of "ergonomic" products to carry out the assessment. This latter practice appears to be fairly specific to the United Kingdom. In other countries, it is deemed necessary to separate the assessment process from the supply of products. The logic of this is obvious and the UK model clearly has potential for the supplier to abuse the assessment opportunity. However, in a competitive market, suppliers know how important it is to develop a relationship of trust with their client. Furthermore, knowledge of what products are available and how they might address an individual's needs can often be more valuable than a "medical" understanding of the condition, placing the vendor in a good position to carry out an effective assessment (Williams and Haslam, 2006). While in no way underestimating the contribution of, for example, an occupational health professional, it is most important that they keep up-to-date with product information as well as their own medical knowledge. To this end, many consultants and occupational health professionals have their own preferred "ergonomic" product vendor whom they contact for specific product knowledge and intervention advice.

Once the assessment has been completed, a report is written and, from this, a product specification is created. At this point, the employer may go to their product supplier of choice or seek multiple quotations (as is required by the UK AtW scheme).

Finally, once the products have been purchased, the goods are delivered, set up, and configured for the user, who is also trained in their use. This final stage is crucial to ensure maximum benefit for the user.

Customer feedback from this process typically demonstrates very satisfied clients. The nature of each user's needs means they often have a lot to gain from such interventions and this "ergonomics at the coalface" approach is usually highly effective.

DEPARTMENTAL

The scenario is very different in departmental applications. Following are some typical examples of the same:

In an ideal world, employers should start to think about ergonomics in the home as part of the preparation for introducing home working. Identifying, at the very least, a laptop stand with a separate keyboard and mouse seems obvious. More enlightened employers ensure that home work areas are assessed and needs such as desks, chairs, fire extinguishers, and lighting are all considered. All too often, however, the conversation starts after the homeworker programme is already in place.

Similarly, an employer issuing tablet computers may give no thought to the ergonomics of their use and, if so, the first time it is highlighted is when the occupational health team starts to get complaints from users about neck problems.

A hospital may find that personnel in the sonography department are going off sick with shoulder and neck problems and decide to increase working hours for the remaining personnel in order to shorten appointment delays and meet performance targets. As might be expected, this simply results in more sonographers taking time off because of sickness. Only then does the employer examine the ergonomics of the sonography process.

These examples are all typical of the sort of departmental issues that arise. Regrettably, the call for human factors and ergonomics (HF/E) intervention is all too often when things have gone wrong rather than at the planning and designing stage. Under these circumstances, a retrofit solution may well have to be a compromise.

Barriers to early interventions are often caused by silo cultures within organizations and interested parties not being involved early enough. For example, home worker rollouts are often initiated by the estates or premises department and, while HR and IT are always involved, occupational health, well-being, or health and safety personnel are either not asked to contribute, or their recommendations are seen as costs rather than investment. When the business model for such a project focuses on short-term cost savings, it is very difficult to make the case for ergonomics investment in the present, to save significant absenteeism and presenteeism costs in the future.

By contrast, when involved early enough, the "ergonomic" product supplier can add value at all stages of a project rollout. Working with the occupational health and/or health and safety team and/or their in-house HF/E professional or external consultant, possible ergonomics or posture issues can be identified early and addressed. In a holistic approach to the rollout, HF/E considerations will also be included. Psychosocial factors for remote workers will be accommodated in training and coaching, alongside the issues of changes to traditional management techniques.

ORGANIZATIONAL

For large projects, ergonomics is nearly always discussed but interested parties tend to be more selective about when and how it comes into the conversation. Some architects and designers love to see rows of uniform desks and chairs and they resist the ergonomist's desire to disrupt this with greater adjustability and flexibility. Often, "the look" is a critical consideration and the best looking chair or monitor arm is not always the most "ergonomic" product.

In recent years, bench-desking has predominated. This concept is very cost-effective because, instead of individual desks, multiple workstations share a common framework. However, this also leads to rooms full of identical desks at identical heights with, usually, no scope for personalization. While some designs allow limited adjustment, this concept is particularly unsuitable for tall users who are obliged to use a desk which is too low for them and cannot be raised. Very often, these workstations are also used in hot-desking environments where identical layouts result in compromise for many users, as reported in much recent research and comments in the media (e.g., Landau, 2014).

To address the need for different types of work area for different types of work activity, designers incorporate multiple zones—including breakout and touchdown areas, quiet zones, meeting rooms, pods, "phone boxes," canteens, and bistro-style settings in addition to the traditional work desk. The comparatively new concept of "activity-based working" is a development of this model. However, in all but the most carefully considered designs, there are areas that fail to attract an appropriate amount of use.

Providers of enterprise installations are, generally, very different operations from the specialist businesses dealing with individuals. The levels of HF/E knowledge within the former, while variable, are usually much lower and, typically, have been learned from the furniture manufacturers. This therefore leads to inconsistencies in the "ergonomics message" received by employers.

AESTHETICS AND EASE OF USE OF "ERGONOMIC" PRODUCTS

In the past, good "ergonomic" products in the workplace were perceived as industrial or agricultural in appearance, but good aesthetic design can now be compatible with good ergonomics. For example, manufacturers of input devices (keyboards, mice, etc.) now aspire to the "Apple look and feel," even incorporating this into their packaging.

Most recently, ease of use has made significant progress. While intuitive adjustment and setup is an essential part of good design, there has often been a compromise in the past. For example, the most adjustable chairs have not necessarily had the most accessible controls. Now, clearly labeled, easily (and instantly) accessible adjustments are a priority and growing numbers of products achieve this.

Sit–stand working has been a particular area in which simple, obvious, and instant adjustment has become the norm. It is also an area where employers naturally turn to a specialist vendor rather than their general furniture supplier. It is common for organizations to want to start with a small trial programme, often with adapter units on existing desks rather than replacing legacy furniture with the full sit–stand product. In such circumstances, the specialist supplier's understanding of the HF/E issues and awareness of different products will provide the optimum solution and such suppliers can also give guidance about managing the trial process.

OTHER PARTIES IN THE SUPPLY OF "ERGONOMIC" PRODUCTS

It is more than 20 years since the original Health and Safety (Display Screen Equipment) Regulations 1992 in the United Kingdom, and the way we work and interact with different types of computers have changed dramatically in that time. In an effort to manage the risk of work-related upper-limb disorders (WRULDs), a plethora of computer keyboards, mice, trackballs, touch pads, and other pointing devices have appeared. Many of these are now available through the IT distribution network and, because they plug into a computer, employers often order them through their IT department. The culture and operational methods of the IT channel are very different from the ergonomics community and product knowledge, where it exists at all, is usually confined to the technical specification of the products. Typically, therefore, "ergonomic" product vendors have developed ways of working with the IT channel to provide products through the most efficient route but also to provide the necessary support required to ensure that users get the right products and knowledge to use them.

THE INTERNET

Many "ergonomic" products can be purchased on the Internet and many millions of words have been written about workplace ergonomics and "ergonomic" products in blogs and online articles. Although there are many practitioners providing high-quality, interesting, and informative content, the scope of someone's online influence is not necessarily proportional to their knowledge or the objectivity of their musings. Many "ergonomic" product vendors are active on social media and comment on or respond to such articles where a more informed perspective is required.

CONCLUSION

There is a substantial demand for "ergonomic" products and a strong cohort of capable and knowledgeable organizations to deliver them, together with the training, support, and customer service necessary for optimal interventions. However, the understanding of ergonomics, its role, and its benefits is inconsistent, and many outside the ergonomics community have only a superficial understanding of what ergonomics entails and what it can achieve.

Individuals with disabilities or musculoskeletal problems have probably benefited most from an ergonomics intervention and probably have the greatest understanding of "what is ergonomics?" At the other end of the scale, personnel involved in major enterprise installations may be dependent on information from furniture manufacturers' representatives, which may be neither impartial nor very informed.

REFERENCES

Landau, P. 2014. Open-plan offices can be bad for your health. [Online]. *The Guardian*. 29 September. Accessed March 4, 2016. Available at: http://www.theguardian.com/money/work-blog/2014/sep/29/open-plan-office-health-productivity.
Williams, C. and Haslam, R. 2006. Ergonomics advisors—A homogeneous group? In: Bust, P.D. (ed). *Contemporary Ergonomics 2006*. London, UK: Taylor & Francis, pp. 117–121.

24 Human Factors and Ergonomics Practice in User Experience

Demonstrating Value in a Fast-Growing Field

Lisa Duddington

CONTENTS

PRACTITIONER SUMMARY

User experience (UX) is a fast-growing discipline that is widely recognized. The field is unregulated and, as demand has grown, there has been a decrease in the level of skill and experience of the average practitioner. In digital projects, there are often limited budgets and time constraints. It is rare that a practitioner is able to use all the tools and techniques of the ideal user-centered design (UCD) process. Also, unfortunately, research is often the first thing to be downsized or removed. This has the effect of reducing the connection with the discipline of human factors and ergonomics (HF/E). As we move away from mouse control to touch, gesture, voice control, and virtual reality, there will, however, be a greater demand for the skills of HF/E.

INTRODUCTION

"UX" is a relatively new term that is most commonly used to describe HF/E as applied to digital platforms and the interactions that users have with digital interfaces, such as websites, software, and apps. There is little agreement on the precise definition. While UX is mostly seen as the interaction a user has with digital platforms, many professionals are now recognizing that the experience a person has with a company goes much further than this. Having a good user experience has little impact if your other touch-points do not also provide a good experience. So some agencies now prefer to use the term "customer experience" to account for the need to focus on the 360-degree experience a customer has with a brand. For example, if a customer has a poor experience when calling your helpline or visiting your retail store, it does not matter how good your website is, they are unlikely to return and are likely to tell others about their negative experience.

SHIFT FROM USABILITY TO UX

UX has always existed but under different terminologies: human–computer interaction, computing and psychology, usability, systems analysis, etc., with elements of HF/E. When I started working in this field, I was a usability specialist, and a large part of a professional's role then was educating others, building relationships, and gaining respect, not in the least to have their recommendations actioned.

UX is an umbrella term that pulls together everything that might affect how a person behaves and feels when they interact with a product or service. The term gained increasing popularity from the mid-2000s onwards. It had greater appeal than usability, and controversially, in 2012, the Usability Professionals Association (UPA) changed their name to the User Experience Professionals Association.

The term UX (as it is more often known) has had both positive and negative impacts. Positively, it has had a huge impact on awareness of both UX and usability. For a professional, there is less time spent on educating now and more on carrying out tasks. There are many more practitioners, events to attend, and books available to read. There is also a greater respect for your opinion, as engineers now understand the value and benefits it brings to them. In fact even developers have begun to change their job titles to include UX—you can now find "UX Developers."

As UX is such a broad term, it has attracted people from lots of different industries. A team of UX specialists might consist of people with backgrounds in psychology, marketing, visual design, information architecture, development, and many more. I find this a huge positive. When a team is made up of people with different sets of skills it can be stronger and more creative. However, the broad appeal of UX is also its downside. It is often misunderstood as a profession that is easy and quick to learn. As a natural response to the sudden demand for UX professionals, the entry level has been lowered and many people have successfully sidestepped careers from other fields, often with little or no retraining. The result is a substantial difference in capability between practitioners and it is very difficult for employers and buyers of UX to distinguish these differences.

IMPACT OF THE RECESSION

Before the recession, professionals working in the digital industry would be split into specific roles: usability specialist (who would also conduct user research), information architect, interface designer, interaction designer, graphic designer, specification writer, prototype developer, and so on. The recession coupled with the rise in popularity of UX had an impact on these roles. For the most part they were replaced by one job title: UX designer. This person was to encompass all of the above skills. It is impossible for one person to do all of the aforementioned jobs well and the industry has already started to get back to prerecession times, once again splitting jobs into different specializations.

RESEARCH IS TOO OFTEN SACRIFICED

There are often limited budgets and time constraints on UX projects. This means that choices have to be made with regards to which tools to use. Despite increased awareness of the importance of UCD, many companies will cut out research when budgets are limited; therefore the actual user-centered part is sacrificed, and one can argue that it's not actually UX if no users are involved. Many are also still fearful of negative feedback if they undertake research with the end user. They feel very comfortable with UX designers and mistakenly believe that a UX designer will be able to identify and fix all issues within their own designs. Of course, in reality, a UX designer is not the end user and their work needs both user input and validation.

Within the e-commerce sector, there is a rise in the use of remote unmoderated research. Although this is the cheapest method of research, it leaves the company vulnerable to asking leading questions, bias in analysis, and missing the richness of information that is acquired through a two-way conversation with the end user. However, there are many companies who do value user research and see the benefits it has within their business, from helping in concept creation to optimization, and these are the ones who benefit from deeper user insights.

THE USER-CENTERED DESIGN PROCESS

Companies usually fall into two camps when innovating: those that practice research-driven design and those that practice design-driven innovation. In research-driven design, ethnographic research is conducted to analyze current behavior and experience to identify unmet needs and gaps for product development. Design-driven innovation starts with design before research is conducted—the design drives the research direction as opposed to the research driving the design direction. There is no right or wrong approach.

UCD is an iterative process and each project will need a different approach depending on what the client knows already about their users and previous work undertaken. Typically, at the start of a project, stakeholder interviews will be conducted. Personas will be created based on existing knowledge, then validated through user interviews. Time will be spent on user scenarios, creating optimal user journeys, and designing the information architecture.

A period of initial concept research and design will be undertaken and this early feedback is used to narrow down concepts to a few that are preferred and most commercially viable. These are then mocked-up into low-fidelity sketches or wireframes so that early UX feedback can be gathered to influence the design direction. The design then progresses iteratively, using short user-testing sessions to feed into the design and provide validation. In the Web and digital fields, the pace is very quick and research often needs to be prepared, conducted, and reported within a week (and often in as little as a day) so that there is little delay in the project.

In recent years, the preferred wireframe format is in the form of low-fidelity prototypes using software such as Axure. These are quick to create and can be both tested and edited during user research, therefore enabling changes to be made and tested instantly, without needing to wait for another round of research. They are also responsive—the designs can be tested on a real physical device, i.e., mobile phone, thereby giving increased validity to the research. They are very realistic and users often assume they are the real thing and not a mock-up.

Importantly, user research should be conducted throughout the design process, not just at the start and end. The addition of interaction and visual design can create new issues that are important to catch as soon as possible. Particular focus should always be placed on the top-user journeys and form-filling, as these are key areas where customers feel lost.

CASE STUDY: SONY ERICSSON

At Sony Ericsson, I worked within the R&D department in the research and design of smartphones. I oversaw the user interaction and usability of both the hardware and software, as well as aiming to create a seamless flow between the two. My job title was "Usability Specialist." Having completed a Masters in Human Factors and Ergonomics, I was responsible for both the user interface and product hardware usability.

Typically, a project would start with a concept developed in-house. Several alternative hardware designs would then be sketched up by the industrial design team. At this point, I would review the sketches to ensure that the hardware was supportive of platform software requirements (e.g., to ensure that vital controls were present) and identify any potential ergonomic issues the user might have when using the device. Designs would then be reworked and turned into block model prototypes through 3D printing.

At this highly confidential stage, I would test the designs with internal employees with specific focus on the comfort and feel of the device in the hand, the distance among buttons, the topography, and the tactile feel of buttons and navigation controls, mimicking the top use cases to identify potential for errors (such as unintentionally pressing the keylock button while taking a photo). The prototypes would progress with additional functionality, such as a real keypad with tactile feedback, until they eventually became early working prototypes that could be used as functioning devices. At each stage we would conduct expert reviews. These reviews were based on checklists we created to ensure best practice and consistency in our user experiences. Our product designers were industrial designers with little knowledge of ergonomics and no involvement with end users so we would work very closely with them. This involved knowledge sharing such as creating requirement documents to help educate and guide them to create products with considered ergonomics and good usability. Prototypes would be tested with users to identify any usability and ergonomic problems, then we would begin our push for changes—this was often the hardest part as each team wants the best for the product from their perspective. For example, industrial designers often wanted a sleek design, whereas those in marketing wanted a standout feature.

The other half of my role was to oversee the usability of the user interfaces. As our software was designed by specialist user interaction and interface designers, they would often seek my help and advice on usability or request user research to inform their designs. I would work across all stages of the product development life cycle, from feedback into initial concept user interfaces to benchmarking the top use cases of final working devices across internal products and against competitors. Daily work consisted of reviewing designs and prototypes created by the designers, suggesting changes and working with them to improve the user interaction, assessing new features/functionality/technology for usability and UX, and conducting ongoing user research.

User research would consist of planned larger pieces, to investigate and analyze mobile user behavior, to smaller scale research to quickly feedback on usability for a design. Often designs would be put through several rounds of user testing and design until they were perfected. Our preferred method of research to investigate real user behavior and attitudes was ethnographic—we would interview users and observe their mobile interaction within their home environment. It gives the research greater context and meaning, not to mention that users are much more relaxed and their behavior more natural. Eye tracking would have been incredibly beneficial but unfortunately it was not possible to eye track mobile devices back then. It is a technology that I now frequently use to analyze mobile user experiences.

Finally, part of my role involved looking towards the future of UX. I was heavily involved in pre-concept phases, testing, and feeding back on new technology that was being considered for future devices. To give you an example, gesture-based interaction and control are becoming mainstream only now, but I was conducting research into gestures back in 2005 to assess things like which gestures people found to be the most instinctive and natural to control the functionality and features within their mobile device. Ten years later, I am now starting to see some of these same gestures that I identified and defined in smartphone devices.

CURRENT AND FUTURE UX TRENDS

UX has a close relationship with technology and the digital world. As such, the landscape is constantly adapting to change. Some of these changes are as follows:

- *PET design.* PET stands for persuasion, emotion, and trust. It involves the application of psychology to design. This encourages the user to feel an emotional connection, to increase trust in the company/product/service, and to persuade them to action. PET has gained popularity in e-commerce design, whereby the purpose of PET is to increase the likelihood of a visitor making a purchase.

- *Progressive reduction.* Rather than provide the same interface from the first use onwards, progressive reduction is a technique whereby the user interface adapts to the user's knowledge over time (see Figure 24.1). When the user is a beginner, this technique recognizes that he will need more help and as he becomes familiar with an application, he will need less help.
- *Flat design.* The current trend in design is "flat design." One of the popular flat design button styles is "ghost buttons" that sit upon a background image. A ghost button is a button that consists of just an outline and no infill. It has poor usability as it does not really look like a button you can press. It is a classic example of form over function.
- *Wearable tech.* Wearable tech with small screens and in some cases, no screens, will pose greater challenges for UX designers who are used to designing for Web and mobile. The smaller the screen, the more difficult it is to design well.
- *Digital and physical experiences become one.* The Internet of things (IOT) is blurring the lines between the digital and physical world but there is still a huge disconnect between the two. Designers will need to focus on creating seamlessuser experiences across online and offline platforms, treating them as one.

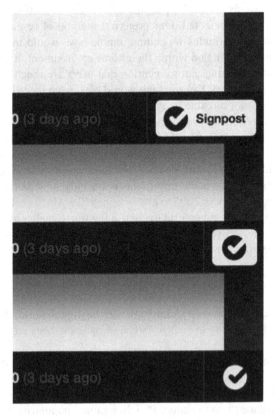

FIGURE 24.1   An example of progressive reduction—as the user learns the purpose of the button, he no longer needs to read the text, so the control changes.

THE FACTORS THAT REALLY LEAD TO SUCCESS OR FAILURE

Simply including UX in your process does not guarantee that your product or service will be a success. There are other factors that impact the ability of your UX to succeed or fail.

First, the success of any UX activity will rely on buy-in for UX in-house, especially from the management team. The ideal situation is to have UX embedded in the whole company culture or at least someone on the board or management team who believes in UX. This will ensure that your work and recommendations are considered at a level where they will make a difference and it will ensure you get the budget you need to conduct enough research to inform your UX work. An approach that often works is to make user research more visible so that people can see the difference between their opinions and how real users behave. Invite people to attend research or showcase interviews on a television screen in the coffee area.

Second, research is often the first thing that is removed from projects when budgets are limited. However, with no research there is an increased risk of failure upon launch and increased need for rework. What you should aim for is "fail fast and fail often" in the early stages because failing here is cheap (and often called Lean UX). This also brings you confidence and peace of mind as it ensures that by the time you go to market the concept and early prototypes have already been tested, your product, website, software, or app is more likely to be a success. The initial illusion of saving money can often turn into spending much more later down the line, to fix problems. If you do not include users in your UX process then you cannot really call what you have done "UX." User feedback and iteration is crucial.

Third, the competency of the designers and researchers is crucial to the success of a project. As UX is a service, the end deliverable and outcome can only be as good as the person conducting the work. There are people practicing under the term "UX" who are not qualified or experienced. While for a website this might just mean that you lose sales, for safety-critical software and medical apps it can be a real danger, with potential in putting lives at risk.

HUMAN FACTORS PROFESSIONALS AND UX

UX practitioners can vary considerably in their skills and this largely comes from their background before entering UX. Some will have HF/E and psychology backgrounds and will be strong with research methods, while others may have previously been developers strong on prototyping designs and understanding technical limitations. This is what makes a UX team very interesting—everyone could be called a UX designer but they could all have very different specializations. Therefore, depending on the company and their view of UX, HF/E practitioners may find themselves working in a team with other highly qualified HF/E specialists or they may find themselves working with designers, developers, and marketers (all under the job title of UX).

The multitude of ways in which people arrive in UX can also pose a problem for HF/E practitioners, whose specific value may not be evident at first to clients and employers. I first started my career as a usability specialist at a time when you needed an undergraduate degree, and ideally a postgraduate degree to gain employment;

I entered the field via a degree in computing and psychology, and a Master's in HF/E. This has now changed considerably and it is much easier to enter the field of UX. HF/E professionals considering entering the field of UX may find that their skills are not fully appreciated and are underused, and may have to clarify the benefits to performance and well-being that a background in HF/E brings. However, I believe this will change in the future as technology advances and challenges us to design for very small interfaces and utilize gesture control.

Nonetheless, after 11 years as a practitioner, I love working in UX because every day I face different difficult challenges and I feel a great deal of satisfaction knowing that I have made people's lives a little better. I also learn something new every day—the digital world is very fast paced and as digital technology and software changes so rapidly, it is imperative to keep up-to-date every day with technological news.

CONCLUSION: THE FUTURE OF UX

The digital world and how we interact with it is changing rapidly. Touch interfaces and gestures are already commonplace. In the future, our interactions with technology will change further. We will be more reliant on voice control and gestures. People who used to work in Web design and transitioned easily to UX design will find it much harder to make the transition to designing for challenging interactions that have little or no interface. It will push people to understand physical human limitations and cognitive capacity. I believe there will be a much greater need for the skills of HF/E.

UX is a fast-growing industry and one that is still finding its feet in many ways. The lack of regulation, lack of sufficient investment within many companies for UX activities (particularly research), and a larger but lower skilled workforce pose the greatest threats.

Highly qualified HF/E professionals may find themselves devalued within the current UX climate, but as technology advances and interaction paradigms change, there will be a much greater need for HF/E skills within this profession.

25 Human Factors and Ergonomics Practice in Web Engineering and Operations
Navigating a Critical Yet Opaque Sea of Automation

John Allspaw

CONTENTS

PRACTITIONER SUMMARY

Web engineering and operations is a nascent, fast-growing, and largely untouched discipline from a human factors and ergonomics (HF/E) standpoint. The domain is ripe with complexity and ambiguity, due to the dynamic nature of the systems and networks involved, across both geographic and geopolitical boundaries. The traditional effects, challenges, and trade-offs brought by the use of automation are amplified to a great extent, and there is an unmet need for HF/E support. The field is unique in many ways, but primarily in that the operating environment is quite opaque and that operators are also frequently the *designers* of their technical systems. There are no singular overarching regulatory, standards, or policy-making body for these services. Instead, there is a myriad of overlapping local and regional regulations for various layers of the services, such as telecommunications, privacy, content, and commerce.

INTRODUCTION

As Internet-connected services continue to weave their way into the fabric of modern life and business, the design and operation of the supporting software systems has largely gone unnoticed, and to a large extent unstudied by the HF/E community. The teams engaged in this work are also quite invisible to the general public consumers of these services. Well, until Google or Facebook is experiencing an outage, in which case their work can become quite visible (Judge, 2014; Tsidulko, 2014).

The fact is, HF/E research and practice is effectively absent from this domain, and this chapter should be viewed as a call to HF/E practitioners to consider it a "greenfield" domain. In light of this, this chapter will describe the WebOps field, and its challenges and opportunities, and introduce some of the HF/E-related work that has been done.

LIKE OTHER DOMAINS, BUT DIFFERENT

On the one hand, there are a number of characteristics of Web engineering and operations that look quite familiar to other domains studied by HF/E researchers and practitioners. In this domain, we see business-critical systems providing functionality to the (mostly) commercial audience. We see individuals and teams working in high-tempo and competitive markets, which have the potential for a myriad of technical, cultural, and organizational challenges. Much like military, aviation, and healthcare environments, it is awash with people coping with complexity and time pressure on a daily basis. The design of the technical environment (alerts, dashboards, and controls) has similar challenges faced by HF/E researchers elsewhere, and just as incidents can occur in most safety-critical domains, outages or degraded performance can occur that affect business continuity at a cost of millions of dollars, or have unintended consequences that spread to non-Web domains, such as the loss of availability of electronic medical records (Laffel, 2011).

But on the other hand, this domain is quite different, and this chapter aims to lay out some of those unique qualities and what challenges they present to the HF/E community.

AUTOMATION ISN'T PART OF IT. IT *IS* IT.

Computer science is the primary discipline driving the *design* of software, and has a rich history. In a one-sentence summary, many would agree that computer science is essentially the art of "efficiently implementing automated abstractions" (Guo, 2010).

All facets of the work are steeped in automation, which is necessary for the scale of operations (large organizations such as Google, Amazon, and Facebook, for example, operate hundreds of thousands of servers across the globe). But automation also exacerbates some of the difficulties that engineers can experience in operations, such as feedback or fault-tolerance mechanisms bringing on "cascading" failures. This double-edged sword of automation is familiar to HF/E specialists. However, the study of it *is* unfamiliar to software engineering professionals, generally.

In a recent book by Nicholas Carr, *The Glass Cage*, he suggests that this is due to how software engineers are educated:

> Software programmers receive little or no training in ergonomics, and they remain largely oblivious to relevant human-factors research.

> **(Carr, 2015, p. 159)**

I agree with the sentiment of this generalization.

The language of the domain can reveal just how tacitly automation-rich it is. Typically, an engineer who would execute a command on a thousand servers at once would refer to the action as "manual" because the triggering of the parallel command streams across a fleet of servers was performed by a human. An "automated" action is generally referred to as something that was triggered by *other* software once some performance or time-related criteria that the software has satisfied.

BUT THIS IS NOT JUST COMPUTER SCIENCE

There are many who argue that the discipline should be seen as a superset of segments of computer science and software engineering (Deshpande and Hansen, 2001). The design of software is largely governed by *specification*. How is it expected to work? What benefit is it expected to provide? Specification of software in this way typically speaks about its *intended* functionality, not its actual operation, which can be quite different for a number of reasons. This is a reasonable way for designers to approach writing programs initially, but as development progresses, software often doesn't do what we think it does, for many reasons. It is straightforward to test that it confirms to a specification, but it's difficult to test for all of the contexts that the software will be executed in. No testing methodology or practice completely prevents bugs in software. This is due to "corner cases," "edge cases," and "boundary cases" that were not considered by the author of the code. In this way, there is always a gap between *software-as-imagined* (by the author) and *software-as-operated* (by the user).

AN OPAQUE OPERATING THEATER

The Internet is effectively an *open* system, in terms of systems theory. John Doyle has referred to the Internet as simultaneously both "fragile" and "robust" (Doyle et al., 2005) and the applications that operate on its individual nodes follow that characteristic in terms of availability, performance, and scalability.

The routers and switches that make the global Internet work can and do fail often. Servers that contain content for websites and other increasingly critical services (official government statements and policies, payments processing, bank transfers, electronic medical records, etc.) go "down" for various reasons (hardware, software), and more frequently than most of the public realize.

Yet the network's architecture and protocols allow it to stay working in the midst of those individual component failures, for the most part. How a service that is delivered copes with those failures depends on how not only the application is designed, but also how it is prepared for eventual known failure modes. From an HF/E perspective, this preparation for failures can manifest in many ways: Alert design and anomaly response (Woods, 1995), operator overload/underload during outage scenarios, and diagnosis as it happens in distributed teams.

Software that is delivered on the Internet is typically developed with that use case in mind, whether we are talking about a mobile application on a phone or a table or on a laptop running a Web browser. This means that the characteristics differ from "standalone" or single-user programs, in a few ways:

- It is expected to be "always on"—working 24 × 7, globally.
- The software and infrastructure delivering the software needs to support thousands if not hundreds of thousands of concurrent users at any given time.
- A given site or application can (and increasingly does) depend on other network-connected services, which in turn may be connected to others, all of which are typically independently designed, owned, and operated. An example of this may be an e-commerce website, which may rely on external services whose functionality, availability, and performance are not within its control. The software itself is built of many components: Databases, networks, applications with varying logic, and security controls, all connected and communicating across different protocols, some of which are standardized and known in the public domain, some of which are proprietary.

This heterogeneity and interdependence can (not unsurprisingly) make reasoning about breakdowns and faults by an engineer quite difficult, and the opportunity to make a bad situation (say, an outage) worse (say, by corrupting data permanently) can be easier than one might think.

RESOLVING DISTURBANCES AND OUTAGES REQUIRES *DYNAMIC FAULT MANAGEMENT,* NOT SIMPLE DIAGNOSIS

Because of the opaque and dynamic nature of the systems involved, diagnosis and response to a service outage (or security breach, or performance degradations, etc.) can be quite different from the simple troubleshooting of a broken component.

The interdependencies and maze of diverse connected paths mean that teams of engineers need to work together to reason about the anomalies and signals associated with them at the same time they form a plan for recovery. There has been work in other domains in this area, and I believe it's promising to start with that lens.

> In dynamic fault management, intervention precedes or is interwoven with diagnosis.
>
> **(Woods, 1995)**

> Disturbance management is not simply the aggressive treatment of the symptoms produced by some fault. It includes the continuing and purposeful search for the underlying fault(s) while at the same time struggling to preserve system viability and, especially, how these two lines are coordinated given time pressure, the possibility of bad outcomes and the need to revise provisional assessments as new evidence comes in over time.
>
> **(Woods, 1994)**

> Disturbances may be transparent or opaque, on a 'data-driven', physical level (i.e. display properties or surface level features) or on a 'resource dependent', cognitive level (i.e. owing to characteristics of the human information-processing system).
>
> **(Shorrock and Straeter, 2006)**

Fault-tolerance patterns can cascade … compensating controls (timeouts, retries, etc.) all can have surprising results that are difficult to detect and/or trace, even if you were looking for it.

USERS ARE ALSO THE DESIGNERS

In this domain, the operators of systems are very frequently also the designers. This is in stark contrast to what you might find in other domains. Take, for example, air traffic management software systems that comprise radar displays and other alphanumerical interfaces for air traffic controllers: These are not typically designed or developed by the controllers themselves. To be sure, controllers aren't modifying the systems in question while they are in operation. In many Internet-delivered software organizations, this is the norm.

Why would this distance between user and designer matter? In the air traffic control example, it means that the exploration of how the software is being used needs to be explicit—namely, trained observers watch how controllers use the software and logs of the application are used to infer what functionality needs adjustment or redesign to improve the experience for the controller.

In the Internet services example, engineers who are doing their work may find a need for new functionality to assist them, and they'll simply write that new functionality into the application themselves, sometimes on the spot.

This feedback between user and designer in the same individual is largely tacit and interfaces can evolve at a rapid pace as a result, which can be both a blessing (in the case that the changes are advantages and intuitive for a broad number of users) and a curse (in the case that the change solves a very narrow use case for the author, but adversely affects use cases needed by others).

VARIETY OF OCCUPATIONS

In this field, there are a number of specialist roles. The most common are engineers who specialize in a particular part of the software "stack," such as network, security, database, application, and datacenter/hardware technicians.

There are also generalist teams of engineers who go by many names, such as "site reliability engineer" or "operations engineer" or "production operations engineer"—but at a high level, these roles have as their core skills both systems administration and software development.

The difference between these roles and software developers working on features or customer-facing products is that they largely build and operate the foundational software on which user-facing features sit. For example, a shopping cart or checkout feature used to purchase items from an e-commerce site would depend on payments processing and disbursement systems. Another example might be a photo-sharing feature, which would depend on software that crops, resizes, and stores image files.

This boundary between *infrastructural* and *product* engineering can be blurry, and there is no objective or standard dividing line between the two. The implication for this fuzzy boundary is the continual adaptation and exchange of tacit knowledge among these groups as they map out what teams will take on what parts of the work, and why. This is a ripe area for exploration for HF/E practitioners, because this continual negotiation of responsibility may have significant impact on the design of tools and applications.

PROCEDURES AND COMPLIANCE

For the most part, engineers are both the consumers and authors of procedures in the form of "runbooks," which are context-specific responses to common issues or tasks. In many cases, the runbooks exist in wiki documents and like systems, so they can be referenced and updated as needed.

Unlike other domains, Web engineering and operations is not subject to a large or general corpus of industry or government regulations, outside of the same sort of standard telecommunications rules that govern telephone traffic. There are commerce-related regulations or standards for compliance that do, however, put constraints on the designs of systems. These pieces of regulation (and those similar to them) are focused on the particulars of what the technology is doing, and not around the professional standards of the discipline, as you might find in healthcare or military environments.

BLAMELESS POSTMORTEMS

Despite not having industry-wide standardization or formal agreement on procedures, the community has taken a rather progressive approach to learning from accidents and untoward events. In such an automation-rich environment, there is a strong temptation to use concepts and terms as "human error," mirroring other domains that use them. Inspired by work done in HF/E and taking a systems-thinking approach to the topic (e.g., Dekker, 2002; Woods and Cook, 2003), software organizations have taken this perspective to their world in the shape of "blameless postmortems" (Allspaw, 2012; Zwieback, 2013; Parsons, 2014).

These are (effectively) facilitated debriefings that result in documentation of an event that places importance on the perspectives given by people familiar with the event, as well as placing actions and decisions in the context in which they happened.

As a result, this approach to learning from an accident (such as an outage or data loss event) looks like a mixture of cognitive task analysis (Crandall et al., 2006; Allspaw, 2014) and an after-action review (US Army, 1993). It can yield rich data on how work actually happens in an organization (as opposed to how it is *expected* to happen by management) in the forms of both technical artifacts (logs, dashboards, etc.) as well as narratives around goals, tasks, and constraints that might not otherwise be available explicitly.

From a cultural and social standpoint, organizations that take this approach give engineering staff support for giving details about mistakes that they've made without fear of retribution or punishment. From an HF/E perspective, it means that there is an abundance of avenues to explore.

WHERE HF/E PRACTITIONERS CAN HELP

An HF/E practitioner would be engaged in software operations for a number of reasons. The first challenge is to better understand how engineers *reason* about the systems that they've designed, when they're not functioning the way they're expected. From my experience, some of the questions an HF/E practitioner may look to answer might be the following:

• How could displays of data be made more useful in diagnosing incidents and events?
• How can controls of the systems and subsystems be designed in order to provide the necessary feedback to the operator, especially in light of the opaque interconnectedness of these systems and the risk of unwanted downstream effects on other systems?
• How might alerting systems be designed in order to provide the receiver with more context and assistance to respond to it while still balancing the signal to noise ratio?

A veteran HF/E researcher or practitioner might point out that the above questions are familiar ones. They would be correct. The issue with this domain is not that the questions are entirely novel or exotic; it's that the context for them *is* novel to research.

This practice environment isn't a control room in a power plant, but it certainly shares some similarities. These practitioners aren't mission control engineers, but they certainly share some similar perspectives. Same with weather forecasting, hazardous materials (HAZMAT) operations, air traffic management (particularly air traffic safety electronics personnel; see http://www.ifatsea.org), and professional team sports. It's likely that finding these similarities is a straightforward exercise.

Finding the differences in practice is what needs validation and definition, and that is where I think more HF/E research and practice is needed.

ADVICE FOR HF/E PRACTITIONERS LOOKING TO EXPLORE THIS DOMAIN

As mentioned before, there is not a lot of HF/E research or practice happening in this domain, so the areas of focus for someone coming into an organization are frankly wide open. Having said that, it might be useful to have a high-level map of the territory.

- If the "pitch" of HF/E is to learn about how engineers do their work and take those lessons and turn them into either more efficient processes or design of technical artifacts, most organizations would welcome an outsider known as a *researcher*.
- There may exist a potential pitfall to be aware of. HF/E practice can be confused with user experience (UX) practice. By and large, consumer Web applications have UX groups studying how users of their product use it, reason about it, find (tacitly or explicitly) gaps in the functionality, etc. These groups do use methods and approaches in their work that mirror most of those in HF/E research, but they tend to lean more toward product design as an outcome. Different organizations view the UX role differently, so HF/E practitioners should work closely with those groups to draw the parallels and contrast as sharp as they can.
- This domain is (unlike other software-driven domains, such as financial trading) quite open to sharing narratives and being studied. In the past few years, there has been a groundswell of conferences and practitioner-led events (e.g., O'Reilly Velocity) and publications that demonstrate a vigorous desire to know their own work and share it with others. Secrecy is not a default position in the field. It can be found, but it's not the default.

CONCLUSION

Web engineering and operations is a domain that needs exploring by HF/E practitioners. While software development and programming are related disciplines that have been studied, they are different activities with different constraints and goals. The *operation* of software on such a massive scale, such as the Internet, provides challenges for organizations and individuals that lie outside the boundary of their own companies, so effort must be made to discover the foundational dynamics of how people work in this (relatively) nascent industry. The economic and geopolitical influence that this industry has on modern society is too great to ignore.

REFERENCES

Allspaw, J. 2012. Blameless postmortems and a just culture. 22 May. Accessed April 13, 2015. Available at: http://bit.ly/1zZLJVj.

Allspaw, J. 2014. Velocity conference New York 2014: Complete video compilation. 18 September. Accessed April 13, 2015. Available at: http://bit.ly/1FMyv1B.

Carr, N. 2015. *The Glass Cage*. New York, NY: Random House.

Crandall, B., Klein, G.A., and Hoffman, R.R. 2006. *Working Minds: A Practitioner's Guide to Cognitive Task Analysis*. Cambridge, MA: MIT Press.

Dekker, S.W.A. 2002. Reconstructing human contributions to accidents: The new view on error and performance. *Journal of Safety Research*. 33(**3**), 371–385.

Deshpande, Y. and Hansen, S. 2001. Web engineering: Creating a discipline among disciplines. MultiMedia, *IEEE*. 8(**2**), 82–87.

Doyle, J.C., Alderson, D.L., Li, L., Low, S., Roughan, M., Shalunov, S., et al. (2005). The "robust yet fragile" nature of the Internet. *Proceedings of the National Academy of Sciences of the United States of America*. 102(**41**), 14497–14502.

Guo, P. 2010. What is computer science? Efficiently implementing automated abstractions. February. Accessed April 3, 2015. Available at: http://pgbovine.net/what-is-computer-science.htm.

Judge, A. 2014. Google down: World's largest search engine suffers outage. 4 July. Accessed April 13, 2015. Available at: http://bit.ly/1FmBJGp.

Laffel, G. 2011. Health care in the cloud: A 'case study of what not to do'. The health care blog. 2 May. Accessed April 15, 2015. Available at: http://thehealthcareblog.com/blog/2011/05/02/health-care-in-the-cloud/.

Parsons, C. 2014. Are you operationalizing lessons learned? 17 January. Accessed April 13, 2015. Available at: http://knowledge-architecture.com/blog/2014/01/17/are-you-operationalizing-lessons-learned/.

Shorrock, S.T. and Straeter, O. 2006. Managing system disturbances: Human factors issues. In: Karwowski, W. (ed.). *International Encyclopedia of Ergonomics and Human Factors*. Second Edition. London: Taylor & Francis, pp. 2180–2185.

Tsidulko, J. 2014. The 10 biggest cloud outages of 2014. 24 December. Accessed April 15, 2015. Available at: http://bit.ly/1AFLfrA.

US Army. 1993. A leader's guide to after-action reviews. *Training Circular*. TC 25–20. September. Washington, DC: Department of the Army, pp. 25–20.

Woods, D.D. 1994. Cognitive demands and activities in dynamic fault management: Abductive reasoning and disturbance management. In: Stanton, N. (ed.). *Human Factors in Alarm Design*. London: Taylor & Francis, pp. 63–92.

Woods, D.D. 1995. The alarm problem and directed attention in dynamic fault management. *Ergonomics*. 38(**11**), 2371–2393.

Woods, D.D. and Cook, R.I. 2003. Mistaking error. In: Hatlie, M.J. and Youngberg, B.J. (eds.). *Patient Safety Handbook*. Sudbury, MA: Jones & Bartlett.

Zwieback, D. 2013. *The Human Side of Postmortems*. Sebastapol, CA: O'Reilly Media.

26 Human Factors and Ergonomics Practice in the Construction and Demolition Industry
A Dynamic and Diverse Environment

Daniel Hummerdal and Stuart Shirreff

CONTENTS

PRACTITIONER SUMMARY

This chapter outlines dimensions that may be relatively unique to the construction and demolition industries. Each aspect provides a challenge and an opportunity where human factors and ergonomics (HF/E) practitioners can make a difference in these fields. To demonstrate this, we have provided two case studies of how our work has brought about a different way of understanding and acting in relation to work.

HUMAN FACTORS AND ERGONOMICS IN THE CONSTRUCTION AND DEMOLITION INDUSTRY

Buildings and structures may appear to be unified or relatively well-defined, but the process of putting these together, or taking them apart, is not equally coherent. A diverse set of complex, dynamic, and sociotechnical challenges emerge and combine during various stages of the process. As such, both construction and demolition phases are subject to a unique mix of conditions that sometimes make work difficult, and this is where HF/E practitioners can make a difference.

This chapter outlines industry-specific issues and provides examples of how HF/E practitioners can contribute to making workplaces more efficient, productive, and safe. As such, we hope that readers will gain some understanding of what it is like to work as an HF/E practitioner in the construction and demolition industry, but we also hope to raise some critical questions about practice in this industry.

The information is primarily reflective of our respective work in Australian construction (Daniel Hummerdal) and in the United Kingdom demolition industry (Stuart Shirreff). However, the sector is diverse in technology, risks, and organizational solutions. As such, this chapter aims to provide a concise snapshot of what it may be like to work in these industries.

WHAT MAKES THE INDUSTRY UNIQUE?

OLD PROCESSES BUT NEW TECHNOLOGIES

Construction and demolition is about putting buildings and structures together mechanically, and taking them apart. The process has probably not changed much since first moving out from the core dwellings. That said, with advances in technology and materials, both construction and deconstruction techniques have changed dramatically, and not always for the best.

For example, putting timber window lining in once involved a tape measure, hand saw, small jack-plane, and a hammer (and if you were really lucky a portable radio). Today, thanks to the requirement of timescales for property development delivery, the same task would be carried out with a portable electric saw, an electric planer, and probably a pneumatic nail gun. Because of these tools, up to three times more work may be done in a day, but this comes with an increase in dust exposure, hand-arm vibrations, and noise levels (and of course no radio because you wouldn't be able to hear it). While tools and technology may have developed to improve some performance goals, these developments have often not taken the (accumulated) impact on workers into account.

DIVERSE AND DYNAMIC WORK CONDITIONS

Construction and demolition are characterized by highly diverse and dynamic work environments. Clearly, the construction or demolition of a housing area involves a different set of risks and challenges than, for example, the construction or demolition of a gas compression station or section of plant within a live chemical site. Furthermore, on each site, work is subject to different sources of variability and subsequently different requirements. For example, roofing is more sensitive to rain, temperature, and wind, compared with an indoor activity. Similarly, housekeeping and manual handling intensifies around deliveries and at certain times of day. The point is that both construction and demolition are subject to high degrees of variability between and within jobs, over the short and longer terms. What was once an active site with lots of people and equipment may, over time, become an increasingly remote work site with small teams with little support. A site that at one stage required assessment of earth moving processes may later need to refocus on working at height. This transient nature of work requires both foresight and ongoing adaptation of strategies, tools, and resources, with significant HF/E implications.

FRAGMENTATION

In terms of the life cycle of a building or plant, construction and demolition are recipient stages of "upstream" decisions and actions, notably during the design phase. Additionally, with heightened levels of complexity, cost-efficiency, and innovation, a project's need for specialization and differentiation increases. As a result, the process from design to demolition is often dispersed among a wide range of specialized contractors and subcontractors, working together simultaneously or consecutively, for shorter and longer periods.

As a fragmented industry, there are large gaps between different stakeholders in construction and demolition. This makes it difficult to access and convey information between organizations. While the need to integrate knowledge through consultation stages and between stakeholders is often recognized by both stakeholders and regulators, the disconnect in both time and space makes it challenging to access, and convey, information across organizational and process gaps. For example, "safety in design" workshops are sometimes held to bring different stakeholders together, but the downstream actors may not be known nor have people available, and changes that occur during the construction phase may not be captured and conveyed effectively onto later stages, such as demolition.

The result is a mosaic of organizations and discontinuities. A few examples are as follows:

- Constructors and demolition contractors may have limited access to the designers.
- If drawings are issued by a different engineering contractor than the one carrying out the work, the gaps create lags in updates and coordination of changes.
- Assumptions about standard work practices may not extend beyond a particular organization.
- Different safety standards may create confusions and frustrations.

PEOPLE

Although some trades in construction and demolition require prior formal education, substantial on-the-job training, or a license to operate, many work opportunities do not require prior education, or only entry-level training. As a result, these activities attract people with a practical preference and a "can do" attitude. Sometimes, however, the lack of formal education can mean that some workers have a relatively low level of literacy. This can be a problem for activities that frequently involve the transport of dangerous goods or the operation of high-risk equipment, as these often rely on text-heavy documentation to ensure "the one right way."

This example is one of many opportunities to develop more user-friendly procedure material. Not only can materials be written in a more accessible way, but also there is an opportunity to illustrate work instructions and procedures with photos and illustrations of desired and undesired states.

OPEN ENVIRONMENTS

Both construction and demolition take place in unique geographical settings that are open, and often urban, with limited space. This inevitably creates a need to manage the interfaces between a project and its surroundings. For example, urban settings require: Interactions with nearby residents to manage issues around noise and dust; traffic management for deliveries and large machinery and equipment, with consequences for the public; and the space to lay down material, which may be limited.

Remote locations create their own set of people issues, requiring workers to live away from their families for extended periods. Such situations increase the risk from traveling (especially when driving long distances to and from the location), and introduce social challenges. In Australia, construction companies are increasingly concerned with calls for help in dealing with mental health issues, such as depression and anxiety, which seem to increase with living away from home over longer time periods (Lifeline, 2013).

EVOLVING RISKS AND REGULATIONS

Today, construction in the United Kingdom contributes to approximately 30% of all work fatalities (HSE, 2014). In Australia, the corresponding number was near 20% in 2011 (Safe Work Australia, 2014). The construction sector itself is highly regulated within the United Kingdom and Australia, with compliance to specific legislation, such as the Construction Design and Management Regulations (CDM), along with rules and regulations that are challenging both to companies and the bodies responsible for ensuring compliance.

All of this is overseen by several regulators. In the United Kingdom, for example, it is overseen by the Health and Safety Executive and their specialist units including the Office for Nuclear Regulation (ONR), the Hazardous Installations Directorate (HID), but also by the Environment Agency and possibly a further specialist organization, certainly when working in the high-hazard environments.

SUMMARY

Construction and demolition activities can be characterized as open, dynamic, and complex undertakings. Successful projects thus require consideration of myriad needs and conditions, many of which are difficult or cannot be anticipated at the design stage. While regulators have done much to increase attention to health and safety in the workplace, many organizations still struggle to find a good fit between people and technology. HF/E practitioners can help organizations not only with mechanisms to understand this gap, but also tools to help close the gap.

Importantly, the process must always be adapted as conditions emerge in the organization of the work. If this is not achieved, people will have to develop practices and solutions that in essence are unknown to the rest of the organization. When this happens there is a risk that the organization does not know enough about what is actually going on, and misses opportunities to capture innovations, as well as addressing practices that entail increased risk and inefficiencies.

WORKING AS A HUMAN FACTORS AND ERGONOMICS PRACTITIONER IN THE INDUSTRY

HF/E is a relatively new field to both construction and demolition, which are used to being governed by the rule-book and project delivery contracts. As industries that are well-rehearsed in the compliance paradigm, the idea of engineering the system to better fit humans sometimes runs counter to traditional thinking. The ability to adapt and to associate alongside the workforce is something that has been lacking in the construction and demolition industry for a long period of time. HF/E is sometimes met with skepticism and people may consider it too much of a "warm and fuzzy approach" to be meaningful. However, HF/E is increasingly recognized as an alternative perspective that can bring about change as a result of looking more holistically at problems and solutions. Significantly in the United Kingdom, the large-scale construction project for the London 2012 Olympic Games provided an ideal opportunity to develop and apply previous and new HF/E principles and techniques to present alternatives to the more traditional normative rules-based system for construction in the United Kingdom (HSE, 2012). This construction project was hailed as a great success, with no fatalities and an injury rate below the norm. When other attempts to solve a problem fail, organizations seem more likely to turn to the HF/E professional for ideas and advice. Over time, the early inclusion of the HF/E perspective becomes integral to better understand people at work.

CASE STUDIES OF TRANSLATING IDEAS INTO PRACTICE

A central challenge for many organizations is to ensure that decision-makers are aware of what is actually going on in their organizations. Various reporting schemes are traditionally put in place to provide communication between the management and frontline operations. Such schemes, however, tend to be focused around unwanted conditions and outcomes, such as incidents, near-misses, and hazard reports. Not only does such a negative focus risk introducing a reactive culture, but it may also miss what it is that the people need in order to achieve desired outcomes. Two examples of how both of us have implemented a new way of keeping the discussion about work alive, even when everything seems to be going well, follow.

ALFA—Ask, Listen, Find Out, and Act

Ask, listen, find out, and act (ALFA) is a process used by Theiss (a construction company) with the purpose of improving work conditions. ALFA describes the basic building blocks of the process. Through a series of professionally facilitated focus groups (10 participants in each group, each session lasts about 90 minutes), the goal of the Ask and Listen phase is to elicit real examples of when work is difficult, or otherwise highlight improvement opportunities. This is where having an HF/E perspective is helpful; to follow-up and dig deeper in order to understand what's intervening and hindering work from flowing smoothly. The conversations are recorded and the information transcribed. Any personally identifying information is removed. Separate focus groups are held with workforce, supervisors, and managers to enable an open environment to share experiences.

The examples elicited from the focus groups are subsequently used in a "Find Out" workshop. A representative cross-section of the project members collaboratively analyze and organize the information. The result is a prioritized list of issues the project can address to improve outcomes. The "Find Out" workshop takes around 4 hours to complete.

The "Act" phase of the ALFA process is carried out by an Improvement Team, consisting of selected project members in a position to coordinate action development, implementation, and follow-up. They meet regularly to review and, if necessary, modify the action priorities identified in the earlier stages of ALFA. Once an issue has been "closed out," the team starts to work on the next highest priority on the ALFA Action List. Throughout this process, the project manager remains in charge of what initiatives will be accepted and put into practice.

In one of Thiess' projects, 15 focus groups ("Ask and Listen" sessions) were organized. In total, 172 project employees participated in the sessions. Just over 1000 minutes of recorded conversations resulted in 305 examples of operational experiences.

During the subsequent "Find Out" session, project employees (approximately 25 people) identified 14 patterns of work that helped or hindered work. These patterns were organized into key challenges for the project to explore, and prioritized through a voting process among the workshop participants.

As the ALFA process allowed the project to identify its own struggles, creating ownership for addressing the issues was less of a challenge than if an outsider would hand over a list of findings. The ALFA process became a starting point for a continuous improvement cycle that almost a year later is active and contributes to keeping the discussion about work alive. Benefits from engaging the workforce in these types of conversations include increased understanding of project complexities, improved opportunity for innovative solutions, financial savings, and increased worker satisfaction.

Positive Reporting

At KDC (demolition business), we wanted to generate more information about what was going on in the company, but not only around unwanted situations. The idea was that we could increase the organizational intelligence by inviting a different set of reports coming through our systems.

One of these interventions was the rebranding of "Near Miss Reports" to "Site Safety Observations" (SSOs). Near miss cards are a designed reporting system for minor nonconformities or where there was a potential for something of greater severity to have happened. Many companies use such systems with a variety of names, such as abnormal events or unforeseen or unexpected occurrences. As these names suggest, these reports typically do not lend themselves to positive reporting. They do work to an extent and KDC are actively involved in a variety of these systems with their current client base, but they do not sit well within the reframed remit of the safety team. Unfortunately many clients request a specific number of reports on a monthly basis, which can lead to the creation of reports simply to comply with performance indicators, rather than to enhance the safety and health of those on site. With this type of reporting, it is also extremely difficult to obtain workforce engagement as many see it as whistle-blowing—something associated with punishment rather than support.

KDC recognized the need to change the title of the report cards to better match the true role of the process. The problem was exemplified in October 2008 when the raising of a near miss card on a project in a chemical site in Scotland instigated a full investigation by the client's safety team. The near miss was, in fact, a minor occurrence: A fence panel had blown down overnight due to adverse weather conditions, but the site was still secure as no night working was being carried out. However, due to the term "near miss" being used, the site safety protocols required a full investigation.

Around the same time, a trend was recognized that while many near miss cards were received (approximately 70) fewer than 8% were positive. To move from a negative approach to a positive approach, the decision was taken to rename the procedure and cards to SSOs. This title is deliberately neutral and as such allows for both positive and negative observations to be raised in equal standing. Furthermore, operatives seem to accept that a "safety observation" can be positive and are consequently more forthcoming. In 2009, there were 84 SSOs, of which 12 were positive; still relatively few but almost double the previous year. This initiative is still led primarily by the safety team and other senior managers and directors. It is carefully monitored and figures are collated monthly, but no targets are set in regard to quantities required. Instead, trends are monitored and good practice is promoted.

Various aspects of the SSOs helped to make this a success:

- They were devised as an opportunity to record any or all safety observations both positive and negative on an operational site or environment.
- The simple checklist allowed for identification of where any observation might have an impact on the work in hand.
- The opportunity to remain anonymous eliminated fears of finger-pointing or blame.
- The behavioral section and peer review sections allowed for the company safety team and management to give feedback and act on the observations.
- Ultimately the SSO formed part of the "open loop" risk management system adopted by KDC and gave the operatives a voice on day-to-day safety.

In 2013, the safety team recorded 631 SSOs for the business, of which 81 were positive. As a result of the work to embed the SSO system, trends can now be reviewed. This also allows for practical selection of personal protective equipment, which can be monitored and reviewed, and feedback can be given to the work force as well as senior managers.

CONCLUSION

The construction and demolition industries provide HF/E practitioners with varied, dynamic, and diverse challenges. As fields are traditionally governed by command and control methods, a central challenge for HF/E practitioners is to understand and bridge the gap between work as it is supposed to happen according to plans and procedures (work-as-imagined), and work as it actually happens (work-as-done). We have illustrated the development and implementation of two initiatives that intend to address this gap—ALFA and the SSOs. By broadening the way the organizations understand what is actually going on, improvement conditions can be identified and addressed—something which may be impossible to detect with an auditing, monitoring, and enforcing of what should be going on.

Although HF/E may initially be met with skepticism, the development of practical interventions that deliver both quick wins and long-term improvements are likely to help make HF/E an established way of understanding and improving work. The success of HF/E practitioners in these domains is partly predicated on the ability to explain why new ways of doing things are needed—in particular to senior management. As such, knowledge of HF/E principles and models is but one part of making progress in this field. Without the ability to work near and with those at the frontline of operations, HF/E would be just another top-down management tool.

REFERENCES

Health and Safety Executive (HSE). 2012. Pre-conditioning for success 2012. Characteristics and factors ensuring a safe build for the Olympic Park. RR955. London, UK: HSE. Available at: http://www.hse.gov.uk/research/rrpdf/rr955.pdf. Downloaded March 26, 2015.

Health and Safety Executive (HSE). 2014. HSE statistics on fatal injuries in the construction industry in Great Britian 2014. Accessed December 6, 2014. Available at: http://www.hse.gov.uk/statistics/industry/construction/construction.pdf.

Lifeline, WA. 2013. FIFO/DIDO mental health research report 2013. Perth, Australia: Lifeline. Accessed March 24, 2015. Available at: http://www.lifelinewa.org.au/download/FIFO+DIDO+Mental+Health+Research+Report+2013.pdf.

Safe Work Australia. 2014. Key work health and safety statistics, Australia, 2004. Canberra, Australia: Safe Work Australia. Accessed December 6, 2014. Available at: http://www.safeworkaustralia.gov.au/sites/SWA/about/Publications/Documents/841/Key-WHS-Statistics-2014.pdf.

27 Human Factors and Ergonomics Practice in Agriculture
The Challenges of Variety and Complexity

Dave O'Neill and Dave Moore

CONTENTS

PRACTITIONER SUMMARY

Agriculture globally encompasses an extremely wide range of activities, jobs, people, and technological settings, and there can be confusion about where the boundaries of this sector lie. Technologically, the degree of agricultural mechanization ranges from the use of human muscle and energy using very basic hand tools, through to automated systems requiring little human interaction except for occasional control operations. The workforce is unusually difficult to define, especially because the farm is normally a home as well as a workplace. There are marked seasonal fluctuations in labor demand, which may be met in a variety of ways. Accordingly, it is a costly and often elusive sector to study. The potential areas of human factors and ergonomics (HF/E) engagement are numerous; they have worldwide relevance and could affect up to half the world's population. Highly participative methodologies have consistently proved to deliver the most successful interventions. For HF/E practitioners to gain credibility in farming communities, we have to offer opportunities for wealth creation (or money saving), faster working (or time saving), or greater comfort (or reduced workload and drudgery).

INTRODUCTION: THE CONTEXT FROM AN HF/E PERSPECTIVE

In the exploitation of natural resources, the production of food is probably the most ubiquitous. According to the Food and Agriculture Organization (FAO) of the United Nations, about half of the world's population is engaged in food production (FAO, 2013), from small-scale subsistence, predominantly, to large-scale commercial enterprises, which contrast strongly with the heavily labor-intensive practices of subsistence farming families. Small-scale subsistence farming dominates because of the number of families in Africa, Southeast Asia, and South America engaged in growing food crops mainly for family consumption.

Agricultural production systems vary from moderate to very complex, irrespective of how technologically advanced the countries are. Subsistence farmers tend to employ complicated risk-management strategies and grow a variety of crops on small, often dispersed plots of degraded soil, and erratic rainfall. The logic is that if one or two plots fail to produce a good yield, other plots will be successful and the family should not starve. On the contrary, commercial farmers would favor monocropping practices and concentrate on one or two (arable) crops. Such farmers would have access to large stretches of good quality land and a range of mechanized equipment for land preparation, crop establishment and protection, maybe irrigation, and harvesting.

The use of human labor and ingenuity is common across all farming systems, although much mechanical and electronic assistance may be available. The more power-assisted equipment is used, the more dangerous, isolated, and machine-paced the occupation of farming tends to become. Nevertheless, it can be dangerous without any powered equipment if working at a height or handling livestock. Machine use may also be associated with less task variety and a shift in the ratio of the dynamic to static muscular work being done. The long-term potential for the gradual onset of musculoskeletal or respiratory disorders is almost always present. There is considerable scope, therefore, for ergonomics interventions as described in the following sections.

The aim of mechanization is to make available better tools and equipment, principally to increase productivity (of labor and land) and decrease production costs. According to Rijk (1999), there are nine stages to the agricultural mechanization process (see Table 27.1). As subsistence farmers progress toward the status of commercial farmers, they have to move down the list on the left-hand column of Table 27.1. Moving down from one stage to the next requires many cultural, social, economic as well as technical factors to be in place. Without the appropriate ergonomic inputs to facilitate the steps, progress will be inhibited and the whole process will be delayed, especially in moving through the earlier stages.

TABLE 27.1
Nine Stages of Agricultural Mechanization

Key Feature	Brief Description
Hand tools	Simple devices (originally sticks and stones) used to increase labor productivity. Hand tools are still important even in highly mechanized agriculture.
Draught animal power	Animal muscle power is substituted for and complements human muscle power.
Stationary power substitution	Engine power is substituted for human and animal power in stationary operations.
Motive power substitution	Engine power is substituted for human and animal power in field operations. It is particularly beneficial for power-intensive operations such as plowing.
Human control substitution	The equipment control functions are allocated to machinery elements (depending on complexity).
Adaptation of cropping practices	Machines are becoming dominant and cropping practices may change according to the characteristics of the machine, rather than machines facilitating traditional practices.
Farming system adaptation	Farming methods as well as cropping practices are modified to achieve the economies of scale needed to justify investment in machinery.
Bioadaptation	Engineering improvements are near their limit so characteristics of crops and animals offer better opportunities to raise productivity.
Automation of agricultural production	Minimal human intervention: information from sensors used to initiate and control events and processes.

Source: After Rijk, A.G., *CIGR Handbook of Agriculture Engineering*, Vol III, ASAE, St Joseph, MI, pp. 536–553, 1999.

SUBSISTENCE AGRICULTURE

Subsistence farmers have very little cash, no credit-worthiness, and have to depend on poor and often unreliable natural resources (e.g., soil, water/rainfall). They are to be found mainly, but not exclusively, in industrially developing countries. Typically, their farming practices are risky (Chambers et al., 1989) because of their need to grow a range of local staple crops to ensure their families are fed in the event of a failure of one (or more) of the crops. Their cropping practices tend to be extensive in several small plots but, nevertheless, be limited to areas of up to 3 hectares (but often a much smaller area in total). Mechanized assistance for land preparation, crop care, and processing is severely limited so manual labor predominates (see Figure 27.1). The greatest constraints on crop production and productivity are land preparation and weeding; the former because of the energy required and the latter because of the time required.

FIGURE 27.1   Land preparation with a hand hoe.

Typical levels of work intensity range from around 200 W at about 30% VO_{2max} (i.e., 30% of the maximum physical work effort) for simple weeding tasks to around 330 W at about 50% VO_{2max} for digging in hard soil. Maintaining a VO_{2max} of about 25%–30% over a standard working day is regarded as moderate hard work and levels exceeding that are unlikely to be sustainable.

In subsistence farming, the most urgently required ergonomics interventions are those aimed at relieving work (energy) demands and reducing time spent (and the associated drudgery) on the most time-consuming tasks; weeding and crop processing (at the household level). Transportation is also an important activity that can be both energy intensive and time-consuming, depending on the local circumstances, such as distance to crop gardens and markets, water collection points, and wood (fuel for cooking).

These interventions address the first two stages on mechanization shown in Table 27.1. Poorly designed, maintained, or even inappropriate tools can lead to inefficiencies and waste limited human energy. For example, in land preparation, any efficiency shortcomings of the hoe make the work particularly arduous at this level of energy expenditure. To maximize productivity, it is essential to ensure that the hoe is fully suited to the task.

The harnessing of animals to apply their strength and energy in carrying out the basic farming operations is another mechanization option. Table 27.2 shows typical work rates of working animals with human parameters shown at the bottom of the table.

This shifts the main human function from that of supplying the mechanical energy needed to perform the task to that of controlling it. That is not to say that the task of controlling animals may not be strenuous—it depends on their training and other circumstances such as the state of the soil and the condition of the implements (for soil preparation and weeding). Nevertheless, the human energy expended in controlling draught animals gives a better return in terms of area cultivated and crops managed than it would for hand hoeing or manual weeding.

SMALL-SCALE COMMERCIAL AGRICULTURE

This sector may also be dominated by family-run units but the main difference from subsistence farming is that these enterprises aspire to operate in the cash economy and rarely rely on bartering for the provision of labor or other inputs (e.g., nonfamily members being rewarded for their labors by being allocated a proportion of the yield

TABLE 27.2
Work Capabilities of Various Animals and Man

Animal	Average Body Mass (kg)	Approximate Draught Capability (N)	Average Speed (m/s)	Power Developed (W)
Oxen/bullock	500–900	600–800	0.56–0.83	560
Cow	400–600	500–600	0.70	340
Water buffalo	400–900	500–800	0.80–0.90	560
Light horse	400–700	600–800	1.00	750
Mule	350–500	500–600	0.9–1.00	520
Donkey	200–300	300–400	0.70	260
Camel	450–500	400–500	1.10	500
Man	60–90	300	0.28	75

Source: Campbell, J.K., *Dibble Sticks, Donkeys and Diesels: Machines in Crop Production*, International Rice Research Institute, Manila, Philippines, 1990.

after harvest). The business approach also means that crops are taken to market and sold directly within the community when surpluses permit. Such marketing activities usually require a greater reliance on transport than subsistence farming and involve establishing relationships with dealers or intermediaries.

To be successful, small-scale commercial farmers must have a greater labor and land productivity than subsistence farmers and so depend on a higher level of mechanization, although not such a high level as their large-scale commercial counterparts. These farmers, almost by definition, are better resourced than subsistence farmers and will usually have access to draught animal power and, usually, some form of engine power. They are also likely to be farming on better quality land and may have access to irrigation. They may use natural fertilizer (dung, manure) but also purchase inorganic fertilizer and crop protection chemicals for spraying.

Commercial farmers use more machinery than subsistence farmers but it is not necessarily engine powered. For example, they can generally afford hand-powered crop-processing machines for postharvest operations, often used for cutting, and therefore they will be exposed to greater risks of injury. There are many cases of farmers (or agricultural workers) losing fingers or limbs, because such equipment is often inadequately guarded. Pedal-powered threshers are also popular for farmers who cannot afford the engine-powered versions. Figures 27.2 and 27.3 show a manual threshing operation and the use of a pedal-powered thresher, respectively.

The pedal-powered thresher increases labor productivity by a factor of about 5. It also significantly reduces the drudgery of the task. An engine-powered thresher would further increase productivity by a similar factor (depending on the size of the engine and, hence, the throughput).

Productivity gains can also be achieved by irrigating crops. Although engine-powered pumps are the commonest means of achieving this, it is also possible to deliver water to the growing crop using a treadle pump (see Figure 27.4). One advantage is that treadle pumps have very few moving parts to go wrong, and with no

FIGURE 27.2   Manual threshing.

FIGURE 27.3   Pedal-powered thresher.

dependence on fossil fuel or electricity there are very few maintenance needs. The major disadvantage is that these pumps depend on an operator to be working whenever a water flow is required.

Small-scale commercial farmers tend to be operating at stages 2–5 shown in Table 27.1. They are able to benefit from greater access to stationary, nonmobile, powered equipment because it is becoming relatively cheaper than it used to be, especially with the availability of more second-hand units. It is an attractive investment because the productivity gains are immediately apparent. A similar situation applies to mobile-powered machinery, particularly two- and four-wheel tractors. These are generally more expensive than stationary units but provide the farmer with much greater flexibility as they can meet all the farmers' needs:

FIGURE 27.4   Operation of treadle pump.

- Draught power for tillage and other field operations
- Transport (but not high speed)
- Stationary power source, provided a power-take-off assembly is fitted

The use of powered equipment almost inevitably introduces health and safety hazards. These include noise and vibration, emissions from internal combustion engines, and the presence of high electrical voltages and currents. All of these can cause physiological damage, plus the risk that the engine can run out of control. This can present a significant threat in the case of mobile machinery, where not only the operator but bystanders also may be at risk.

The use of agrochemicals is attractive because these represent another means of raising productivity. The use of pesticides can cause greater harm than fertilizers because they are designed to destroy organic matter, albeit unwanted matter. Therefore, they have to be properly contained and carefully targeted at whatever they are intended to destroy. Human exposure during pesticide application tasks must be avoided. Tasks include mixing, loading into sprayers, the spraying operation itself, and tending to crops soon after they have been sprayed. An unintended consequence of the increased use of pesticides is a rise in the incidence of poisoning, both unintended and suicides.

There is a wide range of ergonomics interventions appropriate to small-scale commercial farming. Most of them are primarily concerned with safety and well-being because the productivity concerns have been addressed through the appropriate stages of mechanization summarized in Table 27.1. Ergonomics requirements include the guarding of rotating parts and (sharp) blades. Some inventiveness may be required if the material fed into a processing machine is bulky and a large opening is necessary. Also, the controls must be designed and positioned to prevent, or minimize the risk of, incorrect operation and be within the strength capabilities of the users to operate. Any displays indicating how the machine is functioning must be clearly legible, unambiguous, and easy to understand. These control and display criteria apply to mobile as well as stationary machines but they are likely to be more complex on mobile machines, so the ergonomics criteria may be more difficult to meet. In cases of poor ergonomics design, it may be necessary to redesign the human interface. Cultural variables must also be taken into account (e.g., switch position for "off" may be up or down) and adaptations may be required on important equipment. With the increasing use of pesticides, protecting the operator and ensuring that he or she is using spraying equipment correctly is of utmost importance. Normal working clothes can offer a degree of protection but whenever possible, personal protective equipment (PPE) should be used during all handling of chemicals. Ergonomics interventions to emphasize and reduce the dangers of pesticide poisoning include the following:

- Affordable, light, and protective clothing to reduce discomfort factors, providing appropriate training on symptoms and treatments of poisoning and natural methods of pest management
- Illustrating instructions using icons and symbols that are culturally recognized and socially sensitive so that they can be properly understood
- Providing safety and health information in the users' own language(s)

LARGE-SCALE COMMERCIAL (INDUSTRIAL) AGRICULTURE

Large-scale commercial farming is highly mechanized and characterized by an abundance of engine-powered equipment, electronic sensors, and displays offering decision-making guidance, and a small labor force. Farming activities are undertaken using a wide range of complex machines. Notwithstanding these complex machines, and the human role being predominantly that of a controller, work in the agricultural industry is still more physically demanding than that in most other (production) industries. In the progression from small-scale to large-scale (and high input) farming, the most obvious changes are the increased levels of technology in the equipment, the further reduction in human energy expenditure, and the further increase in information processing required by the operator. The pursuit of yet higher levels of productivity leads to greater use of agrochemicals and, often, sophisticated irrigation systems.

The need to work with animals, often with body weights and strengths exceeding those of a typical human, sets agriculture apart from most other industries. The difficulties of handling and generally managing farm animals, from the shearing of sheep to the milking of cows, create ergonomic problems specific to agriculture. Irrespective of the degree of mechanization or automation, equipment designers must consider not only the human–machine interface but also the human–animal interface and, preferably, the animal–machine interface, in pursuing complete system compatibility. Milk production is the principal enterprise for many farmers and the activities involved pose wide-ranging ergonomics problems. These include milking the cows (particularly equipment and workplace design), monitoring their health and productivity (identifying and responding to relevant indicators), and designing livestock housing to minimize human, as well as animal, health risks. The design of milking parlors must minimize postural problems, promote safe and efficient work routines, and provide an acceptable thermal environment. Satisfying all these demands within the prevailing economic constraints generally involves ergonomics and engineering compromises. Monitoring health and productivity becomes more difficult as herd size increases and can be facilitated by information technology, but care has to be taken such that technology is "user-friendly." Robotic milking, which is now a reality although not yet very widely adopted, reduces problems associated with the workplace, but the greater isolation of the herdsman places greater reliance on effective information systems to facilitate the husbandry of individual animals.

To avoid the increased information processing demands exceeding the capacity of the typical operator (agricultural worker), it may be appropriate to introduce some (or complete) well-designed automation to the monitoring of machine performance (e.g., spray application) or mechanized processes (e.g., grain drying). Computers and information technology in various forms are now common on farms and the interfaces and software packages that enable their use must be "user-friendly" for farmers. Figure 27.5 shows a typical United Kingdom tractor with a large spray boom attached (inset shows some of the controls in the cab).

The modern agricultural tractor is the icon of commercial farming just as the hand hoe, or maybe animal-drawn plough, is the icon of subsistence farming. The tractor offers an interesting and instructive ergonomics case study.

Tractors present a complex array of ergonomics issues: (1) workplace layout—the design and location of displays and controls, especially considering lines of sight outside the cab being safe and suitable for both field work and road use; (2) protecting the driver from noise and vibration; (3) protection from overturning; (4) providing an environment of satisfactory air quality; and (5) avoiding thermal stress. For a summary of some of the ergonomics research and development done to ameliorate the effects of noise, vibration, overturning, air quality, and heat stress, see Chisholm et al. (1992).

FIGURE 27.5   Modern United Kingdom tractor with spray boom.

The adopted method of protecting against these dangers and stressors, by enclosing the driver (operator) in a cabin, can introduce additional ones, so careful design is required to achieve the best (and affordable) compromise. For example, enclosing a driver in a strong cabin (incorporating rollover protection), and with a suspended seat, offers some protection in the case of an overturn, protection from whole-body vibration and, to a certain extent, from noise. However, it reduces the field of view, attenuates tactile feedback from the machine, and inhibits communication—all potential safety issues. Protecting the driver from excessive noise levels is achieved mainly by enclosing the driver in the cab but, as large areas of glass are needed to allow the driver good vision outside the cab, the cab can be a solar radiation trap, thus making it uncomfortably, or even dangerously, hot for the driver. A climate control system would, therefore, be usually recommended for the cab. Opening the cab windows or doors is not necessarily a solution as this may cause the noise limit to be exceeded or allow air pollutants such as dust or chemical spray into the cab. Similar considerations apply to more specialist vehicles such as combine harvesters and sprayers.

The highest levels of mechanization described in Table 27.1 are found in large-scale commercial farming. Most of the ergonomics interventions relate to health and safety issues but there is also scope in the development of new equipment to improve human performance, optimize the design, and "user-friendliness" of the human–machine interface in terms of both hardware and software. The health and safety issues, in addition to those referred earlier in relation to tractors and other vehicles, include the following:

* Hand-arm vibration from power tools
* Handling of large or untrained animals
* Entering hazardous situations (e.g., oxygen-depleted atmospheres in grain stores)
* Slip and fall injuries associated with working on poor surfaces, ladders, roofs, or in trees
* Diseases communicated from animals (zoonoses)
* Sickness or diseases from exposure to organic/inorganic dusts, agrichemicals, and sprays
* PPE against harmful environmental agents
* Injuries from inadequate machine guarding
* Environmental noise (annoyance) from fixed agricultural plant (e.g., grain dryers)
* Work practices, design, or organization considering social issues (e.g., isolation)

AGRICULTURAL CONTRACTORS AND OTHER "LABOR UNITS"

A consistent issue for those conducting studies or attempting to formulate interventions including policies and regulations is the lack of clarity over who is, and isn't, a part of the workforce. Table 27.3 gives a sense of the diversity of commonly encountered types of position found even on small farms over the course of a year.

TABLE 27.3
Employment Types Observed in the Agricultural Workforce

Position	Description
Land owner	The owner may also be the sole source of labor on the farm. At the other extreme, the owner could be the government or an absentee landlord who doesn't even live in that country, takes no active role, and can be very difficult to contact to discuss interventions or to prosecute.
Salaried managers	A position found in agribusinesses. Executive position that doesn't involve getting the hands dirty.
Managers	Generally lives on the property with day-to-day responsibility for supervising other staff on behalf of the owner. May also do physical work on smaller properties.
Permanent laborers	Staff in full-time positions with the usual terms and work conditions for permanent employees in the given country. Will generally receive training and other benefits. Paid wages with extra for overtime.
Seasonal laborers	Typical of harvest time operations for a particular crop lasting up to 6–8 weeks. May be on a fixed term labor-only contract or simply on an understanding of availability. The farm provides all the major equipment.
Casual/day workers (independent)	Typical of harvest time operations. Labor-only with no formal contract but put through the farm books and paid according to the rules.
Casual/day workers (labor hire)	As above but the farm may have a formal contract with a company or individual gangmaster to provide labor as required on a day-to-day basis dependent on weather and condition of crop.
Casual—cash in hand	Labor-only. Less frequently called and not put through the farm books for tax and occupational insurance purposes. The worker may get more in their pocket, but also has no protection in the event of injury on the job.
Migrants (legal)	Often used in picking tasks for perishable crops that require a large labor force for short periods. May operate via a special visa system that allows people from specific neighboring nations to enter specifically to work on one farm for a defined period.
Migrants (illegal)	As above but without any protection regarding minimum pay rates and conditions. Despite the negative press they attract, the harvesting operations in many countries would be unsustainable without a substantial undocumented labor force. Exposed to abuse.
Contractors	Often highly skilled and they bring their own tools and machinery. May be used for specialist tasks, or as supplementary capacity where a task has to be completed faster. Their annual workload frequently lacks the variety that can provide some protection from overuse issues for full-time workers. Can include haulage to market.
Contractors (testing)	Conduct testing of soils and other lab-based services (agronomists, veterinary tests). Frequently work in isolation as sole operators.
Students	Holiday jobs. Exchange programs. Placements from agricultural colleges. Relatively inexperienced and often working in isolation.

(Continued)

TABLE 27.3   (CONTINUED)

Employment Types Observed in the Agricultural Workforce

Position	Description
Family members	Can be in any of the above roles. Plus that of unpaid and underage laborer without training or industrial representation.
Sharemilkers and cooperative members	Some countries have ownership options whereby the workers are also buying equity as they earn. Can complicate decision-making and hierarchy of control
Ex-farm workers	A group rarely considered that includes retired farmers who still help out on the family farm, and ex-workers who remain in the labor pool and may be called. In addition, there are unemployed or underemployed people in the rural communities with long-term health issues caused or substantially aggravated by farm work.

This wide range of employment types adds significantly to the complexities of applying ergonomics in the agricultural sector. The challenges of understanding and dealing with the sociotechnical and socioeconomic issues in carrying out successful ergonomics interventions make working in the agricultural sector exceptionally fascinating as elucidated in the key challenges for HF/E practitioners in the agricultural sector in the following sections.

ACHIEVING SUCCESSFUL HF/E INTERVENTIONS

The diversity of circumstances and scenarios makes generalizations or recommendations regarding methods of limited value. However, there are certain approaches and techniques (see Table 27.4) that have delivered results and that the HF/E practitioner should understand.

The conflicts between health, safety, productivity, and comfort as implied in the third example are common in the agricultural sector. It is particularly noticeable in the use (or otherwise) of PPE where the discomfort of wearing extra clothing in hot conditions, the annoyance of restricted movement, or the frustration of not being able to communicate normally can result in the protection being abandoned. Persuading farmers to adopt safer systems of work is always a challenge irrespective of where they work. In commercialized farming, which is less "informal" and there is a greater opportunity of their being imposed or adhered to, there is a wealth of good advice from agencies such as the (UK) Health and Safety Executive (see, e.g., www.hse.gov.uk/agriculture/). Although commercial farming may be the target, the principles of safe systems of work apply to any agricultural practice.

The most likely areas of activity for ergonomists working in the agriculture sector are summarized in Table 27.5.

It is important for the HF/E practitioner to understand the difference between good and bad interventions. We assume our ideas fall into the first camp, and that is not always true. Simplistic campaigns are usually insulting and can do more harm than good. The hardest things to change are where it would detract from what the workers themselves like about the work, it changes the balance of interpersonal

TABLE 27.4
Effective Approaches and Methods for Successful Interventions

Approaches and Methods	Narrative
Bottom-up as well as top-down	High-level epidemiological data are essential for identifying trends, emerging threats, and high-risk situations for detailed investigation. However, the understanding to interpret the data and develop interventions that are effective has to come from bottom-up studies.
Participative intervention and implementation design	Just because something is a good idea doesn't mean to say it will be taken up. There may be very good reasons why it is not, including costs of upkeep or unfavorable system integration.
Targeting performance, health, and safety simultaneously	Micro- and small-business research repeatedly shows the importance of working with the hierarchy of priorities adopted by the sector. Business survival ranks ahead of minor ailments, and hunger next winter certainly takes priority over possible respiratory problems in 30 years. This is especially true for casual workers who optimistically expect to be into better work soon.
Focus on recurrent scenarios, not single factors	Acknowledge the complexity of incident events and their interacting networks of causal and contextual factors.
Farmer-to-farmer surveys	The response rate of community-based phone surveys where both parties are from the rural community is reported to benefit significantly from having subject matter experts at both ends of the line.
Heuristics/"rules of thumb"	Studies on equipment before finalization of design, including evaluations of the suitability for various countries to which it will find its way as a new or used machine.
Business census	In countries that run a systematic full or sampled census, questions of interest to HF/E practitioners can provide very valuable data. Ireland is the best example with health and safety questions included at regular intervals in the annual survey of 1000 farms.

prestige in some way, it is counter to the immediate interests of the larger landowners, or it involves people already highly exposed who can't afford even a small failure because of the inordinate consequences.

KEY CHALLENGES FOR HF/E PRACTITIONERS IN THE AGRICULTURAL SECTOR

DIVERSITY OF THE TARGET POPULATION

The target population comprises all who work in the myriad of farming activities, formally or informally, as family support (young or old, 9–90, because that is the reality), as casual labor (rewarded financially or "in kind"), as members of a gangmaster's team, or as the millions who grow crops to survive. Many spend their working life outside systems (e.g., social security) that offer protection (e.g., migrant laborers, subsistence farmers, and even sole-charge semiautomated jobs).

TABLE 27.5
Potential Areas of Human Factors/Ergonomics Engagement in Agriculture

Area	Examples
Hand tool design	Optimizing performance and ease of use.
Working practices	Risk assessment and designing safe systems of work.
Manual material handling and load system design	Both dead and live loads. The design of packaging (sacks and containers) bulk-handling systems and stacking areas. Training and awareness of risk multipliers.
Labeling	Clear marking of hazardous substances, especially pesticides. Systems approach essential for managing/controlling risks.
Stock handling	Isolating and controlling individual animals for treatment. The design of handling systems including compartmentalization of trailers to minimize shifting of live loads in transit. Training and awareness regarding herd psychology.
Vehicle design	Access, control/display issues, rollovers, working safely with attachments, and power takeoffs.
System integration	Optimal design and selection of vehicle/stock/appliance combinations to avoid dangerous or ineffective mismatches.
Design for safety	Optimizing new plant and (processing) equipment for fast and safe maintenance, repair, and blockage clearing.
Gradual onset problems	Long-term health issues related to dust, chemicals, noise, and other insidious health threats, low pay, and insecurity.
Automation	Robotics, use of global positioning systems, laser accuracy, automated growth measurement.
Environmental factors	Operating in extreme climatic, economic, and psychosocial conditions.
Organizational design	Workforce performance and cost-effectiveness across seasonal fluctuations and from year to year.
Costs of injury and disease	Quantification of direct costs and indirect costs—including extended family and community impacts.
Dissemination/extension	Understanding social, economic, and cultural factors in the spread of proven beneficial ideas in small and medium businesses in the rural community. Stakeholder analysis and risks in microbusinesses; minimizing constraints to encourage adoption.
Learning	Barriers to intervention uptake in the rural community. Literacy, learning difficulties, English as a second language.
Globalization and climate change	The impacts of both on the viability of small owner-operated farms.

RESISTANCE

Because every farm and every farming situation is specific, few farmers are prepared to accept that an outsider can help them do things better. Thus, most farmers resist interference. They mainly work for themselves or in small- to medium-size enterprises (SMEs), who don't employ HF/E practitioners, and don't give up their time readily. Farmers are practical people who are used to improvising and prioritizing.

They intuitively know that things rarely have a single cause, because the day they come unstuck, they are doing something they have done thousands of times before. Furthermore, farms are more scattered than other businesses, and so it takes longer and costs more to get there, and there is no guarantee that a visit will yield any useful information (if the farmer has another priority).

This resistance pervades the farming system. Farm owners (and managers in particular) actively tend to resist regulation. In some cases, their businesses can be too precariously balanced to accommodate poorly conceived and overgeneralized directives.

Funding is a particular problem. On the spectrum of HF/E dollars invested, the agricultural workforce sits at the extreme opposite end to military pilots, who benefit from a human factors integration (HFI) program, or those controlling safety-critical high-energy installations. A situation may have to be recognized at a national level (i.e., a political embarrassment) before an ergonomist gets involved. Given the well-recognized under-reporting in this sector, the actual scale of a problem would normally be far larger than officially recognized. When people die in ways that interest the media, and are reported often enough that questions are asked at the governmental or UN level, then funds may be allocated to look into it. Unlike miners or the victims of bomb attacks, farmworkers die one at a time. Rarely, but crucially, funding comes from coalitions of health research funders concerned about long-term gradual onset health issues. Because of the scale of the issues, such exercises are politically scrutinized. Also, bodies representing landowners have a keen interest in managing public perceptions, as customers want to believe in a healthy and fair supply chain.

NEED TO COLLABORATE

This is not uncommon for ergonomists but in the agricultural sector there is a strong possibility that other professionals may have had no exposure to HF/E. These disciplines may include agronomists, anthropologists, economists, sociologists, soil scientists, toxicologists, veterinarians, etc. There is a particular overlap between HF/E and social anthropology or applied sociology; the unit of interest is a rural population, not just a workforce.

Where engagement is possible, the participatory approach (with focus groups, etc.) is generally the most successful. Although it is time consuming (and therefore expensive), it is very rewarding (for all involved) and the surest way to achieve an intervention that survives and is sustained. The key is to work with the stakeholders (having been carefully identified) to develop their ideas so they have ownership and the community is motivated to succeed. The HF/E practitioner has to keep that wider system in mind, while also facilitating the improvement of very basic tools and processes.

LACK OF CURRENT AND RELEVANT RESEARCH AND DEVELOPMENT

Ergonomists often complain about the lack of funding for implementing improvements or the lack of recognition at the start of the design process but, in the case of small-scale farmers, there seems to be almost no funding or current HF/E research (or development) to improve their circumstances. The very few exceptions to this would be the work of the Agricultural Engineering Section at FAO, the International Labour Organization (ILO, another United Nations agency) with a small section interested in agriculture (see, e.g., Niu and Kogi, 2012), and the Bill and Melinda Gates Foundation (see, e.g., www.3d4agdev.org).

The more mechanized farming systems have benefitted from R&D, mainly during the 1950s until the 1990s, when further investments in agricultural engineering development seemed to become unfashionable. In this period, there were significant advances in (1) agricultural vehicle cab safety and comfort, (2) livestock handling, (3) crop establishment and protection, (4) harvesting equipment, and (5) in-field and on-farm processing and storage. Most of these areas have received some ergonomics attention but could have benefitted from more, particularly with regard to safety and health.

The paucity of (recent) references reflects the very low level of R&D activity; the country supporting the greatest activity is probably India with small-scale commercial farming being the main target (see, e.g., http://www.icar.org.in/en/node/909). A wide range of agricultural R&D was covered in a conference sponsored by the International Ergonomics Association in 2007 (see Khalid, 2007). Much of the information presented in the keynotes and the overviews remains up to date.

Another question is "who benefits from R&D?" If better tools improve productivity, market mechanisms adjust the piece rate (payment by output) accordingly. Also, most HF/E study tools are designed initially for and by other sectors, for example, military and petrochemical. The default tools for farming need to be better aligned with, and developed for, SMEs specifically.

CONCLUSIONS

There are many areas HF/E can make a significant difference to the well-being and performance of agricultural workers. These will vary depending on whether the agricultural context is subsistence farming, small-scale commercial farming, or large-scale commercial (industrial) agriculture. The variability of the agricultural workforce also poses challenges that are greater than most other industries. Some recommended approaches have been outlined, but the nature of the engagement is likely to be influenced primarily by the level of mechanization (sophistication).

Engagement with farmers by those outside their immediate community is notoriously difficult. Nevertheless, there are three ways of getting the attention of farmers, by addressing what could be described as their generic needs. Our experience suggests that these are often, in priority order: (1) offer something to make them richer (or save them money), (2) offer a way of saving time (via faster working methods or practices), and (3) offer a means to reduce their workload or drudgery (via easier to use or more comfortable equipment). Despite the potential impact on performance and human well-being, agriculture remains relatively untouched by HF/E. If it were possible to overcome the scarcity of resources and difficulties of engagement, more attention to this area could have worldwide benefits for people, populations, business, and food supply.

REFERENCES

Campbell, J.K. 1990. *Dibble Sticks, Donkeys and Diesels: Machines in Crop Production.* Manila, Philippines: International Rice Research Institute.

Chambers, R., Pacey, A., and Thrupp, L.A. (eds.). 1989. *Farmer First: Innovation and Agricultural Research.* London: Intermediate Technology Publications.

Chisholm, C.J., Bottoms, D.J., Dwyer, M.J., Lines, J.A., and Whyte, R.T. 1992. Safety, health and hygiene in agriculture. *Safety Science.* 15(4–6), 225–248.

Food and Agriculture Organization of the United Nations (FAO). 2013. FAO Statistical Yearbook 2013. Accessed January 12, 2015. Available at: http://issuu.com/faooftheun/docs/syb2013issuu/19.

Khalid, H.M. (ed). 2007. AEDeC 2007: *Proceedings of Agricultural Ergonomics Development Conference.* 26–29 November 2007, Kuala Lumpur, Malaysia.

Niu, S. and Kogi, K. (eds). 2012. *Ergonomic Checkpoints in Agriculture.* Geneva: International Labour Office in collaboration with the International Ergonomics Association.

Rijk, A.G. 1999. Agriculture mechanisation strategy. In: Stout, B.A. and Cheze, B. (eds). *CIGR Handbook of Agriculture Engineering*, Vol III. St Joseph, MI: ASAE, pp. 536–553.

Part IV

Communicating about Human Factors and Ergonomics

Part IV of the book considers issues associated with communicating about human factors and ergonomics (HF/E), at all levels and in various forms. This part of the book has four chapters.

Chapter 28 (Nigel Heaton and Bernie Catterall) explores communication with senior management in organizations. Nigel and Bernie describe ways to improve communication with senior teams to support them in articulating HF/E issues and begin to deliver improvements.

Chapter 29 (David M. Antle and Linda L. Miller) considers engaging participants in HF/E. David and Linda discuss how the engagement of key employees allows for more appropriate and versatile HF/E interventions.

Chapter 30 (Don Harris) turns to writing as an HF/E practitioner, and provides practical advice for practitioners.

Chapter 31 (Dom Furniss) takes a comprehensive look at public and social media engagement for HF/E practitioners. Dom discusses his experience of outreach and networking, including the variety of approaches to public and social media engagement.

28 Communicating with Decision-Makers
Getting the Board on Board

Nigel Heaton and Bernie Catterall

CONTENTS

PRACTITIONER SUMMARY

Human factors and ergonomics (HF/E) practitioners have the best opportunity ever to communicate with senior decision-makers to get our message to the Boards of companies, but we need to adapt our approach to their needs. For instance, the emphasis on risk and risk management within corporations gives us an ideal tool to talk to the senior executive. We must become an expert at talking to the Board in language that they use and are familiar with. As organizations are driven by measureable objectives we need to align our message to these objectives and demonstrate how the HF/E practitioner contributes to the solution. A simple framework that allows organizations to explore their vision, their core values, and the effectiveness of their risk registers provides us with a start-point to work with the senior team to demonstrate how they can articulate what really matters and begin to deliver improvements.

OUR PRACTICE IN CONTEXT: HEALTH AND SAFETY AND HF/E

Although HF/E has a holistic focus on both the operational upsides (e.g., efficiency gains, optimizing well-being) and downsides (e.g., mitigating health and safety risks, avoiding harm), it is the latter that traditionally has brought HF/E to the attention of the Board of many organizations. In our experience with Boards and senior management, we have used HF/E to argue the case for health and safety. Health and safety is just one goal of HF/E, but it is an important function and department for our customers, in a way that HF/E is not.

This chapter, therefore, focuses on our experience of talking to Boards about managing their health and safety risks and how we take an HF/E approach to do this. We aim to provide practical advice about securing management commitment (and ultimately ownership). We argue that at its core, communicating with decision-makers (as with anyone else) is not just about having a message to give, but about tailoring both the content of that message and the method of delivery to meet their needs. HF/E practitioners should be ideally placed for this, with our grounding in participatory approaches and understanding of stakeholder analysis.

TAILORING OUR MESSAGE: UNDERSTANDING WHAT MATTERS TO THE BOARD

Organizations exist to meet their business objectives and, whether in the public or private sectors, at senior levels, these concern a small number of performance measures. In our experience, in the private sector, the primary performance measure is profit—either delivery of shareholder value through profits (for listed companies) or, in private ownership, the delivery of profits to enable the company to continue. In the public sector, the performance measures tend to concern delivering services in a "value for money" way. In both the public and private sectors, money is the main focus. So, if HF/E is to be important to the Board when communicating with them, we must engage with finance.

The senior management team also focuses on a number of subsidiary objectives to deliver the financial objectives. It is in these areas that the application of HF/E principles and knowledge can play a key role. As part of their governance structures, organizations have sought assurance that financial performance is being well managed, and have required managers to report on budgets, spending, and income.

RISK

Many organizations are now requiring managers to report on financial risks and as the organizations mature, they require managers to report on risks per se. These risk-mature organizations employ risk models that break down performance against organizational objectives. It is essential that the HF/E practitioner becomes risk savvy and uses the language of Boards when conveying HF/E messages. While the specific approach to corporate risk management varies widely, five generic objectives are central to organizational success. We can use these as conduits for our message, adopted from the objectives identified in the Turnbull report on risk management, internal control, and related financial and business reporting (Financial Reporting Council [FRC], 2014) and subsequent publications.

OUTPUT

To achieve financial objectives, the operational goals of the company have to be met. These are assessed through measures of output (e.g., of products or services). HF/E, as a discipline, has a long history of influencing output. We are able to ensure that output goals are achievable without excessively compromising the health, safety, and welfare of the workforce. As practitioners, we are often involved with organizations who have focused on output at the expense of the operator. Output-focused interventions often fail to consider the impact of increasing targets on the operator or end user. Whether this is increasing incidence of musculoskeletal disorders (MSDs) on the production lines or stress in the health service, output improvements need to be coupled with the consideration of those who are delivering those improvements and the effects on them and the sociotechnical system as a whole. A typical communication method for health and safety practitioners tends to concern injuries and associated costs. This casts the practitioner in terms of an overhead cost. As HF/E practitioners, we can instead use HF/E interventions before people become harmed. In these cases, we find that the simple footage of jobs before and after intervention and tools that report on MSD symptoms before and after interventions alongside statements from end users can support a powerful message that HF/E interventions improve human work and increase output—safely.

QUALITY

Quality goals are key drivers in many organizations, and often clearly link quality to finance, performance, and people. Again, HF/E can contribute to quality goals in a number of ways, and explaining this to senior teams is important. For example, Yeow and Sen report on a study where a combination of operators' eye problems, insufficient time for inspection, and ineffective visual inspection was generating a yearly rejection cost of US\$298,240. Ergonomics interventions were made to the job and produced savings in rejection cost; reduced operators' eye strain, headaches, and watery eyes; lowered the defect percentage at customers' sites; and increased the factory's productivity and customer satisfaction (Yeow and Sen, 2004).

We also need an understanding of how quality goals, and especially performance targets, impact human performance in a nonproduction environment. In one reported case (see Dubner and Levitt, 2007), teachers were required to hit high pass rates for their students. This goal encouraged the teachers to cheat in order to meet the highest quality goals. The influence of performance targets has affected many areas in public and private organizations (Shorrock and Licu, 2013). As HF/E practitioners, we can look at the wider system implications of quality goals and targets. We need to help the senior leadership teams understand how simple quality targets can distort the system and generate unexpected side effects in the way that people behave. We are able to improve our quality metrics through better inspection techniques, through understanding the links between quality goals and human behaviors, etc.

RESOURCE AND ASSET MANAGEMENT

Organizations need to ensure not only the right tools and equipment for the job but also their availability and ease of accessibility when needed. They also need effective systems to deliver appropriate specifications for the equipment and environment. The procurement, provision, maintenance, and disposal of equipment need to be delivered in a timely and cost-effective manner. HF/E practitioners know the problem of bolting-on HF/E once equipment has been specified and installed. The "Cranfield man" (a person constructed to operate a lathe, rather than a lathe constructed to be operated by a person) (Singleton, 1964) illustrates the need for HF/E in the specification, design, installation, and use of equipment. But this need is not realized automatically. Instead, the HF/E practitioner is often empowered to intervene through legal requirements. For example, within the European Union (EU) there are provisions about the use of equipment where usage is defined in the widest context to include installation, routine use, maintenance, decommissioning, etc. (the United Kingdom's Provision and Use of Work Equipment Regulations [Health and Safety Executive, 1998)—based on EU work equipment Directive 89/655/EEC). At this point, the communication channel is usually in the form of a risk assessment and the HF/E practitioner needs to ensure that the assessment includes ergonomics issues and a clear statement of the cost-benefit for any additional controls or changes. Risk assessments that identify significant risks, where the impact is very high (in terms of cost, reputation, health and safety, etc.), need to be fast-tracked to the Board usually via risk registers that allow for urgent flagging of some types of risk.

PEOPLE MANAGEMENT

People management is a key factor for all successful businesses—be that in selection, training, occupational health, staff retention, disciplinary procedures, injury management and, increasingly, dealing with an aging workforce. Well-being programs have vastly increased in importance over the last 10 years because the value of a fit, healthy, committed, and capable workforce (as opposed to the opposite) is now increasingly recognized at senior management levels. In this case, the main way that HF/E can influence tends to be in terms of job specifications and screening. We can provide advice on how to specify jobs that people can do and the type of screening that needs to be employed for specific jobs. Relatively few organizations are proactive around well-being, but more are recognizing the importance that a wellness program can play in reducing injuries and absence. For example, during the construction of the Olympic Park, as much emphasis was placed on health as on safety (HSE, 2012).

SAFETY MANAGEMENT

The potential reputational damage cost of poor environmental and/or safety performance has increased the interest of most Boards and senior management teams. Whereas 20 years ago ergonomists had to work hard to crowbar this idea into senior management thinking, most senior management teams and Boards are now much more aware of the importance of this aspect of their work. The importance of these issues has been raised by high-profile disasters and their consequences (including increased costs of enforcement action and personal injury claims), together with improved understanding of potential personal and corporate accountabilities for failings. Of course, not all senior management teams and Boards understand this, but their door is, in our experience, much more open to persuasion than it was even a few years ago. Our experience is that organizations invest considerable resources in safety management and contingency planning. Here, the HF/E practitioner needs to work with the risk or safety management team to make the business case for the type of problems that emerge from systems. They will also have a role in helping devise post hoc mitigation plans. However, these often neglect key HF/E issues. Risk management has become the domain of the big consultancy practices, and contingency planning becomes owned by the contingency planning consultant who will spend months analyzing the organization and developing the plan. As practitioners, we can offer insights into the critical people factors that should lie at the heart of effective safety management, and we need to talk to the senior team about the importance of physical design and the importance of designing jobs that people can do.

NOTE ON PERFORMANCE MEASURES

In a scathing indictment of BP, the Baker Panel report noted that managers managed what they were measured against (BP US Refineries Independent Safety Panel, 2007). The implication is that if there are significant risks to the business that are not represented in key performance measures, then those areas of risk in those parts of the business will not be managed. For better or for worse, organizations are driven by measurement. We might disagree with the adage that you can't manage what you can't measure, but in reality the measures that organizations impose on individuals and groups drive behavior. Indeed, external drivers such as share price or even the behavior of shareholders will fundamentally affect the way in which organizations are run. As long ago as the 1920s in Dodge versus Ford, the courts ruled that the needs of the shareholder outweigh the needs of the worker (when Ford attempted to give its workers a pay rise) (Bakan, 2005). We must use measures that show that our objectives have been reached. So if there are no measures of HF/E performance, then it is unclear what managers are trying to achieve.

We also know that performance measurement per se has become an objective in itself. Organizations will seek to meet the requirement to measure performance, independently of what the measure tells them about the organization. So we need to be mindful of the influence that HF/E performance measures may have on behaviors, of the organization, of the team, or of the individual (managers and other staff members). HF/E performance measures are not, of themselves, going to improve performance but there is a common management mindset that you can only manage what you can measure. As HF/E practitioners we need to address this problem and propose measureable aspects of HF/E performance that will drive improvements. For example, while reducing the number of MSDs might seem a laudable aim, this might be achieved simply through under-reporting. More interesting measures such as number of risk assessments completed and all actions undertaken, evidence of number of staff who have been trained with Kirkpatrick level 4 measures to demonstrate results, are potentially more meaningful.

All these business imperatives have HF/E considerations and we can use them as ways-in for our message. We need to ensure that our message is tailored around the issues that matter to the Board. This is part of understanding our stakeholders, and of course core to HF/E!

TAILORING OUR DELIVERY: USING COMMUNICATION TOOLS THAT BOARDS UNDERSTAND

We have argued earlier that tailoring the message is a key part of how we communicate with Boards. In addition, our delivery methods should be tailored to integrate with artifacts and systems already used by the Board. We have found that a tool kit of communication tools can support the HF/E practitioner and help improve communication with the Board. We provide three examples in the following.

VISION STATEMENT

Organizations increasingly communicate both internally and externally using some overarching vision statement. The idea of a corporate statement of identity has been used by organizations such as IBM ("Think") and John Lewis Partnership ("Never knowingly undersold") to provide an external explanation of something about the organization. Vision statements for health and safety are often in the form of commitments or pledges to prevent harm to employees, contractors, and anyone else. Boards often debate these visions. For instance, what do they actually mean and how will they know that they are effective in driving behaviors at all levels of the organization? These debates tend to be driven from a moral perspective but sometimes have money associated with them. For example, Cemex spent £7 million to invest in a redesign of their vehicles as a result of a campaign started and led by Cynthia Barlow, following the death of her daughter under the wheels of a Cemex lorry (see White, 2000). The vision became one that Cemex would no longer kill cyclists, as it had done on many occasions previously. So far, they have achieved their vision, and according to a recent report, Cemex has become the first company in the United Kingdom to operate and trial the new Mercedes-Benz Econic tipper truck, which has been designed to reduce the risk of hitting cyclists and pedestrians.* HF/E practitioners can help organizations to set and commit to visions that will improve the lives of those affected by the organization. These vision statements can reflect the dual HF/E goals of improving performance and

---

* http://www.theconstructionindex.co.uk/news/view/cemex-drivers-gets-clearer-view-with-new-20-tonne-econic--tipper.

well-being. Most telling, Cemex's vision was backed up by the company spending £7 million to achieve the vision; it has been able to walk the talk.

CORE VALUES

To support the vision statement, organizations develop explicit core values. These are both inward-facing and outward-facing. Core values tell employees how they are expected to behave and they explain to everyone else what to expect. Many organizations have spent considerable time and money in developing these values. Organizations such as Du Pont have developed a wide range of behavioral norms for their core values, such as the right of anyone to challenge anyone else if they witness a behavior that is unsafe, or the belief that safety on the job is only half the story. The core values ought to drive behavior, though as has been demonstrated in the courts, Du Pont as an organization has been associated with some appalling health and safety problems (see Blake, 2015).

The UK Fire Service has developed three core values that communicate to fire fighters and the public how fire fighters behave when confronted with a fire. This allows fire fighters to risk their lives in highly controlled ways to save lives but prohibits the taking of risk when lives or property are already lost. We have worked with Boards to develop simple core values that communicate expectations and behaviors. We know that individual Board members need to internalize the core values and live them. It is no good saying that you have values for health and safety then dumping tons of contaminated waste. The key message is that people drive values through their beliefs and demonstrate them through their behaviors. For example, core values might be as follows:

- Nothing we do is worth dying for.
- Everyone has the right to go home in the same state at the end of the day, physically, psychologically, and emotionally, just a bit more tired.
- Everyone is required to intervene and be intervened upon if an act is perceived as unsafe.

RISK REGISTER

Organizations mostly use risk registers as tools to demonstrate compliance with governance goals or as assurance tools. It can be debated as to how effective these registers are, but they provide an excellent opportunity for the practitioner to communicate with the organization about HF/E risks, the resources needed to manage these risks, and the success in managing the risks. HF/E practitioners must be knowledgeable about the various risk registers in use and able to use them to help achieve HF/E goals. Simple communication tools such as the X-Y-Z model of communication can be very effective. Here, a risk is described as an adverse event that might occur to prevent the delivery of an objective. The grammar for this description is: For a specific objective, the risk that X occurs because of Y resulting in Z, where X is an adverse event, Y is a single plausible reason for the event to occur, and Z is a single plausible impact. Thus, in fewer than 30 words we can outline any risk concern. We can support that with the much more detailed assessment of risk, but at a top level we will have got our item onto the register.

CONCLUSIONS

We have never had a better opportunity to communicate with senior decision-makers. The emphasis on risk and corporate risk management, controls assurance and assurance processes, have provided HF/E practitioners with numerous hooks to talk to senior teams. To do this effectively, we have to communicate in the right way, using the words that senior managers look for and using a language that makes clear how we are aligned to the broader business needs. As organizations are driven by objectives and the need to meet them, we need to help provide managers with assurance as well as deliver measureable improvements. We need to be risk savvy and understand how risk management processes work. Organizations often develop visions, core values, and risk registers. All of these can be influenced by the HF/E practitioner. Our abilities to understand systems, people, and work give us a unique insight into how organizations can get better. We have a message that many organizations are open to: We can help you improve both by helping with loss control, but much more significantly by designing out certain risks and improving the overall corporate management of risk.

REFERENCES

Bakan, J. 2005. *The Corporation: The Pathological Pursuit of Profit and Power*. London: Constable & Robinson Ltd.

Blake, M. 2015. Welcome to beautiful Parkersburg, West Virginia: Home to one of the most brazen, deadly corporate gambits in U.S. history. Accessed March 4, 2016. Available at: http://highline.huffingtonpost.com/articles/en/welcome-to-beautiful-parkersburg/.

BP US Refineries Independent Safety Panel. 2007. Report of the BP US Refineries Independent Safety Panel. Accessed March 4, 2016. Available at: news.bbc.co.uk/2/shared/bsp/hi/pdfs/16_01_07_bp_baker_report.pdf.

Dubner, S.J. and Levitt, S.D. 2007. *Freakonomics*. London: Penguin.

Financial Reporting Council (FRC). 2014. Guidance on risk management, internal control and related financial and business reporting. Available at: http://www.icaew.com/en/library/subject-gateways/corporate-governance/codes-and-reports/guidance-on-risk-management, accessed 1 September 2016.

Health and Safety Executive (HSE). 1998. Provision and use of work equipment regulations. Available at: http://www.hse.gov.uk/work-equipment-machinery/puwer.htm, accessed 1 September 2016.

Health and Safety Executive (HSE). 2012. Safety culture on the Olympic Park. HSE Research Report RR942. Available at: www.hse.gov.uk/research/rrpdf/rr942.pdf, accessed 1 September 2016.

Shorrock, S. and Licu, T. 2013. Target culture: Lessons in unintended consequences. *HindSight*. 17, 10–16.

Singleton, W.T. 1964. A preliminary study of a capstan lathe. *International Journal of Production Research*. 3(3), 213–225.

White, F. 2010. Cycling death: UK mom takes on global giant and wins. August 25, 2010. Citizen Action Monitor. Accessed March 4, 2016. Available at: https://citizenactionmonitor.wordpress.com/2010/08/25/cycling-death-uk-mom-takes-on-global-giant-and-wins/.

Yeow, P.H. and Sen, R.N. 2004. Ergonomics improvements of the visual inspection process in a printed circuit assembly factory. *International Journal of Occupational Safety and Ergonomics*. 10(4), 369–385.

29 Engaging Participants in Human Factors and Ergonomics

David M. Antle and Linda L. Miller

CONTENTS

PRACTITIONER SUMMARY

Human factors and ergonomics (HF/E) has a tradition of working with stakeholders and end users within a participatory approach. Employee/end-user engagement allows for better comprehension of specific nuances and intricate details of the workplace/environment, project context, and interactions between various elements in the system. It also allows for solutions and outcomes that are more appropriate, effective, and acceptable for the employees/end users. Engaging employees/end users and fostering various levels of participation require consideration of key roles for those involved, assurance that ethical obligations are met, and facilitation/training by HF/E practitioners. Engagement levels can evolve over time, with progression toward sustained capacity for HF/E throughout a project/enterprise. Establishing common HF/E practices, procedures, and language across levels of a project/enterprise can be achieved through employee/end-user engagement.

BACKGROUND

HF/E provides a range of theories and approaches to understand work, and inform the development, or redesign of physical, organizational, psychosocial, and/or cognitive parameters of work. Approaches for HF/E practitioners are demonstrated throughout this text, and in most instances the driving force for HF/E initiatives comes from the practitioner. However, work is a complicated array of interactions between physical, psychosocial, and organizational factors. While we are able to measure and categorize some of the physical and cognitive elements, understanding their interplay with other constructs requires more intimate knowledge of the work. Too often, guidance on HF/E is driven primarily by quantitative studies and evaluations in laboratories. However, in most cases, laboratory studies and their outcomes do not match the reality of industrial work (Buckle, 2010), and often "control" key factors that affect real-life work (Wilson, 2000). The scientific principles themselves may have applicability, but it is important to adapt the knowledge to fit the needs of the organization, production outcomes, and the employees (Carrivick et al., 2005).

To understand the organizational and psychosocial elements within a workplace, and ensure that suggested adaptations fit the workplace and its employees, HF/E practitioners need context-specific knowledge. This can only be achieved by harnessing the knowledge of individuals at that workplace. For this reason, HF/E has a tradition of working with stakeholders and end users. "Participation" of workers in ergonomics projects became a model for assessing ergonomics and human factors issues, using an approach commonly termed as "participatory ergonomics" (PE).

Although managerial involvement is a key element to program success in any HF/E initiative, this chapter will review engagement of frontline workers within a participatory approach to HF/E projects. We review the development of such approaches, outline key considerations when planning frontline worker engagement, and review strategies to achieve this engagement. To close the chapter, several projects that outline these principles are presented. Much of the background material for this chapter and the cited case studies are extrapolated from work focused on reducing work-related musculoskeletal disorders (WMSD) in various workplaces. However, the general principles of employee participation and engagement can be applied to any HF/E project.

DEVELOPMENT OF PARTICIPATORY MODELS

Participatory models for investigation and intervention evolved in health research as a means of doing research with communities, rather than on communities. This approach was in contrast to traditional research and intervention approaches that are typically detached from the larger community (Tandon et al., 2007).

The fundamental concept of participatory approaches requires that the intended beneficiaries of the research be directly engaged in the project (Cargo and Mercer, 2008). Participatory approaches in health research continued to evolve in the literature over the last 40 years using a variety of analogous terms, including: community-based participatory-action research, participatory rural appraisal, empowerment evaluation, community-partnered participatory research, cooperative inquiry, dialectical inquiry, appreciative inquiry, decolonizing methodologies, participatory or democratic evaluation, social reconnaissance, emancipatory research, and forms of action research embracing a participatory philosophy (Cargo and Mercer, 2008).

The central idea of these approaches requires that community members, organizational representatives, and practitioners/researchers are each involved in all aspects of the project to enhance comprehension of the social and cultural dynamics of the community, ensuring that knowledge is adapted and interpreted in light of these elements (Israel et al., 1998; Tandon et al., 2007).

The participatory approach to research and intervention gradually became part of HF/E. Traditionally, ergonomics was taken to enterprises with a consultancy approach, where external experts provided advocacy after a short period of observation and then left the workplace to implement solutions on their own. Unfortunately, the ergonomic information provided in this manner is not always effective at reducing problems; the solutions may not be adapted and applied to meet the reality of the workplace and the requirements of the end users (Carrivick et al., 2005). Because workplace issues involve physical, psychosocial, psychophysical, and organizational factors (Theberge et al., 2006), an external consultant, on their own, is not able to assess all of these factors in investigations and conceptualization of solutions. The ability to gain perspective on multiple factors requires social interaction and knowledge exchange between ergonomists and key stakeholders of an enterprise (Guérin et al., 2006). Recognizing the need for these interactions, HF/E evolved.

Social models of ergonomics and empowerment of enterprise stakeholders were developed in France over the last 40 years from the work of Allain Laville and François Daniellou, among others (Guérin et al., 2006). In the 1980s, quality circle experiences in Japan and participatory workplace design processes in Northern Europe and North America resulted in the development of social interaction and empowerment approaches to ergonomics that shared the same social interaction elements (Rivilis et al., 2008). Eventually, this work led to the development of participatory ergonomics as a term to describe participatory and social models of interventions and research throughout Asia, Europe, Canada, and the United States in the 1990s (Rivilis et al., 2008).

TYPES OF EMPLOYEE ENGAGEMENT

Ensuring strong participatory approaches requires careful consideration of how you will involve partners and other stakeholders. The type of participation can be somewhat variable, but it is important that the participants share power in decision-making to ensure the process remains participatory (Macaulay, 1999).

Participatory approaches can involve moving results into action among community stakeholders in two ways:

1. Gathering information from them, informing them of outcomes, reviewing information, and discussing implementation strategies. In HF/E, this involves an expert-driven approach, where the ergonomist solicits direct input from various stakeholders and employee groups to help identify work components pertinent for analysis, and they may gather stakeholder ideas for solutions to the identified issues (Guérin et al., 2006).
2. Involving participants directly in the entire intervention process (Israel et al., 1998). In HF/E, this is accomplished by developing a stakeholder/ end-user team, led by a professional HF/E practitioner, and made up of employees or their representatives, managers, and health and safety personnel (Rivilis et al., 2008). The team undergoes training in HF/E concepts, and takes a role in the selection of projects, conduct of interventions, assessment of issues, development of solutions, and the implementation of changes (Rivilis et al., 2008).

The expert-driven approaches are normally required on entry into a new community so that trust can be developed and elements for larger projects can be put into place. Ultimately, though, the team approach helps engage participants throughout an entire intervention process and helps encourage long-term sustainability.

Achieving the team approach empowers participants to influence decision-making for their own aspirations in the future (Bradbury and Reason, 2003). This process is termed "capacity-building" in participatory approaches (Cargo and Mercer, 2008; Crisp et al., 2000; Hawe et al., 1997) and it is important to the success and sustainability of ergonomics and human factors projects. Teams can have the capacity to deal with some issues and champion continued support for the projects and interventions. The HF/E practitioner will likely always be required to lead the process and provide facilitation and direction, but capacity-building allows stakeholders to influence the work itself, the workplace, health activities, etc.

KEY CONSIDERATIONS FOR DEVELOPING
FRONTLINE WORKER ENGAGEMENT

Participatory models in HF/E (as well as in broader health forums) hold many advantages. However, engagement of participants is not an easy process. Researchers and practitioners of HF/E often struggle to develop the socially constructed partnerships associated with a successful intervention (Vézina and Baril, 2009). While there are models and guidelines available in PE research that outline required success factors, there is limited information on exact strategies to facilitate the process and engage participants. Fortunately, we can learn from some lessons taken from the broader participatory research/practice forum, and also rely on some experiences of HF/E advocates.

ESTABLISHING TRUST AND UNDERSTANDING PREVIOUS EXPERIENCES

Trust and respect among the researcher/practitioner and the participants is cited to be critical to ensuring the participatory nature of the project (Bradbury and Reason, 2003; Cargo and Mercer, 2008; Israel et al., 1998). However, developing trust and respect can be complicated when there are unequal distributions of power and control, unequal representation from various groups, disagreement over distribution of resources, and disagreements over methods and timing of activities (Israel et al., 1998). To facilitate the development of trust and respect, Israel et al. (1998) recommend establishing, up front, the following:

1. An agreement over operating norms
2. Identification of common goals and objectives
3. Democratic leadership and representation
4. The presence of a community organizer
5. Involvement of a support staff (in PE, this might relate to the suggestion of developing a steering committee)
6. Opportunities for the researchers/practitioners to explain their roles, skills, and competencies to participants
7. Opportunities to investigate any prior community involvement in similar projects (in the workplace this may include examining past initiatives that were considered successful to determine factors that allowed for the effort to succeed)
8. Opportunities to identify the key members of the community who must be involved

ETHICAL CONSIDERATIONS

There is a lack of literature detailing the specific ethical "norms" to follow when working in PE projects. Therefore, it is useful for those undertaking PE projects to have an expanded knowledge of their ethical obligations. We can gain these insights through some work in broader participatory research forums, where there is a focus on ethical participatory research; for example, Macaulay et al. (1999) noted that participatory research should include strong ethical practices to allow high-quality training, development of infrastructure, and data collection and storage. In 2004, Minkler provided a review of key ethical issues in participatory-action research projects:

- *Community-driven issue selection*—It is important that the topic of the research project be important to the community you are working with. Avoid having the practitioner take complete control of selecting projects.
- *Insider–outsider tensions*—Communities may fear or distrust the presence of outside researchers or professionals. Trust and acceptance should occur naturally over time.

- *Racism and cultural humility*—If the outside researchers/practitioners are not from the same racial, ethnic, or cultural background as the community partners there can be misunderstandings and perceptions of mistreatment or racism toward project partners.
- *Participation and its limitations*—When choosing representatives to take part in projects these individuals may not represent the entire community. Care must be taken when making generalizations about the community based on the representatives. Furthermore, participation should not interfere with the normal "flux" and activities of the community.
- *Sharing findings and getting to action*—It is important to have memorandums of understanding and up-front agreement of what can and cannot be reported within the community and in outside publications, conferences, reports, etc.

Understanding Principles of Capacity-Building

The team approaches in PE attempt to develop an internal capacity for ergonomics within an enterprise. However, the theory behind capacity-building remains vague in PE literature. Fortunately, borrowing from broader participatory research helps to orient our understanding of capacity-building within a PE project. For example, Crisp et al. (2000) outlined four main approaches to capacity-building:

- *Bottom-up organizational approach*—This term describes the need to provide knowledge and information to the lower levels of a community (or organization) so that well-trained individuals can continue to work after initial funding and investment stop, and reduce the need for outside consultants. Crisp et al. (2000) suggested that careful consideration must be taken to select the training process and individuals who will be trainees. The organization must ensure that there are opportunities for individuals to practice their skills in a practical setting.
- *Top-down organizational approach*—In this model, the training programs are initiated only after conversations with stakeholders and staff during regular decision-making processes. This helps to mobilize resources (both personnel and nonpersonnel) that can contribute to building capacity.
- *Partnerships*—This approach highlights the partnership between researchers and participants, where they partake in regular communication, exchanging knowledge in a two-way flow. The participant gains "expert" knowledge and the researcher gains an understanding of contextual and community factors related to the project.
- *Community organizing approach*—This is the most ambitious approach because it involves working with the community to have them take part in solving their own health issues. The difficulty associated with this approach is that community participants must gain the skills and organizational structure required to maintain/develop procedures to address issues, make informed decisions, resolve conflicts, and share information.

It is likely that each of these approaches is needed throughout the life of a project/ program, with the first three approaches helping to develop the skills and structure required to leave behind a "community organizing approach" to sustain the work after the research project has ended. These approaches are similar to those described in PE literature (Rivilis et al., 2008; St-Vincent et al., 2000), but formal descriptions of the capacity-building theory are absent in PE.

Hawe et al. (1997) suggested that the development of administrative units related to a project, steering committees for projects, action plan developments, and development of resources for use within a community are all evidence of capacity-building during a participatory project. Hawe et al. (1997) elaborated that, ultimately, capacity-building is a means of ensuring sustainability of the benefits of a project (knowledge and activities) and improved capacity for the community to solve problems, long after a project has ended.

OUTLINING A PROCESS TO ENGAGE FRONTLINE WORKERS

Employees, as the end users of any designed intervention, must be able to support the intended changes to work. If they are not comfortable with the changes (from both prevention and production perspectives), the intervention will not be effective and may even be rejected. In addition, participation can help ensure that their issues are heard and their information/feedback can play a critical role in the development of effective workplace changes. It is important, however, to carefully consider *who* should be involved, and what their role should be.

ENGAGING PARTICIPANTS USING AN EXPERT-DRIVEN APPROACH

First and foremost, individuals who work in selected intervention areas should be given an opportunity to become involved in the project. If this is to come in the form of an expert-driven participatory approach, the best method is to gather the HF/E practitioner's initial assessment of the work (perhaps even video recordings of the work) and then review the information with key employees to understand key contextual elements of the work. This is referred to in the French language as *autoconfrontation*, "where a worker is invited to explain his/her work ... the narration gives a picture about the actual work process but may also go deeply into the cognitive content of the work, going from the concrete to the abstract. *Autoconfrontation* is a way to enrich the understanding of the problem and to analyze its determinants. In doing so, the solutions begin to take form" (Kuorinka, 1997). Validation of the results with the employee through *autoconfrontation* allows for a better understanding of work and differences in organization and technique among the employees (Mollo and Falzon, 2004). As part of this worker engagement, some key topics that might be explored include (but are not limited to) the following:

* Likes and dislikes with the workstation and/or environment
* Anthropometric considerations
* Psychosocial issues
* Job history and training

- Reasons for the location of tools/objects
- Perceptions of the pacing of work among employees
- Reasons why movements are performed in a certain manner or sequence
- Reasons for differences in technique among workers completing similar tasks
- Perceptions about how worker stature affects how they do their job
- Reasons for inefficiencies or errors
- Opportunities for cooperation and interaction among employees when completing job tasks that might augment or reduce risk

Participants selected to be involved in the project should be representative of the workforce within the area being assessed. This may require reviews/interviews with multiple frontline workers. Selection of the participants should consider the following:

- Seniority and experience levels within the department
- Gender representation (where applicable)
- Ethnic representation (where applicable)
- Anthropometric representation
- Task, environment, and work organization differences

Maintenance and engineering personnel (where these exist) should be intimately involved with the review process as well, because it is often their responsibility to refine and implement recommended changes. In addition, they have useful knowledge about what is feasible and awareness of potential upstream and downstream consequences of changes that might affect worker activities and productivity.

ENGAGING PARTICIPANTS USING A TEAM APPROACH

Establishing a team at a workplace, where capacity is built to enhance projects and build permanence for HF/E programs, is not an easy task. Engaging employees in this process requires careful consideration. While the expert-driven approach will be used by the facilitating HF/E practitioner and the team to gather information from the larger workforce, the "Ergo-Team" that has been trained by the practitioner helps to provide insight and direction in the day-to-day operations of the project.

There is no ideal size for an Ergo-Team; it depends on the size of the company, resources available, and goals of the project. However, those selected to become part of the team should be interested, motivated, committed, and properly trained. While such training is outside the scope of this chapter, a guide to training frontline workers to take part in these initiatives can be found at the PE website via the website of the Memorial University of Newfoundland (2016). This guide and associated tool was developed using first-hand experience, as well as through reviews of research and practice in PE and human factors.

Team members require strong leadership skills, a willingness to collaborate with others, good listening and organizational skills, and good communication skills. They should work well within a group setting and be open to others' viewpoints and ideas. More generally, they should be forward-looking and committed to the project (Thompson et al., 2006) in that they want to improve the knowledge and capacity within the facility in the linked areas of ergonomics and occupational health and safety. In addition, Ergo-Team members should ideally

- Be permanent employees
- Have a broad range of experience
- Have respect for their peers
- Be supportive of change
- Be seen as fair and reliable
- Be able to make a longer-term commitment to the Ergo-Team
- Not have too many additional roles and responsibilities that would be disrupted by the team, or force them to forego team activities (Antle et al., 2012a, 2012b)

The job of the team includes nine main components:

1. Appropriately, ethically, collaboratively (working with supervisors and workers), and systematically identifying HF/E issues (problems or opportunities for improvement) in the workplace
2. Selecting areas for first and subsequent interventions
3. Acquiring an understanding of basic ergonomics principles and tools for job analysis
4. Obtaining voluntary and informed consent from employee volunteers from the area who are willing to participate in an analysis of their jobs
5. Carrying out structured analyses of those jobs with a focus on relevant HF/E issues (e.g., musculoskeletal disorder [MSD] risks)
6. Presenting the results of their job and work area analysis to workers and supervisors in the intervention area and exploring with them the best ways to address the issues
7. Developing a report on their work and recommendations for ergonomic improvements in those areas for submission to the relevant decision-makers
8. Keeping employees informed, on an ongoing basis, about who the members of the Ergo-Team are, what their mandate is, about their activities, and the results of those activities
9. Following up with employees on changes that are made and evaluating those changes (Antle et al., 2012a)

The HF/E practitioner or facilitator will also need to train team members to evaluate, on an ongoing basis, their progress, lessons learned, and the overall results of their intervention. This continuous learning approach will help the team lay the foundations for effective and more autonomous future job analyses and interventions.

CASE STUDIES OF ENGAGEMENT AND PARTICIPATION IN HF/E

The following case studies are presented to demonstrate engagement of employees in HF/E projects. The methods of fostering employee participation, and the type and breadth of employee participation were directly related to the project progression and outcomes.

These cases will demonstrate how an evolution of employee participation levels over several years can lead to investment in more complex and beneficial projects. These cases will also demonstrate the value of employee engagement in identifying specific nuances and intricate details of work. Finally, we will discuss how employee engagement using participative approaches can help to develop a common language and understanding for HF/E processes.

Case Study 1: Progression Of Employee Engagement Approaches In Two Seafood Processing Facilities

A key element to ensure effective employee engagement is to avoid rushing toward a capacity-building team model. This model requires an organizational understanding and commitment to HF/E, which typically requires several years to develop and cultivate. Instead, it is best to first run a number of projects using the expert-driven participatory approach to establish both relevancy and success. This can be demonstrated through the experience at two seafood processing facilities in eastern Canada.

In Plant A, there had been a three-year project history where expert-driven PE approaches were used. This led to a number of successful project initiatives and redesigns to various production lines. Once the three-year project cycle was complete, the plant wished to develop a team approach to allow maintenance of project momentum between visits of the external experts. A team model of PE was implemented to develop this capacity (Antle et al., 2008). Training of the team occurred intermittently over a number of months, and then the team worked with the ergonomics experts to investigate and diagnose problems at the plant, which were primarily related to soft-tissue injuries among production line workers. After engaging peer employees and completing an extensive set of evaluations, the team arrived at a set of solutions to address problems with workstation design and production flow in an area where sections of crab pieces were being processed. The solutions were approved, implemented, evaluated, and adapted to ensure they met production guidelines, improved comfort, and were acceptable by the workforce (Antle et al., 2008). The approach to developing the team showed that a successful intervention could be carried out with the team approach.

A secondary seafood processing plant in a nearby community (Plant B) became interested in the positive outcomes reported by Plant A at a local health and safety conference. Plant B had not had any previous experience with HF/E projects, but wished to adopt the team approach used in Plant A. Ergonomics experts involved in the first project attempted to facilitate development of the team using the same process and sets of tools produced from the experience at Plant A (Antle et al., 2012a, 2012b). During the project, training efforts required much more time investment than in Plant A and an understanding for the ergonomics investigation process was limited. Momentum of the project was limited, and consistency of participants' attendance and engagement for training sessions, meetings, and assessment sessions were absent. While employees on the team were enthusiastic for their own involvement, they did not support involvement of their peers in the assessment and design of the recommendations. Interviews and reviews with employees were therefore limited in terms of employee engagement at the frontline, despite the best efforts of the ergonomics experts to intervene. No intervention or changes could be enacted by the conclusion of the project.

In both plants, the teams had enough momentum to complete training and begin an intervention. At Plant A, the Ergo-Team was able to function with greater independence and engagement of the larger workforce was accepted as a key step in the process. This demonstrates capacity development and showed promise for continued intervention activity. In Plant B, the team members showed commitment to the training and enthusiasm for the project. However, their inexperience with "participatory" models of assessment and intervention slowed momentum and there was resistance to peer employee involvement. This may be related to existing managerial culture.

Based on these experiences, it is suggested that an expert-driven participatory approach be the basis of projects until sufficient experience at all levels with the ergonomic process is built throughout the organization. Without this, capacity-building and a team approach are unlikely to be successful.

CASE STUDY 2: USING EMPLOYEE ENGAGEMENT TO DEVELOP EMPLOYEE JOB TRAINING IN A CONCRETE PIPE PRODUCTION PLANT

To demonstrate the power of employee engagement to advance an HF/E project, consider a request from a concrete pipe production plant. The primary objective was to compile an assessment of various task components that present risks for WMSD, and then identify key knowledge about work organization strategies, procedures, and equipment that can reduce risk of accident and injury.

Employee leaders involved with the project had been previously trained in principles of basic industrial ergonomics. Therefore, they only required specific training on involvement of employees and application of analysis techniques, detailed in a PE tool package (Antle et al., 2012a, 2012b).

The observation and analysis was facilitated by HF/E practitioners, and along with a PE team developed at the plant, employees from various work areas were involved to review the work using the autoconfrontation approach. The employees were able to point out specific components of the work that led to increased strain or difficulty for the various tasks. A list of recommendations was compiled and then reviewed at a meeting that included a larger team of supervisors, employees who complete the various jobs, and upper management.

Ultimately, several key recommendations on policy changes were proposed and accepted for implementation. These included the following:

- Changes associated with team lifting operations
- Introduction of worker rotation between jobs
- Increased spacing of concrete pipe molds to allow for postural variability
- Use of a crane to lift items inside the pipe to prevent manual lifting of heavy objects while on unstable ladders
- Increased cleaning frequency for the platform stair cart's wheels to allow for easier movement

Several engineering and equipment changes were also proposed and implemented. These included the following:

- Stools and ladders with larger platforms to allow improved foot support and stability
- Installation of mechanisms to raise pipes to a higher level when doing rework on the edges
- Use of vibration dampening gloves when using impact hammers to remove mold pieces
- Use of slide hammers to remove mold pieces rather than pry bars

Employee involvement ensured production and employee factors were considered when implementing the suggestions. However, the greatest contribution the participatory approach yielded was the identification of specific nuances in the work techniques that influence the physical efforts. These nuances are often impossible to see during short periods of observation by an HF/E professional. Explanations by experienced employees are required to truly understand intricate aspects of the work. These explanations were later implemented into training modules for new employees. For example, employees noted that novice employees rework and cut out portions of wet concrete by first working at standing height levels, and later complete the task for the areas above the shoulder and/or below the knee. This strategy results in prolonged durations in flexed trunk positions, overhead reaching, and kneeling/crouching. More experienced employees worked in linear patterns from the lower to higher levels. This is beneficial because loosening the lower concrete allows gravity to help with removal from higher levels. This reduces the amount of force output and repetition required for the task. Furthermore, it allows frequent changes in posture from kneeling/crouching, to standing and trunk flexion, to standing, or overhead work. This postural variability can help to prevent overexposure to any one risky posture.

Another example of a helpful strategy identified by employees was the timing of when to remove molds from drying concrete. If an employee removes a mold too soon, it will cause the concrete to collapse and the concrete will be unusable. However, if they wait too long, it becomes excessively difficult to remove the mold with manual efforts. Understanding how to interpret texture, feel, and color as indicators of "readiness" for mold removal are important skills that need to be disseminated to reduce physical demands.

These key strategies were formalized into standardized operating procedures and captured for training of less experienced and new employees.

CASE STUDY 3: ENABLING PARTICIPATIVE ERGONOMICS WITH A COMMON LANGUAGE

In one large multinational company, representatives of the health and safety corporate team identified that MSDs accounted for the majority of reported injuries and associated lost time or restricted duty. The organization determined that a number of facilities had started their own PE approach, but soon identified that to share knowledge, strategies, and resources they needed a common language to maximize their efforts. In each location, teams were developed using local experts to help develop and drive each PE initiative. All large capital projects (related to engineering changes) required corporate approval to move forward. In a number of cases, projects were rejected without clear justification for the decision, frustrating local facility PE teams. To help advance the efforts of local facility PE teams, the corporate health and safety team realized that not only did a common method need to be developed to analyze problems, but a common and consistent presentation procedure was also needed to move the project forward at a corporate level.

In an effort to develop common methods and strategies for PE projects, employee leaders were asked to participate in the identification and development of common ergonomics education, training, and tools for all facilities in the organization. Education and training was delivered at all sites for PE teams including WMSD risk factors, employee engagement techniques, and intervention strategies. Additionally, all engineering and facility health and safety staff were educated on tools and methodology to advantage capital project proposals that addressed ergonomic concerns to corporate level for approval. This initiative, though time-intensive, resulted in a significant number of projects being approved and implemented across the multiple divisions of the organization. Having a common language and set of tools that could be used by PE teams allowed for communication to remain consistent, and allowed corporate levels to better comprehend the return of investment potential for each initiative across the company. Ultimately, this allowed for a fair and consistent comparison with approving projects and distributing investment funds.

CONCLUSIONS

Engagement of frontline employees is critical to the comprehension of work and to the development of meaningful assessments of policy, equipment, and worker strategy needs. In addition, by relying on participatory models it is possible to attain more accurate descriptions of the work determinants, better refine solutions, and ultimately improve buy-in and acceptance of the intervention among employee groups.

However, the development of participatory methods is complex. Evolution of the program toward a capacity-building "team" approach requires a significant time investment. Throughout any participatory project, it is important to ensure proper engagement of the employees, and identify the traits required when selecting key stakeholders to involve in a project.

Continued work to explore methods for employee engagement is required. Differences in industrial contexts and guidelines to take these elements into consideration are the keys. While the information presented in this chapter is intended as a starting point, each workplace and the employees who work there are very different. Interactions between physical, psychosocial, and organizational elements will never be exactly the same between two workplaces, even if they are within the same industry or company. Engagement of key employees allows these complexities to be less of a confounding factor in intervention design and allows for a more appropriate and versatile intervention.

REFERENCES

Antle, D.M., MacKinnon, S.N., Molgaard, J., Neis, B., Vézina, N., and McCarthy, M.A. (eds.). 2008. Stakeholder perceptions of participatory ergonomics success factors. In: *39e Congrès Annuel Association de Canadian Ergonomistes 2008*, October 5–8, 2008, Château Cartier, Gatineau/Aylmer, Québec, Canada.

Antle, D.M., Neis, B., Vézina, N., McCarthy, M.-A., MacKinnon, S.N., and Molgaard, J. 2012a. *How to Implement an Ergo-Team Approach to Participatory Ergonomics*. St. John's, NL, Canada: SafetyNet Centre for Occupational Health and Safety Research.

Antle, D.M., Neis, B., Vézina, N., McCarthy, M.-A., MacKinnon, S.N., and Molgaard, J. 2012b. *Implementing an Ergo-Team Approach to Participatory Ergonomics: Training Workbook and Tools for Ergo-Team Members*. St. John's, NL, Canada: SafetyNet Centre for Occupational Health and Safety Research.

Bradbury, H. and Reason, P. 2003. Action research. *Qualitative Social Work*. 2(2), 155–175.

Buckle, P. 2010. 'The perfect is the enemy of the good'—Ergonomics research and practice. Institute of Ergonomics and Human Factors Annual Lecture 2010. *Ergonomics*. 54(1), 1–11.

Cargo, M. and Mercer, S.L. 2008. The value and challenges of participatory research: Strengthening its practice. *Annual Review of Public Health*. 29(1), 325–350.

Carrivick, P.J.W., Lee, A.H., Yau, K.K.W., and Stevenson, M.R. 2005. Evaluating the effectiveness of a participatory ergonomics approach in reducing the risk and severity of injuries from manual handling. *Ergonomics*. 48(8), 907–914.

Crisp, B.R., Swerissen, H., and Duckett, S.J. 2000. Four approaches to capacity-building in health: Consequences for measurement and accountability. *Health Promotion International*. 15(2), 99–107.

Guérin, F., Laville, A., Daniellou, F., Duraffourg, J., and Kerguelen, A. 2006. *Comprendre le travail pour le transformer: La pratique de l'ergonomie*, Third Edition. Lyon, France: ANACT.

Hawe, P., Noort, M., King, L., and Jordens, C. 1997. Multiplying health gains: The critical role of capacity-building within health promotion programs. *Health Policy*. 39(1), 29–42.

Israel, B.A., Schulz, A.J., Parker, E.A., and Becker, A.B. 1998. Review of community-based research: Assessing partnership approaches to improve public health. *Annual Review of Public Health*. 19, 173–202.

Kuorinka, I. 1997. Tools and means of implementing participatory ergonomics. *International Journal of Industrial Ergonomics*. 19, 267–270.

Macaulay, A.C., Commanda, L.E., Freeman, W.L., Gibson, N., McCabe, M.L., Robbins C.M., et al. 1999. Participatory research maximises community and lay involvement. *British Medical Journal*. 319(7212), 774–778.

Memorial University of Newfoundland. 2016. Participatory ergonomics. Accessed March 28, 2016. Available at: http://www.participatoryergonomics.mun.ca/.

Minkler, M. 2004. Ethical challenges for the 'outside' researcher in community-based participatory research. *Health Education & Behavior*. 31(6), 684–697.

Mollo, V. and Falzon, P. 2004. Auto- and allo-confrontation as tools for reflective activities. *Applied Ergonomics*. 35(6), 531–540.

Rivilis, I., Van Eerd, D., Cullen, K., Cole, D.C., Irvin, E., Tyson, J., et al. 2008. Effectiveness of participatory ergonomic interventions on health outcomes: A systematic review. *Applied Ergonomics*. 39(3), 342–358.

St-Vincent, M., Toulouse, G., and Bellemare, M. 2000. Démarches d'ergonomie participative pour réduire les risques de troubles musculo-squelettiques: bilan et réflexions. *Perspectives Interdisciplinaires sur le Travail et la Santé*. 2(1). Available at: http://pistes.revues.org/3834.

Tandon, S.D., Phillips, K., Bordeaux, B.C., Bone, L., Brown, P.B., Cagney, K.A., et al. 2007. A vision for progress in community health partnerships. *Progress in Community Health Partnerships: Research, Education and Action.* 1(**1**), 11–30.

Theberge, N., Granzow, K., Cole, D., and Laing, A. 2006. Negotiating participation: Understanding the 'how' in an ergonomic change team. *Applied Ergonomics.* 37(**2**), 239–248.

Thompson, G.N., Estabrooks, C.A., and Degner, L.F. 2006. Clarifying the concepts in knowledge transfer: A literature review. *Journal of Advanced Nursing.* 53(**6**), 691–701.

Wilson, J.R. 2000. Fundamentals of ergonomics in theory and practice. *Applied Ergonomics.* 31(**6**), 557–567.

Vézina, N. and Baril, R. 2009. Apprendre à intervenir: difficultés rencontrées par de jeunes ergonomes et stratégies. Paper presented at the *40th Annual Conference of the Association of Canadian Ergonomists—'Ergonomics—Think it Live it,'* Loews Le Concorde Hotel, Québec City, Canada.

30 Writing as a Human Factors/Ergonomics Practitioner

Don Harris

CONTENTS

PRACTITIONER SUMMARY

The ability to write clear, concise reports is a mandatory skill for practitioners, not an option: ultimately, your report is a major part of what you are paid for. Your work is only as valuable as the clarity of your communication: if people can't understand what you have done and what your conclusions are, then what was the point? Clear communication and engaging with your target audience will make it more likely that you will be engaged for follow-on work. Learning how to write effectively will save you time and effort.

INTRODUCTION

From the outset it has to be said that there is no such thing as a "standard" format for a practitioner report. There are several different basic types of report but every report should satisfy the specific individual requirements of your client. Clients may be internal (in-house) or external, as in a consultancy report. And this introduces the most important principle when writing as a human factors and ergonomics (HF/E) practitioner: you are writing for your client.

Before writing any form of practitioner report five basic questions should be addressed. In no particular order they are as follows:

- What are the objectives of the work?
- Who will be the "consumers" of the report (*and* there may be more than just one audience)?
- What format of report will be required (will several different versions be needed for different readerships)?
- When is/are the various report(s) required?
- Are there any issues of confidentiality (either with regard to the report or material you may reference in your report)?

Furthermore, don't just focus on the end report: what about interim reports, monthly client briefings, and so forth? Although this chapter is almost entirely about "final" reports, these other documents are also important for client engagement, especially in long-term projects. A guiding principle is to "agree before you write"; this refers to timing, content, method, and detail of all communications.

AUDIENCE

One of the greatest challenges when producing any report is writing for the target audience, or more specifically deciding exactly who the target audience is, their knowledge of the issue being investigated and the HF/E principles in question (this is often easier when writing for an in-house client). And while the readers may only have limited HF/E knowledge it is vital that the report never, in any way, "talks down" to them. You can usually assume that anyone reading any version of the report will be in some kind of technical or management position, and will usually be an informed and educated reader.

This illustrates a fundamental communication issue faced by any HF/E practitioner. When writing a journal or conference paper (or even a book chapter…) the target audience is relatively homogeneous: readers will be fellow scientists with a similar background. In writing this chapter I am making a fundamental assumption about the readership; they have a background in HF/E, most probably university educated (with the likelihood of a postgraduate qualification and/or experience); they will be familiar with producing basic, laboratory-style reports in an academic style; they have a career in HF/E, and they are reasonably fluent in the English language (but note that we operate in an increasingly international industry).

Such uniformity of experience cannot be assumed when writing a practitioner or consultancy report, and different versions of your report are likely to be read by people with different backgrounds. If you have to produce a technical report, then it can be expected that these readers will be familiar with the problem being addressed and will have a grasp of the scientific/technical issues being investigated. They may also be familiar with some HF/E issues. Shorter versions of your document may be required for middle/senior management, e.g., executive summaries. It is less likely that these people will have an HF/E background but they will be familiar with the company's requirements. They will have experience in management (and may have originally come from a technical background). If any version of your report reaches Board level, similar assumptions can be made. However, try to make relatively few assumptions about the background of your readership, other than the fact that they are well-informed and intelligent.

You should be able to get a reasonable "feel" for your readership from any feedback to your initial work proposal. Pay attention to the questions that people ask: Did they understand what you were proposing? How many questions did they ask to clarify things? What was the nature of these enquiries? Furthermore, do not be afraid to ask your project manager in the client organization for their advice, particularly if different versions of the report will be required for different parts of the company.

FORMATTING AND STYLE

Large multinational companies often use advanced document control systems. When working for an external client, establish right from the start if there are specific house styles or even file-naming conventions. Check software versions not just for word processing software but also graphics packages and any other specialist software. Some companies have report templates to "help" with formatting! Ask if one of these is available and practice using it. And finally, don't forget the obvious things—like what size paper is required (especially if working for a transatlantic company)!

STRUCTURE

Given that there will be a whole range of potential readerships for your reports, it is essential to maximize readability. Keep all reports as short as possible and well structured. There are golden rules for good structure. Always work from the general to the specific, setting out the big picture first before filling in details. This helps to build a mental framework. There is an old adage when making presentations which applies equally to writing good reports:

* Tell them what you are going to tell them.
* Tell them it.
* Tell them what you told them.

Remember that your readers are human—always put yourself in the position of your audience. They do not have perfect memories, may not be familiar with the subject matter, and may not read your report all at one sitting. Good structure using headings and subheadings to signpost the work helps the reader. The occasional summary at an appropriate point can act as a well-placed aide-memoire (but don't overdo it).

Although most of what follows will revolve around the tried and tested "Introduction; Method; Results; Discussion; Conclusions" format, this is not mandatory. Indeed, this is inappropriate for some of the types of client report. There is no such thing as *the* correct way to structure a report. Don't feel as constrained when writing a practitioner report (in terms of the structure) as when writing an academic paper. There are report structures that work, and structures that don't. And finally, always tell your reader *why* they are being told *what* they are being told and draw the necessary conclusions and inferences based on your findings.

STYLE

Getting just two basic aspects of writing style right can both enhance readability and avoid irritation: (1) getting tenses right (and consistent) and (2) using short, simple sentences.

It is traditional that scientific writing is undertaken in the third person to emphasize scientific impartiality. This is not quite as important in practitioner reports but in general, if in doubt, stick to the third person even though doing this can occasionally lead to clumsy sentence construction. Try to be as clear and direct as possible in getting your message across.

For some, tenses can be much more problematic but there are some simple rules to aid consistency and clarity. When describing work previously published (usually in the Introduction and the Discussion) you are telling the reader what was *done* (past tense). The Method section is simple. You are again telling the reader about something that you *did* in the past, relative to when they are reading the report so again, use the past tense.

The Results section is often best presented mostly (but not solely) in the past tense. You are telling the reader about the results of analyses performed earlier. However, when drawing a reader's attention to a particular result, the present tense is often the one that fits most naturally.

The rules for use of tenses in the Discussion section are the same as those used in the Introduction. When describing work of other researchers use the *past* tense; when describing the results, use the *past* tense. However, when making inferences and explanations based upon your work, use the *present* tense. Imagine you are having a conversation with your reader; by doing this you naturally use the "right" tense for the situation. If in doubt, speak it out loud. You will find that you use the most appropriate tenses for the situation quite naturally. This also helps to avoid overlong, "clumsy" sentences.

With regard to the length of sentences, keep them as short as possible. Whenever you use the word "and" consider splitting the sentence into two. I use a simple rule: if a sentence is three lines in length, then I start to worry that it is too long. If a sentence is over four lines long, then *it is* too long. However, first get the text on the page— *then* edit it. You don't want to lose your train of thought tidying up the English.

Also try to avoid too many levels of embedded clauses. Just one is enough. Use too many and you often have to read the sentence again to remind yourself how it started. Use the active voice as far as possible (this also helps to give the impression that you are writing with some authority). Writing without equivocation also helps to build confidence in your reader that you know what you are talking about. It also makes for snappier, punchier sentences which have a clear point to them and makes your whole report easier to digest.

Finally, always remember that HF/E is an international discipline and you may be writing for an audience who does not have English as their first language.

FIGURES

When producing client reports it is permissible to liberally illustrate all sections of the report with diagrams and figures (as appropriate). However, make sure that your figures are not over complex, do not contain too many elements, and/or are labelled in a very small font "to fit it all in." Keep figures as simple as possible. A well-drawn figure can make complex interrelationships easy to understand but a poorly drawn figure can create confusion.

While illustrating a report liberally can aid structure and understanding, it is still important to avoid including figures that are simply a waste of space. For example, a histogram depicting frequency counts in just a few categories is far better explained in a simple table.

The main driver for the format of figures still remains the size of the figure on the *printed* page, even if it is produced as an electronic document. The physical page size is always a hard limit. Before preparing any figure assess the maximum area on the printed page. Often the ultimate limiting factor is not the number of lines or the size of the boxes, it is the size of the labels. These *must* be legible so don't just keep making the font smaller to fit it all in.

Unlike academic journal papers, practitioner reports are often reproduced in color and are available electronically. This makes producing attractive, easy to read figures much easier as it is possible to use color coding to designate categories if necessary. However, resizing figures to fit in the page can still be problematic. When using software packages such as PowerPoint®, labels will often fail to resize and/or change to a default font. As a result, after finalizing your figure it is best to convert it to another format, usually Tagged Image File Format (TIFF); Joint Photographic Experts Group (JPEG); Graphics Interchange Format (GIF); or a device independent bitmap (BMP). These make resizing the figures much easier but you cannot edit these files, so make sure that everything is correct. Packages employing vector graphics (e.g., Adobe Illustrator®, Microsoft Visio®, or Corel Draw®) avoid these problems.

TABLES

The presentation of tables is a fine balance between not wasting space and producing a table that is too dense and complex to extract information from it reliably. Again, do not be tempted to present complex tables using a very small font "because you needed to fit it all in." If it won't fit into a single table then consider presenting it in two tables. Presenting the table in a landscape format is another option (but this can lead to formatting problems, especially if using company templates). The format of the tables should reflect the basic questions asked by the client. If necessary their layout and content must complement the analyses performed.

Most publishers' guidelines require the use of a minimum number of horizontal lines to separate major table entries and no use of vertical lines. This is not a bad format to follow. It produces reasonably attractive and readable tables if adequate space is allowed between columns (and rows) to promote readability. Row and column headings in any table should be self-explanatory. It *is not* acceptable to use directly the output from statistical analysis software packages. Furthermore, just because the computer calculates results to eight decimal places, it does not mean that such precision is required. If you use a five-point Likert scale, is it meaningful to report means in such a manner? Finally, table headings should be descriptive and self-explanatory even to a reader who has not read the main text of the manuscript. Units of measure should be included (if appropriate).

TITLE

Don't get clever. The title helps provide a framework for the reader's understanding of everything that follows. It should reflect what you did or is a simple statement of what was produced. Keep it short, simple, clear, and to the point. This may make for a boring title but the reader will know what your report is about.

TYPES OF CLIENT REPORT

There are several different types of report that a client may require and clients often need several different reports. The target audience(s) for each should be ascertained from the start.

Good report writing usually works from the general to the specific. The following is a bit of structure, a framework upon which to hang your understanding of what follows. The first practitioner-type report to be discussed is the full "technical" report; this will be followed by a description of the "nontechnical" report (but just highlighting the differences). The next subsection will look at the "executive summary" followed by the shortest version of the practitioner report—the "quick brief." There are many "shades of grey" in between these types of report. It is by no means a definitive list, but for simplicity we'll just stick to these in this short overview.

TECHNICAL REPORT

All projects will probably require a full technical report. This may either be for the client organization (if it is an external commission) or for your company's internal documentation. This will be the type of documentation most familiar to HF/E practitioners. Its structure usually follows the well-worn path of introductory material providing a background to the project; Methodology describing what was done to investigate the issues; Results containing the data obtained and the output of analyses performed; and a Discussion section interpreting the results and outlining their implications. All reports should finish with a strong Conclusions and Recommendations section.

So far, so obvious. The key differences between an academic document and a technical practitioner report lie in emphasis and detail, particularly in the Introduction and the Discussion/Conclusions sections. However, before dealing with the differences, let's discuss the similarities.

The Method and Results sections (including any details of statistical analyses) are much the same as would be found in a university thesis or research report. The Method section describes how you collected the data. The section often begins with a description of the target sample (e.g., size, demographic requirements, and/or skills and attributes needed). This should be followed by a subsection describing equipment or facilities used to collect the data. Descriptions of questionnaires or structured interview schedules should also be included in this section, as should reference to published, commercially available instruments, or scales which are not commercially published but which are well-known and derived from the open literature. In experimental studies it is helpful to provide an overview of the basic experimental design. This should be followed by a description of what participants did, the trials context, and the relevant parameters applied (for example, starting conditions or task completion criteria). The measures taken (dependent variables) will usually be specified next. Finally round off the Methodology with a section describing the procedure employed, a detailed account of what was done to prepare and execute the data collection.

The Results section has two (possibly three) basic components: (1) a description of the sample (if not already described in the Method section) and (2) the presentation of the data and analysis. Another short subsection may also be required to describe the manner in which the dependent variables were computed from the raw data prior to analysis.

The description of your sample should be done in prose rather than tabulated. The reason for describing your sample is to demonstrate that it is a relevant and representative sample, so only report the variables necessary to do this. Keep it short and to the point.

Any section on the treatment of data, if necessary, comes next. The Method section describes how the data were collected but in some instances the dependent variable(s) used are computed from the measures initially collected. This subsection describes how this is accomplished.

The real "meat" of any technical report comes in the description of the dependent variables and the results of the analyses performed. This will usually involve either the tabulation of the results or their representation in some kind of figure; *do not use both!* Never repeat yourself in any way (this is a very common mistake).

In a technical report, tables are often the best way of presenting detailed accounts of the data. However, they need to be clear. The layout and content of tables should reflect the analyses undertaken. Table headings should be descriptive and self-explanatory if a reader has not read the main text. Any units of measure should be included. Row and column headings should also be self-explanatory.

In general, when reporting statistical analyses avoid including the full output tables from software packages (e.g., full analysis of variance [ANOVA] tables). Results should be reported in the standard abbreviated format in the prose and call out specifically to the appropriate figure or table. Do not interpret the results in the Results section: this can wait until the Discussion.

Reporting the results from qualitative analyses in a succinct manner can be a challenge. Often the Results and Discussion run together so that analysis and interpretation of the data run side-by-side. This is often neater and allows for an easier discussion of the resulting issues. When interpreting qualitative data the emphasis is upon providing explanation for phenomena, not on categorization and quantifying their frequency of occurrence. Just a single observation or comment among many might be the key to the interpretation of all the data obtained. If the qualitative data analysis is being driven with reference to a theoretical framework then each inference drawn should explicitly refer back to the aspect of the model under examination, supported by evidence from quotes or observations. This helps to establish an audit trail, from data to analysis to inference. The judicious use of quotes also brings the content of the section to life for the reader.

As noted earlier, the major differences between an academic document and a technical report lie in the Introduction and Discussion sections. Any technical report describes work that has been undertaken for a specific reason on behalf of an internal or external client. The project brief is a good place to start any report. It should include the objectives and scope of the project and any specific limitations to the work undertaken. Long academic introductions drawn from a comprehensive literature review describing the theoretical context of the work are not what is required in this instance. Nevertheless, you should still be aware of the theoretical background before undertaking the commission: it is just that this material need not appear in your report. Work for any client (internal or external) will usually be undertaken to address a specific issue and some background work may already have been undertaken. The relationship between the work described in your report and work previously done is important to describe but make sure that you are *au fait* with any commercial confidences.

There is a good chance that your report will be published in an electronic format. In this case, you will need to know the reporting requirements and pay considerable attention to any hyperlinks you include. This comprises both internal hyperlinks in the document (e.g., from the contents pages to various sections) and hyperlinks to sources external to the report. These may be electronic materials on public websites or links to materials on the client company's document servers. Make sure security or distribution limitations are clearly established and appropriate controls are in place before you inadvertently release a company's confidential data to the world!

The introduction should end with a clear statement of the objectives of the work and reiterate any significant constraints. Don't raise expectations at the beginning of the document that you fail to deliver on at the end of it!

The Discussion section is the focal point of any report. The whole value of the work is contained in the Discussion and Conclusions. What did you find? What does it mean? What are the implications for your client? This is what they pay for. This section is also your opportunity to sell the quality and the importance of the work that you have just completed. There are two main aspects to the Discussion: (1) interpreting your findings and (2) placing them in the context of the material contained in the Introduction.

In a technical report the practical significance of your results are of paramount interest. But remember that your readers are human. By the time they start reading the Discussion it will be 5,000 words, half an hour, two phone calls, and a cup of coffee since the Introduction. Help them out—give them a gentle reminder of the project brief and how you addressed it.

The emphasis in this section should be on *interpretation*. You should always proffer a reason for the results you obtained, otherwise you are falling into the trap of description and not discussion. Whenever drawing any inference you must always refer to the evidence from which you draw your conclusions (e.g., table, figure, or quotation). In doing this you are establishing a verifiable audit trail to support your reasoning. Do not be tempted to "overinterpret" your results but at the same time never let the reader make their own mind up—tell them what their opinion should be!

Interpreting your results is only half the battle. You also need to place your work within the context of the material in the Introduction. Specifically point out how it complements earlier work, takes it a stage further, or maybe even contradicts it. A good report should come full circle. The mark of a good Discussion is that it provides *explanation*.

As a final note on the writing of the Discussion section, remember that irrespective of whether your work is for an internal or external client, you always need to "sell it." Don't undermine yourself by firing a volley of unnecessary self-criticism at the work you have just completed. This leaves the reader with a poor impression. Be honest about limitations to your study but don't be self-destructive.

Conclusions should relate directly to the aims in the project brief, which is set out in the Introduction. They should be short, sharp, and to the point. It is your expert and professional opinion that the client wants—based on all the evidence—so you need to be clear and direct about this. Again, it's all about coming "full circle" to where you started off. Recommendations can again have two aspects to them: (1) practical recommendations and (2) recommendations about how to follow up the study with more work. End on a high note—create a good, final impression.

Many technical reports commence with an abstract of some form. These are about 150–200 words, one short sentence describing each major section of the report (maybe two or three for the Discussion and Conclusions). The key to a good Abstract is that it gives an overview of what the report is all about. Not every aspect has to be included. Place greatest emphasis on the findings. Along with the title, the Abstract provides a framework for understanding of everything that follows. However, you may want to consider if you need a traditional Abstract. If your report has an executive summary (see later), which is essentially a nontechnical extended Abstract, do you need an Abstract? This is a judgment call, but always avoid unnecessary repetition.

NONTECHNICAL REPORTS

A nontechnical report to a client describes exactly the same work, undertaken and analyzed in the same way as the full technical report, but it is written for a reader without a background in HF/E (or possibly even science or engineering). You must still assume that you are writing for an intelligent reader, though (and this report will probably also get a wider readership than any technical report). It is also fallacious to assume that work contained in a nontechnical report utilizes a simpler methodology or analysis. The work is exactly the same, it is just the way that it is described that is different. Do not fall into the trap of "only presenting simple statistics" in a nontechnical report because "the client won't understand anything else." One job when writing such a report is to make sophisticated methods or analyses easy to understand by nonspecialists. Don't dumb down the science in the report: explain it more clearly in a way suited to your audience.

The introductory section can stay much the same as in a technical report, although you may just want to check for any particular technical terms or other pieces of jargon. Using incomprehensible technical terms doesn't make you look clever. From the viewpoint of the reader (who may be paying for it...) it's just annoying.

You can abridge the Method section considerably as the reader of a nontechnical report is unlikely to want to replicate the methodology. Think of the readership more as A.A. Gill (restaurant critic/consumer of food) rather than Delia Smith (writer of cookbooks/producer of food). They will (metaphorically) want to know that their chicken was roasted but don't need to know the finer details about where the garlic was pushed.

The Results section is the real challenge. Do not reduce everything to colorful pie charts and 3D graphs; this is not necessarily the right thing to do. The right thing to do is present the data in the most appropriate way for the target audience. If graphics are not required then don't use them. Nontechnical reports are descriptions of your work for non-HF/E professionals, not comic books.

Step through the analyses in a logical sequence and describe what each does and why it has been done. For example, you may describe an ANOVA with a covariate as a test to see if there is a real difference between groups but taking into account the effect of another variable and removing its influence from the data. If you are using curvilinear regression, you may want to preface the analysis with something to the effect that "not all relationships between variables are linear." Use a graphic if it makes it easier to understand your results or if it emphasizes the point you are trying to make.

When it comes to the discussion of your results there are a couple of options. In general, the content of the material in the Discussion should be almost the same as in the full technical report. However, again you may want to consider running the results and the discussion together so that their interpretation is presented adjacent to the data supporting it. Again, this is common when reporting qualitative research findings. You may find that it makes your explanations easier to follow for the intended readership. The conclusions and recommendations are much the same as in a full technical report.

EXECUTIVE SUMMARY

An executive summary should be a maximum of 2–3 pages or 1,000 words. It is a nontechnical, extended abstract appearing before a full report. The likely audience for a report of this type will be middle/senior management and it is unlikely that they will have a HF/E background. However, they will be familiar with the company's operations and the issues it is facing. It is also likely that one of the readers of the executive summary sanctioned the work you have completed! This is also the part of any report that is likely to be most widely read. It is worth spending time on.

Start the executive summary with the project brief and just a couple of short paragraphs describing the background material, especially if your work is based on previous work undertaken by the client. Finish the Introduction with a clear statement of the objectives.

Readers of the executive summary will have little interest in the methodology. Simple statements briefly outlining the major features of the participants, equipment, and procedure will suffice. No more than one paragraph should be needed here.

The major component of interest concerns what you found and why. Again, consider running together the main results along with their interpretation. Depending upon the nature of your work, it may not be prudent to try and present more than two or three headline findings. These should relate directly to the project brief and statement of objectives. Keep the message strong and don't equivocate. Use simple, clear graphics if they help get your message across. Banish phrases such as "there is some suggestion that..."; "the data could be interpreted as indicating..." and the word "however"! No references are required. The final part should clearly spell out any implications and recommend a clear course of action. Always try to create a good final impression.

Quick Brief

Originally a quick brief was used to update a senior military officer about an issue although it is now becoming an effective communication tool for informing senior executives.

A quick brief can be thought of as the written form of the one-minute "elevator pitch." Its purpose is to inform senior personnel about what was done, why and what the outcome was. Do not attempt to make more than one main point. The emphasis should be on the implications of the results (and perhaps suggestions for what should be done next). Detail is not required and data should only be presented to reinforce any argument (provided that it can be conveyed simply and clearly).

It should start with a simple statement of the requirement and a *short* description of pertinent background material. Any description of the methodology should be minimal (e.g., "a survey was undertaken..."). Numerical results should only be provided if absolutely required and in the simplest form possible (preferably graphical). The greatest weight should be placed upon describing the implications of the results within the practical context set out in the project brief. Clear concise conclusions should be presented. No references are required. Ideally, the presentation should also be reasonably eye catching if it is to be effective, especially if it is to stand out from other Board papers.

A quick brief should not occupy more than a single side of A4 or exceed more than about 300 words. It should only take two minutes to read and digest the implications of its content.

OTHER DELIVERABLES

In addition to the various reports it is becoming increasingly common to make available other formats of deliverable. These are worth mentioning briefly. Clients may ask for a presentation to accompany any report—not the end of a contract presentation but one to support the report material in subsequent internal (or external) briefings by the customer. These presentations may even take the form of video briefings.

If the client requires the raw data, this should also be presented and documented properly; it is their data! A spreadsheet containing thousands of numbers in rows with incomprehensible column headings is of little or no use to anyone.

TELL THEM WHAT YOU TOLD THEM...

Writing practitioner reports is about doing the simple and obvious things properly. Make sure you know what your client wants and write for your intended audience. This will differ between the types of report. Keep it clear and to the point. Remember the work undertaken was to meet the requirements of your client. The content and presentation of your report will determine how easy it is for them to utilize the results. There is a big difference between "simplifying things" and "explaining complex things clearly." Make sure you are clear on this distinction. A judicious quotation can help to illuminate a complex point. You will rarely go wrong in writing if you continually put yourself in the position of your reader. Remember that it is their report, not yours. And always ask yourself this question: "would I want to read what I have just written?"

FURTHER READING

Harris, D. 2012. *Writing Human Factors Research Papers*. Aldershot, UK: Ashgate.

31 Public and Social Media Engagement for Human Factors/ Ergonomics Practitioners
Outreach, Research, and Networking

Dominic Furniss

CONTENTS

PRACTITIONER SUMMARY

Social media is changing the way people communicate and connect regardless of whether we choose to embrace it. Practitioners and companies can use this to their advantage to increase their visibility and connect to professional networks. We should be connecting with the world, informing people about human factors and ergonomics (HF/E), inspiring young adults to pursue HF/E careers, and impacting beliefs and actions. We need to keep conversations about HF/E alive both within and outside our community. There are different levels of public engagement benefits from the societal level, to the discipline, project, and individual levels.

Outreach work takes time and effort, but there are many forms of public and social media engagement to suit different people. This chapter reviews HF/E engagement, identifies gaps, and provides examples for inspiration via a smorgasbord of approaches.

INTRODUCTION

HF/E has traditionally had little contact with the public as a profession. Most contact has been with industry and university education, with relatively few exceptions, for example, www.ergonomics4schools.com. However, these trends are changing for all disciplines. There is a cultural shift toward more outreach and public engagement, fueled by the potential positive impact on the discipline, the public, and professionals, and afforded by new engagement opportunities through social media and Web 2.0. HF/E practitioners and our discipline can benefit from this new world.

Outreach activities are performed to engage with people and communities who would not normally engage with a particular area of knowledge or application. These activities can inspire the next generation to pursue careers in related disciplines. They can educate and inform people about research and services they might not be aware of, and influence people's beliefs and impact changes in behavior. Outreach activities can inform the zeitgeist and culture of our time. More specifically, outreach activities can be targeted to bring about change for a particular group and purpose, for example, putting ergonomics on the school curriculum or putting human factors on the patient safety agenda.

Public engagement has a long history in the United Kingdom. In 1825, The Royal Institute began running Christmas lectures to introduce a subject to a young audience through spectacular demonstrations. Michael Faraday founded these lectures and gave many of them over a 30-year period. The British Broadcasting Corporation (BBC) televised the first Christmas lecture in 1966, and they are still running today. These lectures were designed to inform and inspire people about science.

The profile of public engagement has risen up the agenda in academia for a number of reasons, in part due to trying to secure the continued support of the public to fund research through taxpayer's money, during the financial crisis. In 2010, UK funders of research joined together to sign the *Concordat for public engagement with research* to define expectations and to start to embed public engagement in academia. In 2013, *Inspiration to engage: concordat for engaging the public with research* outlined successful case studies in public engagement and described the benefits of embedding public engagement in the research culture.

Over the past 10 years or so I have been fortunate enough to work within an environment that has fostered a culture change in public engagement. I worked as a postdoctoral researcher at University College London (UCL) on Computer–Human Interaction for Medical Devices (CHI+MED), from 2009 to 2015. UCL was one of the six Beacons for Public Engagement in the United Kingdom. As a beacon, UCL carried out a number of initiatives and encouraged researchers to get involved in public engagement. Separately, CHI+MED was a large research project that had public engagement written into its deliverables as some of its Principle Investigators were already involved in public engagement work (e.g., see notes on cs4fn later in this chapter). This context allowed me to create and explore HF/E public engagement opportunities. In 2014, I received a UCL Public Engagement Award for my contributions in this area, some of which I outline below. While these activities were in the context of a university setting, the experience applies also to practice more generally.

WHAT MIGHT PUBLIC ENGAGEMENT MEAN FOR THE HF/E PRACTITIONER?

Public engagement knowledge and lessons can be drawn from academia through to industry. Like academics, there are many different reasons for HF/E practitioners to consider exploring public engagement activities, and the reasons are as follows:

- A sense of duty to give back to the community and to society
- To raise their profile, and that of their company and their discipline
- To communicate passion in their subject
- To inform people about their area of expertise
- To inspire the next generation of HF/E researchers and practitioners
- To influence policy makers
- To gain skills and experience for their continuing professional development (CPD)

The discipline as a whole needs more outreach and public engagement; this needs to involve researchers and practitioners in order to represent the fullness of the discipline. Both researchers and practitioners need supportive environments to do this as part of their work, which means it should have an organizational purpose too.

On the industrial side, corporate social responsibility (CSR) refers to activities that contribute to some social good beyond direct business targets. This is most closely associated with the "sense of duty" bullet point above. However, engagement activities have the potential to bring direct benefit to businesses, for example, in raising their profile for recruitment and services and enhancing the skills of their workforce. Porter and Kramer (2006) offer a more mature discussion about CSR and distinguish between two kinds. Responsive CSR depends on being a good corporate citizen and addressing every social harm the business creates. Strategic CSR is selective about social initiatives and aligns them with core long-term business values to create competitive advantage. "The essential test that should guide CSR is not whether a cause is worthy but whether it presents an opportunity to create shared value—that is, a meaningful benefit for society that is also valuable to the business" (Porter and Kramer, 2006, p. 83). Creating shared value for accessibility services, for instance, might involve partnering with a relevant charity to create free online materials, or running an open workshop to understand the needs of local clients.

Public engagement and outreach can take many forms and it can be done on different scales. I highlight some of these for inspiration for companies and practitioners in the following.

DIFFERENT TYPES AND LEVELS OF ENGAGEMENT BENEFIT

Benefits from engagement can be found at different levels: from broad engagement at a societal level, to engagement that promotes a discipline or subject area, to more focused engagement around particular issues and projects, to engagement that impacts the individuals taking part (see Figure 31.1). The distinction between these levels is in some sense artificial, but it helps provide a rough framework for thinking about the scope and aims of engagement work.

BENEFIT AT SOCIETY LEVEL

One of the purposes of public engagement is to create an informed society inspired by different areas of knowledge including science, engineering, and the arts. Engagement work seeking to impact at the society level has a broad reach, for example, partnering with broadcasters and media agencies. An early example is the BBC's broadcasts of The Royal Institute's Christmas Lectures.

People who frequently engage at this level sometimes become celebrity scientists. One of the most well-known today is the physicist Brian Cox, but recognizable faces and other household names include Robert Winston (medical doctor with interest in fertility), Richard Dawkins (ethologist and evolutionary biologist), Patrick Moore (astronomer), Jim Al-Khalili (theoretical physicist), Kathy Sykes (physicist), Mark Miodownik (materials scientist), and Alice Roberts (anatomist). All of these have presented high-profile lectures and television shows like the BBC's *The Sky at Night*, *Wonders of the Solar System*, and *Horizon*.

Along similar lines and a bit closer to home, Dr. Kevin Fong (anesthetist) presented a *Horizon* program titled *How to Avoid Mistakes in Surgery* in 2013. This featured many HF/E themes applied to surgery, including research published by Ken Catchpole (see Chapter 13 on HF/E in healthcare) on translating the learning from how a Formula One pit crew operates and coordinates their work, to how surgery teams should organize their handovers to intensive care (Catchpole et al., 2007). Other high-quality documentaries with an HF/E flavor include BBC's *Piper Alpha: Fire Night* and the US Chemical Safety and Hazard Investigation Board's *Anatomy of a Disaster* (CSB, 2008). Although strong HF/E themes appear in these programs

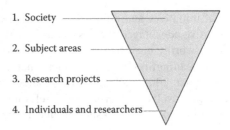

1. Society

2. Subject areas

3. Research projects

4. Individuals and researchers

FIGURE 31.1   Benefits of public engagement at different levels.

their purpose is not to promote HF/E per se, that is, the programs are often focused on the domain issues and safety, and HF/E remains a small and implicit part. Creating programs for national broadcast is time-consuming and expensive. However, modern technology allows us to create cheap films on a more flexible basis, which can be widely distributed through social media. An engaging example of this is a series of YouTube videos released by the Canadian Space Agency. Here the astronaut Chris Hadfield answers from space, questions from high school children. In one fascinating episode he shows how astronauts wash their hands in space (VideoFromSpace, 2013), which has been viewed by millions of people.

At UCL, we have been experimenting with creating YouTube videos for CHI+MED to promote issues around patient safety and medical device design. These include *Man-Machine Nightmares* (Furniss, 2010) and *Microwave Racing* (Furniss, 2010b), which have been viewed by thousands of people online. *Microwave Racing* has won awards from the Human Factors and Ergonomics Society (HFES) and CHI (the flagship conference on human–computer interaction). Online videos have the potential for global reach: we have had requests to use these videos for training in Australia and have heard about *Microwave Racing* being shown in a meeting in the United States and used in classes in Mexico.

Following this we have designed a *digital stories* exercise for MSc students at the UCL Interaction Centre (UCLIC). This acts as a fun ice-breaker in their first week on the course, and gives them skills to use *digital stories* as a cheap way to create videos and get a taste for public engagement (Benedyk and Furniss, 2011; Furniss and Benedyk, 2011). One successful example, *Why Buttons Go Bad* (UCLTV, 2010) was shown at CHI and hosted on UCLTV. It is a great example of what can be achieved with the right ideas and skill sets in only a few hours across one week.

There are numerous examples of interesting ergonomics and human factors–themed videos on YouTube, which can be curated through YouTube channels. These can be hosted personally or through company or department accounts. Our department, UCLIC, has its own YouTube channel to host the various videos produced by the department, which include students' digital stories, public engagement videos, and recorded talks, comedy gigs, and lectures. This functions as a public archive but also facilitates the ongoing engagement with this material, while raising the profile of UCLIC.

Creating open, engaging, and informative videos provides good material for the broader community to use and share. Sharing these materials also helps practitioners and companies raise their profile. For example, from a company perspective, Healthcare Human Factors in Canada created a great video on the poor state of medical device usability called *Oh Shnocks! The state of healthcare technology in '09* (HealthcareHF, 2009). This is a compilation of usability testing highlights they recorded when testing different medical equipment. Other examples include a EUROCONTROL video promoting their SKYbrary information resource (SKYbrary, 2009) and, from a practitioner perspective, Sidney Dekker who has made engaging videos on topics such as just culture, resilience, and why things go wrong. If you are doing a talk it is now relatively easy to record it and share it with the world shortly after—see, for example, *Life After Human Error* from an O'Reilly Velocity Conference (O'Reilly, 2014).

Technology, Entertainment, Design (TED) talks provide one of the most popular and prestigious platforms for online talks. However, searching its database for "ergonomics," "human factors," "usability," and "user experience" does not give many results. There is one by Don Norman on good design, one by Margaret Gould Stewart on user experience, and one by David Kelley on human-centered design. It is obviously easier to locate talks about "design" than HF/E.

TEDx is similar to TED, but locally run and less grandiose. Its conferences are sometimes streamed live online. However, TEDx talks seem harder to search for online. From my department, UCLIC, Yvonne Rogers has spoken on designing technology to keep us engaged with what is important in life rather than becoming digitally detached, and Nadia Berthouze has spoken on detecting emotion through whole-body monitoring. Other HF/E-flavored TEDx talks include Ken Catchpole on medical error, Esther Gokale on posture, and Danny Nou on how ergonomics is used to design for usability and user experience. Sharing these sorts of videos and talks through TED, TEDx, and YouTube is a great way to engage and inform a potentially global audience. There are also different levels of engagement depending on whether you are a professional speaker and thought leader, or just someone passionate about their subject area.

The growing emphasis on engagement has led to broader opportunities within and outside universities to engage with public audiences. At UCL we have participated in lunchtime lectures for the public, both talks to large audiences conducted by lecturers and professors (e.g., Ann Blandford has lectured on *When Technology Design Provokes Errors*; UCL Lunch Hour Lectures, 2011) and talks to smaller audiences conducted by PhD students and postdocs (e.g., I have presented on *The Comedy of (Human) Error*; Furniss, 2012). Speaking to the public has also diversified to include things like science talks in cafes and pubs. Bright Club is an event for academics to perform a funny and informative comedy set for the public. It is outside many researchers' comfort zones, as it was mine, but it was an amazing experience to hold your nerve and think about your subject area in a completely different way, and I am told it was funny too (see *HCI Comedy Set: Bright Club*; Furniss, 2011a). Sarah Wiseman, a colleague, working on CHI+MED, has pursued multiple comedy sets combining science and comedy at, for example, the Edinburgh Fringe Festival and the Greenman Music Festival. These talks are easy to record and share online with a broader audience, for a longer length of time.

Writing can also be a powerful medium to engage a broad audience, even in our multimedia age. Here we are thinking less about technical material to inform a specialist audience and more about materials that transcend the confines of their own subject area, attract new audiences and popular media, and change the way people think. Reading Don Norman's *The Design of Everyday Things* was one of the things that got me interested in studying ergonomics, and I have heard of other students that have been influenced by this text. More recently, Atul Gwande's *Checklist Manifesto* has received a huge amount of attention from popular media, beyond clinicians and those interested in patient safety. It argues for the use of checklists in surgical procedures to help reduce error and has been very influential. Again, there are different levels of engagement depending on your ambition and talent, from grand books by professional writers and thought leaders to magazine articles, blogs, and tweets for us mere mortals, which we will come to later in the chapter.

BENEFIT AT SUBJECT AREA LEVEL

In this spectrum of different levels of impact from engagement, we move down a level from society to discipline. Here we move loosely toward more direct engagement focused on promoting the subject to different audiences. These activities can have some form of activism at their core, that is, not only to excite and inform but strongly focus on bringing about a change in action and policy.

Outside the HF/E area, the Queen Mary University of London's *Computer Science for Fun* (cs4fn, 2016) is a great example of a large focused engagement project seeking to excite students and teachers about computer science (Meagher et al., 2013). cs4fn consists of three main elements: a magazine, a website, and a program of school visits. Each is run in a fun, off-beat style to inspire interest in computer science, and to show its broad application to real issues and interesting problems. The magazine has a print run of up to 31,000 and is mainly distributed across the United Kingdom, but with people downloading the magazine from over 80 different countries. Between 2008 and 2013, the website received over a million visits and hundreds of thousands of PDF downloads. During the same period, members of the cs4fn team gave talks to nearly 20,000 school students on around 270 visits to schools and universities, and engaged a further 10,000–15,000 at science festivals. These levels of engagement contributed to real impact, including a "zeitgeist change, with educational policies now beginning to view computer science as more intellectual and creative than the past association simply with IT" (Meagher et al., 2013).

Professor Paul Curzon, who was the Principal Investigator on cs4fn, also works on CHI+MED, which allowed the projects joint engagement activities (Curzon et al., 2014a). This encouraged cs4fn to branch out to address issues concerned with human factors applied to computing, and CHI+MED benefited from the engagement activities and expertise of cs4fn. For example, there are special articles on the cs4fn website and CHI+MED researchers have taken part in school visits and science festivals. Here, magic shows were used to convey lessons in distraction and misperception that could impact people making mistakes, including doctors and nurses. *The Microwave Racing* video was used in school talks and was also brought to life at science festivals as researchers got school students to race microwaves side-by-side. They used different microwaves to perform the same task as fast as they could (Black et al., 2012). This allowed researchers to talk about the design of the interface to primary school students and to reflect on why usability is so important in reducing medical error. cs4fn has also produced a special edition of their magazine based on CHI+MED research and themes, which has been distributed widely (Curzon et al., 2014b). A subsequent issue has also now had a significant human factors focus.

Engaging with schools is something that has been supported by the Chartered Institute of Ergonomics and Human Factors (CIEHF) for many years. *Ergonomics-4Schools* (http://www.ergonomics4schools.com) is a website that is supported by the CIEHF so school students can find out more about ergonomic themes online.

The CIEHF also supports talks in schools. There is a database of CIEHF presenters who are willing to give talks in schools, should they be requested. A related project, *Designing for Real People*, led by Rachel Benedyk, has contributed to getting user-centered design (UCD) included in the A-level Design and Technology syllabus in the United Kingdom (see Department of Education, 2015). This is a great success, which shows what HF/E-related campaigns can achieve over the long term. However, this work is not without its challenges: without the funding, dedicated staff, and management of something like cs4fn, we are reliant on the dedication and voluntary work from busy people who have a lot of competing demands on their time. Sustainability of these campaigns and projects is a real issue.

Engagement projects at the subject area level can also be targeted at different professional sectors, raising the profile of ergonomics and human factors issues within those sectors. For example, EUROCONTROL's *HindSight* magazine shares safety knowledge, improvements, and helps keeps conversations about safety alive in the air traffic control (ATC) sector, and beyond. The Winter 2013 edition on *Justice and Safety* contains interesting articles on human error and just culture that go beyond ATC (EUROCONTROL, 2013; all articles and issues are also available from www.SKYbrary.aero). In 2009 *HindSight* received the Flight Safety Foundation Cecil A. Brownlow Publication Award for the best aviation safety publication.

In the healthcare domain, in the United Kingdom, the *Clinical Human Factors Group* (CHFG) has played a key role in putting human factors on the patient safety agenda. Martin Bromiley is the founder and current chair. His wife Elaine died from a medical incident, which is detailed in the YouTube video "Just a routine operation" (CHFG, 2014). From an aviation background, he was surprised at how this incident was dealt with, including the lack of human factors learning. The CHFG run talks and workshops on human factors, targeted at the medical profession. They host a website and LinkedIn group, and have achieved significant impact within the healthcare community—including the *Human Factors in Healthcare Concordat* which has been supported by the highest level healthcare bodies in the United Kingdom.

More broadly, World Usability Day (World Usability Day, 2016) happens on the second Thursday in November every year. It has been running since 2005 and has involved 10,000–25,000 volunteers worldwide in the initiative over that period. Every year a different theme is chosen, and volunteers are asked to register an engagement activity to promote issues around usability on this day. This has included companies running open events to raise awareness about usability and the services that are available to improve it. On a similar topic the Interaction Design Foundation (IDF) provides free access to open source interaction design literature from top researchers and professionals in the area (IDF, 2016). The IDF focuses on reaching out to professional designers and students. It has branched out into online courses, mentoring schemes and local groups, but this requires a membership fee to ensure sustainability.

BENEFIT AT PROJECT LEVEL

Public engagement can also benefit specific projects. In healthcare there is a growing recognition that Patient and Public Involvement (PPI) should be included in research. Those involved in PPI sometimes distinguish between engagement, involvement, and participation. Engagement is telling an audience about the outcomes and results of a study that they might not normally come across. Involvement is about getting people to inform research methods, processes, analyses—benefiting from their patients' and the public's perspective on what research topics should be prioritized, how patient information sheets should be worded, how to get broader participation from a community, and so forth. Participation is where people give consent to provide data for research. HF/E is perfect for this as the philosophy of designing systems to be more user-friendly and fit for purpose also applies to the design of research. Users should be involved in research design, participate in the research and design of systems, and be engaged with to make an impact. Other subjects like mathematics, astronomy, and physics do not have this relationship with users. Two recent projects I have worked on have included PPI activities. Exploring the Current Landscape of Intravenous Infusion Practices and Errors (ECLIPSE) has held a PPI workshop to inform research questions about patients' experiences of intravenous infusion therapy and improve the patient information sheets. Learning about carer errors and resilience strategies (CARE-ERRS) has involved a PPI advisory group who codesigned a survey and advised on how to reach out to this community. Both projects have greatly benefited from this specialist involvement.

Citizen science projects are a way of combining research and public engagement. Here the public plays a role in gathering, analyzing, or performing another research role in the project—they become citizen scientists. Haklay (2013) identified four levels of participation in citizen science:

- Level 1—Crowdsourcing, where the citizen scientists gather data
- Level 2—Distributed intelligence where the citizen scientists act as basic interpreters
- Level 3—Participatory science where the citizen scientists participate in problem definition and data gathering
- Level 4—Collaborative science where the citizen scientists define the problem, collect data, and analyze it

Some successful examples of citizen science projects in the United Kingdom include Zooniverse where citizen scientists identify celestial bodies, and the Royal Society for the Protection of Birds' (RSPB) Big Garden Bird Watch, where citizen scientists record the species they see in their garden on a weekend in January. Errordiary is an example of a citizen science project run on HF/E-related themes, such as raising a debate about the ubiquity of error, what it is, and how we should tackle it. Errordiary collects funny, frustrating, and fatal examples of error from the public through Twitter using the hashtag #errordiary and directly through its website. It also collects examples of resilience strategies, using the hashtag #rsdiary, which captures the inventive and informal ways that people develop and adopt to reduce the likelihood of erring. The Errordiary website (see Figure 31.2) is composed of three main parts: (1) a stream of errors, (2) a stream of resilience strategies, and (3) a Discovery Zone where people can engage with stories, video, articles, and games if they want to learn more about the subject.

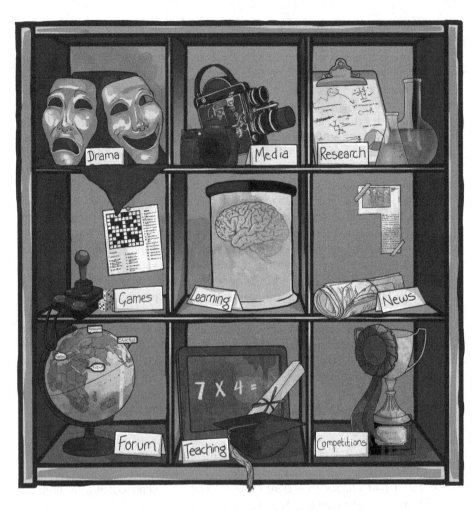

FIGURE 31.2   Picture of the nine elements on Errordiary's Discovery Zone.

Errordiary is a project that brings public engagement, research, and teaching closer together. In terms of public engagement we have engaged with different audiences including people with diabetes, medical professionals, and the public. Hundreds of people have contributed errors and visited the site. In terms of research we have published papers on our novel form of citizen science that we have dubbed a Massive Open Online Diary (MOOD) (e.g., Jennett et al., 2014; Gould et al., 2014); on resilience strategies (e.g., Furniss et al., 2012); and on persuasive games (e.g., Iacovides et al., 2014; Iacovides and Cox, 2014). In terms of teaching we have used the data submitted to Errordiary to enhance teaching on the psychology of human error (Wiseman et al., 2012), and we have developed a workshop structure to inform different audiences about error and resilience (Furniss et al., 2014). Other HF/E citizen science projects and campaigns using what we have dubbed a MOOD model could be conducted to gather public experiences, raise awareness, and debate about HF/E issues, for example, #badHFE, #goodHFE, #UCDgold, #RSI, #poorposture, and so forth.

BENEFIT AT INDIVIDUAL LEVEL

Public engagement can benefit individuals who deliver engagement activities. "By engaging with the public, researchers enhance their communication and influencing skills, raise their personal and institutional profile, enhance the quality of their research by developing new partnerships or incorporating new perspectives, and their confidence and aspirations for the future grow" (Research Councils UK, 2013). In terms of raising your personal and company's profile in a community, this could help your networking and marketability. In terms of skills, Vitae (2012) have developed a skills framework to show how public engagement can impact knowledge, personal effectiveness, research management, teamwork, and communication. From here there is a clear pathway for claiming improvement and professional development from public engagement activities.

SOCIAL MEDIA AND WEB 2.0

Social media has revolutionized the way people stay in touch and communicate. There are now many more ways to create and maintain networks of people, and send messages to individuals and groups far and wide. Web 2.0 refers to the trend toward user-generated content online, for example, through pictures, videos, blogs, and so forth. With these new tools and trends come new benefits and issues. In the following section, I turn to a quick personal review of some of the main tools for social media and Web 2.0. I then conclude by reflecting on the issues and benefits for HF/E practitioners in this area.

A ROUGH GUIDE TO TOOLS FOR SOCIAL MEDIA AND WEB 2.0

Table 31.1 represents a rough guide to tools for social media and Web 2.0. It is personal and incomplete as the landscape frequently changes and the way people use these different tools do too.

ISSUES AND BENEFITS OF USING SOCIAL MEDIA

Social media and Web 2.0 are happening anyway and it is up to us as researchers and practitioners to decide if and how to embrace it. There is a plethora of tools in a changing landscape so I think it is an area to explore and see what suits you. Hopefully the brief guide above might give you some ideas if you are unfamiliar with the area.

One thing that social media and Web 2.0 do is lower the costs of public engagement, open possibilities, and make it more accessible. You no longer have to have the stature of Michael Faraday performing a Royal Institute Christmas Lecture to share your thoughts, interests, and passion with the world.

TABLE 31.1

Rough Guide to Different Forms of Social Media

How It Works	Advantages	Disadvantages
Twitter: Twitter focuses on your network (who you follow and who follows you) and short public messages, called tweets. Tweets are limited to 140 characters. Tweets are public and shared with people who follow you (unless you make your account private). Twitter allows discovery of related tweets on different topics through hashtags (e.g., #errordiary or #rsdiary). Tweets can be retweeted (reshared) by people with different accounts.	• Twitter has great potential for serendipitous discoveries and connections, and flattens hierarchies, e.g., you could talk to celebrities and CEOs. • Hashtags bring people together around topics, events and groups (e.g., at a conference). • Some HF/E professionals and consultancies have used Twitter to good effect: to network, raise their profile, and inform people of their views and activities. • One can manage different Twitter accounts for different purposes, e.g., see @domfurniss, @uclic, @errordiary, @rsdiary, @stevenshorrock, @legoergonomist, @ciehf.	• Twitter can be very transient and searches may be limited. • You can bookmark things by retweeting them and liking them, but these tweets can still be hard to find and retrieve. • Twitter is very public. There are high-profile cases of people ending up in court and losing their jobs through careless tweets. • People who are new to Twitter often try and read all new tweets. It is probably better treated like a cocktail party or busy marketplace where you will hear some things but not others.
Facebook: Facebook works around a profile you create about yourself. You are encouraged to share status updates, pictures, and videos. People are able to connect to each other by becoming "friends." When you log in to Facebook you see updates from all of your Facebook 'friends'. Status updates and media can be shared publicly, with friends, or with friends of friends.	• There is the ability to join different groups on Facebook. These groups can be people with shared hobbies, interests, conditions, and affiliations with different companies, schools, and professional bodies. • With different friends, colleagues, and groups all in one place, conversations can move between social and professional topics, which could promote HFE interests to new audiences.	• Many people use Facebook to stay in touch with friends and family rather than for work. • Lines can easily blur between professional and social circles on social media, e.g., you might become friends with people at work OR you might post about some work activities to friends and family—or the public. This can have implications for professionalism.

Google+: Google+ is similar to Facebook but is Google's version.

LinkedIn: Like Facebook, you create a profile and 'link' to a network of contacts. Unlike Facebook it is marketed for professional rather than personal use.

There is the ability to join different discussion groups that share the same interest, and to follow organizations or individuals.

There are many different features for enhancing your profile and marketing yourself.

Blogging services: Popular blogging services open to all include Blogger and Wordpress. *The Conversation* hosts edited blogs which have more quality control; see Paul Salmon for a HF/E example.

- Google+ offers the ability to create different circles so you can manage which social circles see what more effectively.
- Useful for staying in touch with a professional network over a longer period of time.
- Enables joining different groups that include professional bodies like HFES and CIEHF, different special interest groups like CHFG, and topics of interest like 'Beyond Human Error'. New features like the ability to blog could prove useful.

- The most successful blogs reach thousands of subscribers.
- High-profile blogs can even generate revenue through advertising to their readership.
- Keeping a casual and irregular blog reduces the required effort and expectations.
- Some people use blogs for self-reflection and an archive for their own writing and articles.

- If you know a lot of people it can be difficult to organize everyone you know into appropriate circles (though the same can apply to Facebook).
- LinkedIn can appear quite complicated with different entities like people, groups, companies, and many different features.
- Some features like endorsing people for different skills seem rather dubious. People are endorsed for things they have little expertise in by people who don't know them well.
- LinkedIn is meant to be for people who have worked together, met before, or have some firm connection. However, it is increasingly used for people to amass a huge network of people they don't really know, potentially making one's network less meaningful.
- The time and effort to create and maintain a frequent and good quality blog should not be underestimated.
- There is a long tail of less widely read blogs.

(Continued)

TABLE 31.1 (CONTINUED)
Rough Guide to Different Forms of Social Media

How It Works	Advantages	Disadvantages
Vlogging: Vlogging is like a blog but rather than write you video record yourself speaking to camera and then share it online.	• Popular video sharing services to host vlogging and other video material include YouTube and Vimeo, which could reach out to huge audiences.	• Lectures, presentations, and informative videos might be more relevant to disseminate HFE messages than vlogging per se.
Photo sharing services: Popular photo sharing services include Flickr, Google+ Photos, and Instagram. Here photos can be easily shared online.	• Public engagement possibilities include collecting albums of photos with an HF/E theme. If these photos have the same hashtag or tag they could show up in searches from multiple accounts. For example, searching for #errordiary reveals Errordiary photos.	• Sharing photos through other social media channels, e.g., see @legoergonomist on Twitter, could broaden reach. • It can be hard to keep up a stream of interesting, high quality, and relevant photos and images.
Others: Other noteworthy examples of social media include Tumblr, Buzzfeed, Pinterest, Reddit, and Snapchat. The social media landscape changes and evolves over time. Users need to explore these changes and evolve with them.	• There is great potential in the many different forms of social media that have been developed and are developing.	• Different communities and practices develop around different forms of social media. Time needs to be spent learning these different communities and practices for successful engagement. • There will always be more things available than you can sensibly engage with.

CEO, chief executive officer; HF/E, human factor and ergonomics; CIEHF, Chartered Institute of Ergonomics and Human Factors; CHFG, Clinical Human Factors Group.

Sharing information in such a broad public way can be daunting and caution is advised. Many professional bodies are recognizing this issue and drafting guidance on how their members should conduct themselves professionally online, for example see guidance by the Nursing and Midwifery Council (Nursing and Midwifery Council, 2016). I'm not aware of an HF/E version. People should post responsibly on any social media as sharing can have serious consequences. For example, 17-year-old Paris Brown found herself in the media spotlight when she was appointed as the United Kingdom's first youth police and crime commissioner—journalists dug up inappropriate tweets she had posted years before (between the ages of 14 and 16) and she resigned under the media pressure (BBC, 2013). These stories and others form part of how society is learning to deal with social media.

Although we should be mindful of these risks they are often exaggerated and should not be a problem when dealt with a bit of common sense. There is great value in using social media for public engagement and networking. I have connected to many more people in different ways using Twitter, Facebook, and LinkedIn than I would have done without using them. Indeed, the opportunity to write this chapter came about through my use of social media and involvement in public engagement. Other serendipitous connections include connecting with HF/E researchers and practitioners, patients, and clinicians who share some sphere of interest with me.

In 2015, the National Coordinating Centre for Public Engagement (NCCPE) carried out a survey to investigate the barriers to individual researchers engaging with the public. Their 299 respondents cited a lack of knowledge to do it well (29%), lack of reward and recognition (21%), pressure to publish (17%), lack of funding (16%), and lack of regard of peers (8%) as the main barriers. HF/E practitioners will face similar barriers like lack of time, lack of reward and recognition, and pressure to do core work but we do not know the exact barriers or weightings without the research. Tracking the changing nature of HF/E engagement, and the opportunities and barriers in this area, could be a useful contribution for our community in the long term.

We can collectively make a great contribution through social media by keeping conversations about HF/E alive inside and outside the community. Among ourselves we can share interesting articles, videos, research, and have a dialogue about HF/E issues and concerns. The public nature of this activity means that others can listen in, get involved, and engage as they feel comfortable. Broadly, this can help impact the general zeitgeist about how people think about HF/E, and importantly influence beliefs and actions related to this area.

CONCLUSION

Social media and public engagement are powerful tools for reaching out to impact others inside and outside our community, to inform research, and to network. There are many different tools related to social media and Web 2.0 to help us connect with groups, share thoughts, and generate creative content to spark interest and debate. This chapter has given a broad flavor of HF/E examples of engagement for this area including microwave racing at science festivals, comedy sets in pubs, blogging, tweets, YouTube videos, and public talks.

We use social media and do public engagement because we think it has some impact and value, and we enjoy it. Benefits can be considered at different levels from the societal level to the individual. Impact can be broad and narrow, connecting with a large group of people or only to a few individuals. Impact can also be deep or shallow, creating lasting change in beliefs and behaviors or a passing interest in a subject.

This chapter has covered many different types and levels of engagement, which involve different levels of personal investment, from organized campaigns and extended projects to more casual commenting and sharing on social media. Hopefully, this has inspired further interest in HF/E public engagement and use of social media. By engaging, even in small and simple ways, we will help keep HF/E discussions alive within and outside our community.

ACKNOWLEDGMENTS

The author would like to acknowledge Ann Blandford, Paul Curzon, Steve Shorrock, and Claire Williams for comments on previous versions of this chapter.

This study was funded by CHI+MED, and supported by the UK Engineering and Physical Sciences Research Council [EP/G059063/1].

REFERENCES

BBC. 9 April 2013. Paris Brown: Kent youth PCC resigns after Twitter row. [Online]. Accessed March 16, 2016. Available at: http://www.bbc.com/news/uk-england-22083032.

Benedyk, R. and Furniss, D.J. 2011. Using digital stories to enhance course induction for HCI students. In: *British HCI 2011: Health, Wealth and Happiness: HCI Educators Workshop*. Newcastle-upon-Tyne, UK. Green open access.

Black, J., Furniss, D., Myketiak, C., Curzon, P., and McOwan, P. 2012. Microwave racing: An interactive activity to enthuse students about HCI. In: *The Contextualised Curriculum Workshop at CHI 2012*. 5–10 May 2012, Austin, TX.

Catchpole, K.R., De Leval, M.R., McEwan, A., Pigott, N., Elliott, M.J., McQuillan, A., et al. 2007. Patient handover from surgery to intensive care: Using Formula 1 pit stop and aviation models to improve safety and quality. *Pediatric Anesthesia*. 17(5), 470–478.

Clinical Human Factors Group. 2014. Martin Bromiley: Just a routine operation. [Online]. Accessed March 16, 2016. Available at: https://www.youtube.com/watch?v=GDGMjbm24IM.

CSB. 2008. CSB safety video: Anatomy of a disaster. [Online]. Accessed March 16, 2016. Available at: https://www.youtube.com/watch?v=XuJtdQOU_Z4.

cs4fn. 2016. Computer science for fun. [Online]. Accessed March 16, 2016. Available at: http://www.cs4fn.org.

Curzon, P., Brodie, J., Furniss, D., McOwan, P., and O'Kane, A.A. 2014a. Inspiring future ergonomists by mixing microwaves, magic and medicine. *The Ergonomist*. 527, May, 12–13.

Curzon, P., Brodie, J., Myketiak, C., and McOwan, P.C. 2014b. *CS4FN Issue 17: Machines Making Medicine Safer*. London: Queen Mary University of London.

Department of Education. 2015. Design and Technology GCE AS and A Level subject content. [Online]. Accessed January 18, 2016. Available at: https://www.gov.uk/government/uploads/system/uploads/attachment_data/file/485436/D_and_T_A_level.pdf.

EUROCONTROL. 2013. Hindsight 18–Winter 2013. [Online]. Accessed March 16, 2016. Available at: https://www.eurocontrol.int/publications/hindsight-18-winter-2013.

Furniss, D. 2010. Man-machine nightmares: Chaos buttons, human error and health-care. [Online]. Accessed March 16, 2016. Available at: https://www.youtube.com/watch?v=ifjDWKMNlIk.

Furniss, D. 2011a. HCI Comedy Set: Bright Club (UCL). [Online]. Accessed March 16, 2016. Available at: https://www.youtube.com/watch?v=SeqBRTVaM7w.

Furniss, D. 2011b. Microwave racing. In: *CHI'11 Extended Abstracts on Human Factors in Computing Systems*. Conference on human factors in computing systems. New York, NY. ACM, pp. 497-497.

Furniss, D. 2012. Bite-sized lecture: The comedy of (human) error and cognitive resilience. [Online]. Accessed March 16, 2016. Available at: https://www.youtube.com/watch?v=8_OQXF_Y2go.

Furniss, D.J. and Benedyk, R. 2011. Boundaries and three points on a virtuous circle: Digital stories, public engagement, and teaching students. In: *CHI 2011: the 29th ACM Conference on Human Factors in Computing Systems: Workshop: Video Interaction—Making Broadcasting a Successful Social Media*. 7–12 May 2011, Vancouver, Canada.

Furniss, D., Back, J., and Blandford, A. 2012. Cognitive resilience: Can we use Twitter to make strategies more tangible? In: *ECCE '12 Proceedings of the 30th European Conference on Cognitive Ergonomics*. 28–31 August 2012, Edinburgh Napier University, Edinburgh, Scotland. pp. 96–99.

Furniss, D., Iacovides, J., Jennett, C., Gould, S., Cox, A., and Blandford, A. 2014. How to run an errordiary workshop: Exploring errors and resilience strategies with patients, professionals and the public. In: *The Third Resilience Health Care Net Meeting, Hindsgavl Castle*. 12–14 August 2014.

Gould, S., Furniss, D., Jennett, C., Wiseman, S., Iacovides, I., and Cox, A. 2014. MOODs: Building massive open online diaries for researchers, teachers and contributors. In: *Proceedings of CHI EA '14: CHI '14 Extended Abstracts on Human Factors in Computing Systems*. 26 April–1 May 2014, Toronto, Canada, pp. 2281–2286.

Haklay, M. 2013. Citizen science and volunteered geographic information—Overview and typology of participation. In: Sui, D.Z., Elwood, S., and Goodchild, M.F. (eds.). *Crowdsourcing Geographic Knowledge: Volunteered Geographic Information (VGI) in Theory and Practice*. Berlin: Springer, pp. 105–122.

HealthcareHF. 2009. Oh Shnocks! The state of healthcare technology in '09. [Online]. Accessed March 16, 2016. Available at: https://www.youtube.com/watch?v=WxQLzdLjwp4.

Iacovides, I. and Cox, A.L. 2014. Designing persuasive games through competition. In: *Workshop on Participatory Design for Serious Game Design: Truth and lies at CHI Play 2014*. Toronto, Canada, October 2014.

Iacovides, I., Cox, A.L., Furniss, D., and Myketiak, C. 2014. Exploring empathy through sobering persuasive technologies: "No breaks! Where are you going missy?" In: *2014 BCS Conference on Human-Computer Interaction (BCS-HCI 2014)*, Southport, UK, September 2014. (Abstract only.)

Interaction Design Foundation. 2016. Interaction Design Foundation. [Online]. Accessed March 16, 2016. Available at: https://www.interaction-design.org.

Jennett, C., Furniss, D., Iacovides, I., and Cox, A. 2014. In the MOOD for Citizen Psych-Science. In: *Citizen Cyberscience Summit 2014*. London.

Meagher, L. R., Curzon, P., McOwan, P. W., Black, J. and Brodie, J. 2013. *cs4fn Final Evaluation Report*. London: Queen Mary University of London.

National Coordinating Centre for Public Engagement (NCCPE). 2015. Barriers to researchers engaging. Available at: http://www.publicengagement.ac.uk/blog/barriers-researchers-engaging.

Nursing and Midwifery Council. 2016. Social media guidance. [Online]. Accessed March 16, 2016. Available at: https://www.nmc.org.uk/standards/guidance/social-media-guidance/.

O'Reilly. 2014. Steven Shorrock: Life after human error—Velocity Europe 2014. [Online]. Accessed March 16, 2016. Available at: https://www.youtube.com/watch?v=STU3Or6ZU60.

Porter, M.E. and Kramer, M.R. 2006. The link between competitive advantage and corporate social responsibility. *Harvard Business Review*. 84(12), 78–92.

Research Councils UK. 2013. Inspiration to engage: Concordat for engaging the public with research. Accessed March 16, 2016. Available at: http://www.rcuk.ac.uk/RCUK-prod/assets/documents/publications/ConcordatInspiration.pdf.

SKYbrary. 2009. Land or hold. [Online]. Accessed March 16, 2016. Available at: https://www.youtube.com/watch?v=oDSA7bpchXQ&spfreload=10.

UCL Lunch Hour Lectures. When technology design provokes errors (3 Nov 2011). [Online]. Available at: https://www.youtube.com/watch?v=f8o2ds5ZrV8.

UCLTV. 2010. Why buttons go bad (UCL). [Online]. Accessed March 16, 2016. Available at: https://www.youtube.com/watch?v=X5WXaUQtPRs.

VideoFromSpace. 2013. How to wash your hands in space. [Online]. Accessed March 16, 2016. Available at: https://www.youtube.com/watch?v=9Z2KNDGNnlc.

Vitae. 2012. Public engagement lens on the vitae researcher development framework. [Online]. Accessed March 16, 2016. Available at: https://www.vitae.ac.uk/vitae-publications/rdf-related/public-engagement-lens-on-the-vitae-researcher-development-framework-rdf-apr-2013.pdf.

Wiseman, S., Gould, S., Furniss, D.J. and Cox, A. 2012. Errordiary: Support for Teaching Human Error. In: *CHI 2012: The 30th ACM conference on human factors in computing systems: Workshop: A contextualised curriculum for HCI, 5-10 May 2012*, Austin, Texas.

World Usability Day. 2016. World usability day November 10, 2016. [Online]. Accessed March 16, 2016. Available at: http://worldusabilityday.org.

Afterword

Ergonomics and Human Factors Research and Practice: Lessons Learned from Professor John R. Wilson

Sarah Sharples

CONTENTS

I was lucky enough to be taught and mentored by, and become a colleague of, an inspirational ergonomist, Professor John Wilson. John's work, writings, thoughts, and occasional rants informed the thinking of the editors of this text, as well as many of the authors, and it is an honor to contribute some messages that John felt passionately should be communicated to those who research and practice in human factors and ergonomics (HF/E).

Sadly, John died in July 2013. Since his death almost 3 years ago, many people have contacted and spoken to me and my colleagues at the University of Nottingham and in the HF/E community more widely, and given many examples of how John's work inspired and informed their research and practice.

COMBINE RESEARCH AND PRACTICE

John's work, over a period of over 30 years, encompassed many domains in which HF/E is applied, including manufacturing, retail, product design, virtual reality technologies, and transport (with a particular focus on rail human factors).

Although John worked for his entire career within academic institutions, and since 1983 at the University of Nottingham, like many academics in the field of HF/E he combined his academic research with work as an HF/E practitioner. For many years he combined an academic role with part-time consultancy, and in later years took on a part-time seconded post within the Network Rail Ergonomics team. Combining research and practice was something that John taught those of us who worked with him to use as an explicit strategy in our endeavors. When working with an industrial partner, whether as part of the work directly funded from industry, or within a collaborative project supported by a research organization, we worked on (and continue to work on) real-world contexts, with real-world problems, and real-world users.

Over the years, partly due to external influences such as funding climates and the national economic situation, the work that we tended to undertake within the University moved from shorter "traditional consultancy" activities, such as short-term evaluations of workplaces where there was a design change underway, or where employees had experienced difficulties, to longer partnerships with external organizations. Of course, the nature of work taken on by academics can differ from that undertaken by, for example, an independent practitioner or consultancy (and many academics, like John, also work as independent practitioners in addition to their university duties), but John's approach was to conflate the two activities. In fact, rather than distinguishing between research and practice, he tried to make sure that from our practice, research emerged.

Maintaining a collaborative and supportive link between the different communities of HF/E experts, whether based in academic institutions, independent consultancies, or embedded within private or public sector organizations, is the key to the delivery and advancement of our discipline. This is a particular theme in this book (see, for instance, Chapters 3, 9, and 10). In recent years, U.K. universities in particular have placed a strong focus on "impact"—the way in which fundamental work can be demonstrated to have had an effect on the external context of industry or society. HF/E in general and John in particular, were far ahead of the game when it comes to considering impact. As I'm writing this chapter, I can't think of a single research project that I have worked on, which has not directly or indirectly involved an external stakeholder, whether that stakeholder is a technology developer, a systems or product designer, or an organization. And, of course, almost all HF/E activities involve contact with users, whether they are supporting a participatory design process, providing input into a workplace evaluation, or helping to test a future technology.

John's view was that our role was to ensure that we delivered theories, frameworks, and methods that both advance the discipline of HF/E in general and also provide useful insights and tools that enable those outside the HF/E community to learn and carry out their practice. In this chapter, with apologies to John for anything I've misremembered, misinterpreted, or even willfully distorted in a vain effort to win one of our many lively debates, I will outline some key messages that I think John would like us to have remembered when, as many of us do as we go about our work in HF/E, we wonder "What would John say?"

TAKE A SYSTEMS APPROACH

John was a leading proponent of "systems ergonomics" (e.g., Wilson, 2000, 2014). But, long before the concept of systems ergonomics was articulated in such an explicit manner, John always took a broad, context-led perspective to any HF/E problem that he was investigating. John edited the leading HF/E text *Evaluation of Human Work: A Practical Ergonomics Methodology* (Wilson and Sharples, 2015a)—the first three editions in partnership with his colleague and mentor, Nigel Corlett, and the fourth edition with me (a task that I completed in the year after his death). These writings provide a nice record of the evolution in systems thinking since the first edition, published in 1990. In that first edition of the text, he introduced the "onion model" of Ergonomics. This model really did embody the approach that John took to HF/E research and practice. In all projects, sometimes at the start, but more often, after working on and thinking about the domain for a couple of months, he would sit down with colleagues and start to think about a "framework." As students we sometimes despaired at John's lack of clarity when he explained what a framework actually was, but, thinking back (and perhaps being a little wiser than I was then) I would describe a framework as a representation of the system of interest, the influences and outputs, and often, an opportunity to highlight the fundamental HF/E concepts of interest. A framework enables the integration of research and practice—it highlights the relevant theories, whether physical, such as the causative factors of upper limb disorders, or the cognitive, such as theories around workload or stress, and demonstrated their relevance to the real-world context. It also helped to inform those theories, and probably to highlight their limitations when being applied in a generic and complex work situation. In addition, representing our thinking in a diagrammatic way often informed our methodological focus, and helped us to work together as a multidisciplinary team. I have updated the "onion" model for the fourth edition of *Evaluation of Human Work* (Wilson and Sharples, 2015b), and carry a paper copy of this updated model with me everywhere I go, as I find it is the most useful tool to explain the goals of our involvement in projects and activities to those who are less familiar with the aims of HF/E (Figure A.1).

The systems approach acknowledges that the real world is, like John's office often was, messy (a trait that I have taken on with much aplomb!). But, perhaps like our messy offices, within the muddle of complexity, lie patterns, and our expertise as HF/E practitioners allows these patterns to emerge. Our methods to support knowledge elicitation and to support design and change allow us to embrace the complexity of the system, rather than trying to boil it down to "traditional" input–output cause and effect relationships.

PUBLISH, WRITE, AND TALK

Something that John was very good at doing as an HF/E practitioner was communicating about his work. He wrote books, journal articles, and conference papers, and used these as tools in his practice, rather than it just being a bland "dissemination for dissemination's sake" activity. There are some excellent examples where he has

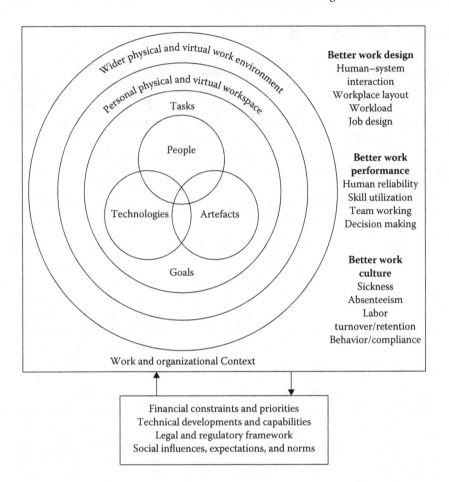

FIGURE A.1   The "onion" model and aims of HF/E.

clearly used a conference paper to collect his thoughts about a certain topic, whether it is about the challenges of establishing systems requirements (Wilson, 2013), or the approaches that he had developed over the years to participating in a program of control room design (Rajan et al., 2005).

John's writing served as a tool for debate. He wasn't afraid to change his views over time, and was always very careful in his writing to acknowledge that while he was taking a particular stance (e.g., on qualitative methodology, or field-based research), there were other stances that were also completely acceptable. Nowadays, academic careers in particular do not tend to allow quite the same amount of time for writing as perhaps they did in the past, due to increased pressures to generate research income, and increased teaching demands, but all practitioners, whether academic-based or not, can benefit from capturing their thoughts and ideas in a written form.

Something that has clearly changed over the past decade is the nature of those forms—we now have blogs and social media available as tools, and journal articles are increasingly openly accessible. We also increasingly find writings about HF/E matters in non-HF/E destinations, such as medical, engineering, technology, or design journals. The interesting consequence of this is that it develops our ability to extend our debate about HF/E issues beyond our immediate workplace colleagues to our international community, and potentially breaks down the barriers between academia and practice.

John was amusingly derisive about the notion of social media (referring to it in one infamous instance as "flim flam"). Establishing methods to ensure that we use such tools to enhance our practice is an essential challenge for our discipline over the next few years (see Chapter 31 by Dominic Furniss). But we must make sure that we do not "throw the baby out with the bathwater." Peer-reviewed papers not only serve as a gold standard within academia for the publication of new theories and data, but they serve as a managed resource for posterity. The fact that we are able to track the evolution of trends and concepts within the literature is valuable and important to maintain as we extend the ways that we communicate. Social media tools, including blogs, are also starting to prove particularly beneficial for communicating the value of HF/E to collaborators and policy makers. It is likely therefore that both researchers and practitioners will need to ensure that their research is communicated through both traditional peer-reviewed channels and social media, with potentially each group biased more to one or the other due to time demands and occupational reward systems.

Communication can enthuse partners and clients of HF/E (see, e.g., Chapters 28 through 31). Being a good communicator is important in many disciplines, but we have some specific challenges in HF/E that make this skill particularly important. HF/E as a discipline has a tendency to use "real-world" terms, such as workload, human error, posture, or force for example, with quite specifically defined meanings. Explaining what "we" mean by these terms is important at the outset of a collaborative project, and in communicating results. Secondly, HF/E still suffers from the tendency to be considered "common sense." Demonstrating the impact of using an HF/E approach on the workplace in the short and longer term is difficult, and requires good communication, along with good practice and clear evidence.

THINK BEYOND THE LABORATORY

John was an extremely keen advocate for conducting HF/E research outside the laboratory setting. His work spanned consideration of physical, organizational, cognitive, and social factors, and he used and developed a number of tools and approaches that had both academic integrity as well as being useful in a practical setting. Notable examples of this are the tools developed for use in the rail context, which spanned factors including the workload (Integrated Workload Scale—Pickup et al., 2005) shift patterns (a subset of the questionnaire tool REQUEST—Ryan et al., 2008) and physical effort (injury risk measurement plan—Muffett et al., 2014).

Not only did this approach leave a legacy of tools, and practices for tool development, but it also had at its heart an "emergent theory approach." Such an approach denotes the practice of allowing the theoretical implications and findings to emerge from the setting, rather than to impose a hypothesis at the outset of collecting data. Rather than testing a specific theory or prediction, this tends to lead toward a much more holistic approach to data collection, and is very sympathetic to systems thinking. This, in partnership with the "framework" concept outlined earlier, allowed the early stages of any project with an external context to focus on capturing the richness of the work environment, and then to think about the specific theoretical relationships that might be explored.

Interestingly, this approach did lead to some debate between John and his colleagues, including myself. The use of an emergent theory approach is very valuable, and has high face validity, as well as serving a practical purpose of ensuring that all are using a common language and understanding of the work situation. But inevitably the framework diagrams and emergent theories tended toward complexity, and with complexity comes challenges in terms of testing and evaluating the theories that have been proposed, as well as in communication of the value of HF/E within and outside the discipline (as perhaps illustrated by the success of easy to grasp concepts such as the Swiss Cheese model of accident causation). At one event that John and colleagues put together to share their approaches with a "younger generation" of researchers, he and his compatriots were, respectfully, described as "irresponsible baby boomers," who had left our generation with a pile of diagrams and concepts, with very little auditable evidence to back them up!

But, while there is a need for this auditable evidence, it is also the case that those diagrams and models that emerged from papers in the 1980s and 1990s are, on the whole, useful. What has changed is the expectation in writing and research to be able to trace the evidence base for theories and propositions. One of my personal favorite papers, which I know John also admired, is the "Ironies of Automation" by Bainbridge (1983). This highly cited paper emerged from the author's own practice while working within control contexts, and her propositions are backed up by reasonably informal observations within the paper, which is written in quite a discursive tone. It would be interesting to consider whether this highly cited paper would in fact be accepted for journal publication today as its form reads like a long blog post or discussion piece (it was first published at a conference in 1982, then in the journal *Automatica* in 1983). This is perhaps where our nonpeer-reviewed forms of publication, such as blogs, and more colloquial forms of communications, such as this book, or the *Evaluation of Human Work* text, serve a particularly valuable role in supporting HF/E practice.

The research theories, frameworks, and methods that emerge from our practice are different from those that we see in disciplines such as experimental psychology, or even some of the specific physiological or biomechanical modelling that our own discipline generates in the lab. It is important that we maintain this practice-led research as an integral part of our discipline, and develop methods that allow us to communicate the science behind this approach. There is of course an important role for structured experimental studies and theory-driven analysis in our discipline, but we must consider these to be part of our portfolio, rather than one particular approach being considered to be superior to another.

BE CONSTRUCTIVELY CRITICAL AND HAVE A CRITICAL FRIEND

I highlighted the importance of communication earlier; with communication comes debate and critique. John had formal activities that particularly encouraged this, most notably in his role as Editor-in-Chief of *Applied Ergonomics* for many years.

The HF/E community is quite small and we support each other in the common goal to evangelize about the value of HF/E in a range of areas. But our passion for the subject can mean that we are sometimes less constructively critical of each others' ideas and approaches than might be helpful.

John was a master of critique without judgment. He would point out the limitations in an argument, without dismissing the argument or position out of hand. I often passed my papers, proposals, and reports to John for review, if he had not been involved in the writing or work. He was very good at looking at the structure of an argument, and thinking about how one piece of work related to the wider discipline of HF/E as a whole. Since John died I am very pleased to have been able to seek out other HF/E professionals from around the world who have similarly critiqued my work, and I would recommend that all practitioners and researchers identify a small group of willing "critical friends" who they trust not to pull any punches when giving (polite!) feedback on work.

One of the values of HF/E is in its variety, in method, practice, and domain. Debate and disagreement is good, and will improve our research and practice alike. The variety in approaches that the HF/E community is able to offer is one of the strengths of our profession. Learning how to conduct and encourage debate respectfully, and with a smile on your face, is critical.

NURTURE TALENT AND COLLABORATE WITH OTHERS

Many ergonomists and human factors specialists, including myself, would almost certainly not be in the profession if it weren't for people like John, who mentor, support, and nurture talent among students and early career researchers and practitioners.

Collaborating with John, as a student, project partner, or colleague, was always an enjoyable experience, if occasionally unpredictable. John wasn't a particular fan of regularly scheduled or formally structured management meetings, but very much valued "corridor conversations" (which often ended up being conversations on a train). Such a meeting would usually result in an indecipherable scrawl on a piece of paper covered in boxes and arrows, which would then form the backbone of our work for months ahead.

But John also knew how to nurture young careers. He was very good at making sure that credit was given appropriately on written papers, and his invited presentations would always be peppered with mentions of his collaborators. He also allowed his younger colleagues to make mistakes, and supported them through the period of recovery from that mistake, whether it had been a mistake in the way a study had been conducted, or the way in which a project had been handled. He realized that we all do make mistakes and rather than hauling someone over the coals, he worked with them to put things right. In fact, thinking about it, he was much more concerned about establishing a resilient research culture than he was about error and blame.

He was also very good at handing over control. One of the hardest things to do when you have built something up from scratch is to hand it over to someone else, and trust them to stamp their own mark on that activity. John had a good mentor himself in this respect, in Professor Nigel Corlett, who founded the HF/E activity at the University of Nottingham and handed the leadership of the activity over to John.

Of course, I'm fairly sure that John cringed at some of the things that people like me (and the editors of this text!) did when he asked us to take over leadership of activities, and probably secretly thought that things would have been done better if he had done them himself, but he realized the value of establishing succession, giving us the confidence to give things a go, and ensuring the resilience of his teams.

It is terribly sad that this resilience has been tested by John's premature death, but I know he would be very proud of the many HF/E professionals around the world who have carried on the work that he would have loved to have collaborated in, and I hope that we have also managed to continue in John's cheerful and supportive spirit.

CONCLUSION

John was one of a kind. Not many internationally leading academics can mix his approachability, intellect, humility, humor, and use of idiosyncratic British phrases and get away with it! He would be enormously impressed by this edited text, and the efforts that the editors have taken, to think so carefully about the issues associated with HF/E research and practice that were so close to John's heart. The advice to take a systems approach; publish, write, and talk; think beyond the laboratory; be constructively critical; and nurture talent and collaborate are lessons that I've taken from my time working with John, but I'm sure there are many other lessons that John's other collaborators and colleagues would have taken, and which add to John's rich legacy.

It is very sad that John is no longer with us to share this text, and join in with its debate, but he would be very proud of this book, and I'm sure would encourage all of its readers to think carefully about how its messages can be used to enhance their own HF/E research and practice. Thanks John.

Professor Sarah Sharples
President of the Chartered Institute of Ergonomics and Human Factors
(2015–2016)

REFERENCES

Bainbridge, L. 1983. Ironies of automation. *Automatica.* 19(6), 775–779.

Muffett, B., Wilson, J. R., Clarke, T., Coplestone, A., Lowe, E., Robinson, J., et al. 2014. Management of personal safety risk for lever operation in mechanical railway signal boxes. *Applied Ergonomics.* 45(2), 221–233.

Pickup, L., Wilson, J. R., Norris, B. J., Mitchell, L., and Morrisroe, G. 2005. The Integrated Workload Scale (IWS): A new self-report tool to assess railway signaller workload. *Applied Ergonomics.* 36(6), 681–693.

Rajan, J.A., Wilson, J.R., and Wood, J. 2005. Control facilities design. In: Wilson, J.R. and Corlett, E.N. (eds.). *Evaluation of Human Work.* Third Edition. London: Taylor & Francis.

Ryan, B., Wilson, J.R., Sharples, S., Kenvyn, F., and Clarke, T. 2008. Rail signallers' assessments of their satisfaction with different shift work systems. *Ergonomics.* 51(11), 1656–1671.

Wilson, J.R. 2000. Fundamentals of ergonomics. *Applied Ergonomics*. 31, 557–567.

Wilson, J.R. 2013. From creation to compliance: Dos and don'ts of negotiating requirements with developers. In: Anderson, M. (ed.). *Contemporary Ergonomics and Human Factors 2013*. Boca Raton, FL: CRC Press, pp. 415–422.

Wilson, J.R. 2014. Fundamentals of systems ergonomics/human factors. *Applied Ergonomics*. 45(1), 5–13.

Wilson, J.R. and Sharples, S. 2015a. *Evaluation of Human Work*, Fourth Edition. Boca Raton, FL: CRC Press.

Wilson, J.R. and Sharples, S. 2015b. Method in the understanding of human factors. In: Wilson, J.R. and Sharples, S. (eds.). *Evaluation of Human Work*, Fourth Edition. Boca Raton, FL: Taylor & Francis.

Index

Printed in the United States
by Baker & Taylor Publisher Services